The Brain from Inside Out

The Brain from Inside Out

György Buzsáki

UNIVERSITY PRESS

OXFORD
UNIVERSITY PRESS

Oxford University Press is a department of the University of Oxford. It furthers
the University's objective of excellence in research, scholarship, and education
by publishing worldwide. Oxford is a registered trade mark of Oxford University
Press in the UK and certain other countries.

Published in the United States of America by Oxford University Press
198 Madison Avenue, New York, NY 10016, United States of America.

© Oxford University Press 2019

All rights reserved. No part of this publication may be reproduced, stored in
a retrieval system, or transmitted, in any form or by any means, without the
prior permission in writing of Oxford University Press, or as expressly permitted
by law, by license, or under terms agreed with the appropriate reproduction
rights organization. Inquiries concerning reproduction outside the scope of the
above should be sent to the Rights Department, Oxford University Press, at the
address above.

You must not circulate this work in any other form
and you must impose this same condition on any acquirer.

CIP data is on file at the Library of Congress
ISBN 978-0-19-090538-5

To my family, with whom everything, without whom nothing.

CONTENTS

Preface ix
Acknowledgments xv

1. The Problem 1

2. Causation and Logic in Neuroscience 33

3. Perception from Action 53

4. Neuronal Assembly: The Fundamental Unit of Communication 83

5. Internalization of Experience: Cognition from Action 101

6. Brain Rhythms Provide a Framework for Neural Syntax 141

7. Internally Organized Cell Assembly Trajectories 165

8. Internally Organized Activity During Offline Brain States 199

9. Enhancing Brain Performance by Externalizing Thought 219

10. Space and Time in the Brain 241

11. Gain and Abstraction 279

12. Everything Is a Relationship: The Nonegalitarian, Log-Scaled Brain 301

13. The Brain's Best Guess 337

14. Epilogue 357

References 361
Index 417

PREFACE

A theory is not to be considered complete until you have made it so clear that you can explain it to the first man whom you meet on the street.
—Joseph-Diez Gergonne[1]

The most complicated skill is to be simple.
—Dejan Stojanovic[2]

Success is the ability to go from one failure to another with no loss of enthusiasm.
—Winston Churchill[3]

As far as I remember, there was only one rule. "Be home before it gets dark." The definition of darkness was, of course, negotiable. My childhood included many animals: turtles, a family of hedgehogs, fish in a toilet tank, pigeons, and barn owls, in addition to our family's cats and chickens. Our pig, Rüszü, and I were good friends. He was always eager to get out from his sty and follow me to our favorite destination, a small, shallow bay of Lake Balaton, in Hungary. A short walk across the street from our house, Lake Balaton was the source of many happy moments of my early life. It provided everything I needed: swimming during the summer, skating in the winter, and fishing most of the year around.

I lived in complete freedom, growing up in the streets with other kids from the neighborhood. We made up rules, invented games, and built fortresses from rocks and abandoned building material to defend our territory against imagined invaders. We wandered around the reeds, losing ourselves only to

---

1. As quoted in Barrow-Green and Siegmund-Schultze (2016).
2. https://www.poemhunter.com/poem/simplicity-30/.
3. https://philosiblog.com/.

find our way out, developing a sense of direction and self-reliance. I grew up in a child's paradise, even if those years were the worst times of the communist dictatorship for my parents' generation.

Summers were special. My parents rented out our two bedrooms, bathroom, and kitchen to vacationers from Budapest, and we temporarily moved up to the attic. Once my father told me that one of the vacationers was a "scientist-philosopher" who knew everything. I wondered how it would be possible to know everything. I took every opportunity to follow him around to figure out whether I could discover something special about his head or eyes. But he appeared to be a regular guy with a good sense of humor. I asked him what Rüszü thought about me and why he could not talk to me. He gave me a long answer with many words that I did not understand, and, at the end, he victoriously announced, "Now you know." Yet I did not, and I kept wondering whether my pig friend's seemingly affectionate love was the same as my feelings for him. Perhaps my scientist knew the answer, but I did not understand the words he used. This was the beginning of my problem with words used in scientific explanations.

My childhood curiosity has never evaporated. I became a scientist as a consequence of striving to understand the true meaning behind explanations. Too often, what my peers understood to be a logical and straightforward answer remained a mystery to me. I had difficulty comprehending gravity in high school. OK, it is an "action at a distance" or a force that attracts a body having mass toward another physical body. But are these statements not just another way of saying the same thing? My physics teacher's answer about gravity reminded me the explanation of Rüszü's abilities given by "my" scientist. My troubles with explanatory terms only deepened during my medical student and postdoctoral years after I realized that my dear mentor, Endre Grastyán, and my postdoctoral advisor, Cornelius (Case) Vanderwolf, shared my frustration. Too often, when we do not understand something, we make up a word or two and pretend that those words solve the mystery.[4]

Scientists started the twenty-first century with a new goal: understanding ourselves and the complexity of the hardware that supports our individual minds. All of a sudden, neuroscience emerged from long obscurity into everyday language. New programs have sprung up around the world. The BRAIN Initiative in the United States put big money into public–private collaborations aimed at developing powerful new tools to peek into the workings of the brain. In Europe, the Human Brain Project promises to construct a model of the

---

4. These may be called "filler terms," which may not explain anything; when used often enough in scientific writing, the innocent reader may believe that they actually refer to a mechanism (e.g., Krakauer et al., 2017).

human brain—perhaps an overambitious goal—in the next decade. The main targets of the China Brain Project are the mechanisms of cognition and brain diseases, as well as advancing information technology and artificial intelligence projects.

Strikingly, none of these programs makes a priority of understanding the general principles of brain function. That decision may be tactically wise because discoveries of novel principles of the brain require decades of maturation and distillation of ideas. Unlike physics, which has great theories and is constantly in search for new tools to test them, neuroscience is still in its infancy, searching for the right questions. It is a bit like the Wild West, full of unknowns, where a single individual has the same chance to find the gold of knowledge as an industry-scale institution. Yet big ideas and guiding frameworks are much needed, especially when such large programs are outlined. Large-scale, top-down, coordinated mega projects should carefully explore whether their resources are being spent in the most efficient manner. When the BRAIN Initiative is finished, we may end up with extraordinary tools that will be used to make increasingly precise measurements of the same problems if we fail to train the new generation of neuroscientists to think big and synthesize.

Science is not just the art of measuring the world and converting it into equations. It is not simply a body of facts but a gloriously imperfect interpretation of their relationships. Facts and observations become scientific knowledge when their widest implications are explored with great breadth and force of thinking. While we all acknowledge that empirical research stands on a foundation of measurement, these observations need to be organized into coherent theories to allow further progress. Major scientific insights are typically declared to be important discoveries only later in history, progressively acquiring credibility after scrutiny by the community and after novel experiments support the theory and refute competing ones. Science is an iterative and recursive endeavor, a human activity. Recognizing and synthesizing critical insights takes time and effort. This is as true today as it has always been. A fundamental goal in neuroscience is identifying the principles of neuronal circuit organization. My conviction about the importance of this goal was my major motivation for writing this volume.

Writing a book requires an internal urge, an itching feeling that can be suppressed temporarily with distraction but always returns. That itch for me began a while ago. Upon receiving the Cortical Discoverer Prize from the Cajal Club (American Association of Anatomists) in 2001, I was invited to write a review for a prominent journal. I thought that the best way to exploit this opportunity was to write an essay about my problems with scientific terms and argue that our current framework in neuroscience may not be on the right track. A month later arrived the rejection letter: "Dear Gyuri, . . . I hope you understand that for the *sake of the journal* we cannot publish your manuscript" [emphasis added]. I did not understand the connection between the content of my essay and the reputation of the journal. What was at stake? I called up my great supporter and crisis advisor, Theodore (Ted) Bullock at the University of California at San Diego, who listened carefully and told me to take a deep breath, put the issue on the back burner, and go back to the lab. I complied.

Yet, the issues kept bugging me. Over the years, I read as much as I could find on the connection between language and scientific thinking. I learned that many of "my original ideas" had been considered already, often in great detail and depth, by numerous scientists and philosophers, although those ideas have not effectively penetrated psychology or neuroscience. Today's neuroscience is full of subjective explanations that often rephrase but do not really expound the roots of a problem. As I tried to uncover the origins of widely used neuroscience terms, I traveled deeper and deeper into the history of thinking about the mind and the brain. Most of the terms that form the basis of today's cognitive neuroscience were constructed long before we knew anything about the brain, yet we somehow have never questioned their validity. As a result, human-concocted terms continue to influence modern research on brain mechanisms. I have not sought disagreement for its own sake; instead, I came slowly and reluctantly to the realization that the general practice in large areas of neuroscience follows a misguided philosophy. Recognizing this problem is important because the narratives we use to describe the world shape the way we design our experiments and interpret what we find. Yet another reason I spent so many hours thinking about the contents of this book is because I believe that observations held privately by small groups of specialists, no matter how remarkable, are not really scientific knowledge. Ideas become real only when they are explained to educated people free of preconceived notions who can question and dispute those ideas. Gergonne's definition, cited at the beginning of this Preface, is a high bar. Neuroscience is a complex subject. Scientists involved in everyday research are extremely cautious when it comes to simplification—and for good reasons. Simplification often comes at the expense of glossing over depth and the crucial details that make one scientific theory different from another. In research papers, scientists write to other scientists in language that is comprehensible to perhaps a handful of readers. But experimental findings from the laboratory gain their power only when they are understood by people outside the trade.

Why is it difficult for scientists to write in simple language? One reason is because we are part of a community where every statement and idea should be credited to fellow scientists. Professional science writers have the luxury of borrowing ideas from anyone, combining them in unexpected ways, simplifying and illuminating them with attractive metaphors, and packaging them in a mesmerizing narrative. They can do this without hesitation because the audience is aware that the author is a smart storyteller and not the maker of the discoveries. However, when scientists follow such a path, it is hard to distinguish, both for them and the audience, whether the beautiful insights and earthshaking ideas were sparked from their own brains or from other hard-working colleagues. We cannot easily change hats for convenience and be storytellers, arbitrators, and involved, opinionated players at the same time because we may mislead the audience. This tension is likely reflected by the material presented in this book. The topics I understand best are inevitably written more densely, despite my best efforts. Several chapters, on the other hand, discuss topics that I do not directly study. I had to read extensively in those areas, think about them hard, simplify the ideas, and weave them into a coherent picture. I hope that, despite the unavoidable but perhaps excusable complexity here and there, most ideas are graspable.

The core argument of this book is that the brain is a self-organized system with preexisting connectivity and dynamics whose main job is to generate actions and to examine and predict the consequences of those actions. This view—I refer to it as the "inside-out" strategy—is a departure from the dominant framework of mainstream neuroscience, which suggests that the brain's task is to perceive and represent the world, process information, and decide how to respond to it in an "outside-in" manner. In the pages ahead, I highlight the fundamental differences between these two frameworks. Many arguments that I present have been around for quite some time and have been discussed by outstanding thinkers, although not in the context of contemporary neuroscience. My goal is to combine these ideas in one place, dedicating several chapters to discussing the merits of my recommended inside-out treatment of the brain.

Many extraordinary findings have surfaced in neuroscience over the past few decades. Synthesizing these discoveries so that we can see the forest beyond the trees and presenting them to readers is a challenge that requires skills most scientists do not have. To help meet that challenge, I took advantage of a dual format in this volume. The main text is meant to convey the cardinal message to an intelligent and curious person with a passion or at least respect for science. Expecting that the expert reader may want more, I expand on these topics in footnotes. I also use the footnotes to link to the relevant literature and occasionally for clarification. In keeping with the gold standards of scientific writing, I cite the first relevant paper on the topic and a comprehensive review whenever possible. When different aspects of the same problems are discussed by multiple papers, I attempt to list the most relevant ones.

Obviously, a lot of subjectivity and unwarranted ignorance goes into such a choice. Although I attempted to reach a balance between summarizing large chunks of work by many and crediting the deserving researchers, I am aware that I did not always succeed. I apologize to those whose work I may have ignored or missed. My goals were to find simplicity amid complexity and create a readable narrative without appearing oversimplistic. I hope that I reached this goal at least in a few places, although I am aware that I often failed. The latter outcome is not tragic, as failure is what scientists experience every day in the lab. Resilience to failure, humiliation, and rejection are the most important ingredients of a scientific career.

ACKNOWLEDGMENTS

No story takes place in a vacuum. My intellectual development owes a great deal to my mentor Endre Grastyán and to my postdoctoral advisor Cornelius (Case) Vanderwolf. I am also indebted to the many students and postdoctoral fellows who have worked with me and inspired me throughout the years.[5] Their success in science and life is my constant source of happiness. Without their dedication, hard work, and creativity, *The Brain from Inside Out* would not exist simply because there would not have been much to write about. Many fun discussions I had with them found their way into my writing.

I thank the outstanding colleagues who collaborated with me on various projects relating to the topic of this book, especially Costas Anastassiou, László Acsády, Yehezkel Ben-Ari, Antal Berényi, Reginald Bickford, Anders Björklund, Anatol Bragin, Ted Bullock, Carina Curto, Gábor Czéh, János Czopf, Orrin Devinsky, Werner Doyle, Andreas Draguhn, Eduardo Eidelberg, Jerome (Pete) Engel, Bruce McEwen, Tamás Freund, Karl Friston, Fred (Rusty) Gage, Ferenc Gallyas, Helmut Haas, Vladimir Itskov, Kai Kaila, Anita Kamondi, Eric Kandel, Lóránd Kellényi, Richard Kempter, Rustem Khazipov, Thomas Klausberger, Christof Koch, John Kubie, Stan Leung, John Lisman, Michael Long, Nikos Logothetis, András Lőrincz, Anli Liu, Attila Losonczy, Jeff Magee, Drew Maurer, Hannah Monyer, Bruce McNaughton, Richard Miles, István Mody, Edvard Moser, Lucas Parra, David Redish, Marc Raichle, John Rinzel, Mark Schnitzer, Fernando Lopes da Silva, Wolf Singer, Ivan Soltész, Fritz Sommer, Péter Somogyi, Mircea Steriade, Imre Szirmai, Jim Tepper, Roger Traub, Richard Tsien, Ken Wise, Xiao-Jing Wang, Euisik Yoon, László Záborszky, Hongkui Zeng, and Michaël Zugaro.

---

5. See them here https://neurotree.org/beta/tree.php?pid=5038.

Although I read extensively in various fields of science, books were not my only source of ideas and inspiration. Monthly lunches with Rodolfo Llinás, my admired colleague at New York University, over the past five years have enabled pleasurable exchanges of our respective views on everything from world politics to consciousness. Although our debates were occasionally fierce, we always departed in a good mood, ready for the next round.

I would also like to thank my outstanding previous and present colleagues at Rutgers University and New York University for their support. More generally, I would like to express my gratitude to a number of people whose examples, encouragement, and criticism have served as constant reminders of the wonderful collegiality of our profession: David Amaral, Per Andersen, Albert-László Barabási, Carol Barnes, April Benasich, Alain Berthoz, Brian Bland, Alex Borbely, Jan Born, Michael Brecht, Jan Bures, Patricia Churchland, Chiara Cirelli, Francis Crick, Winfried Denk, Gerry Edelman, Howard Eichenbaum, Andre Fenton, Steve Fox, Loren Frank, Mel Goodale, Katalin Gothard, Charlie Gray, Ann Graybiel, Jim McGaugh, Michale Fee, Gord Fishell, Mark Gluck, Michael Häusser, Walter Heiligenberg, Bob Isaacson, Michael Kahana, George Karmos, James Knierim, Bob Knight, Nancy Kopell, Gilles Laurent, Joe LeDoux, Joe Martinez, Helen Mayberg, David McCormick, Jim McGaugh, Mayank Mehta, Sherry Mizumori, May-Britt Moser, Tony Movshon, Robert Muller, Lynn Nadel, Zoltán Nusser, John O'Keefe, Denis Paré, Liset de la Prida, Alain Prochiantz, Marc Raichle, Pasko Rakic, Jim Ranck, Chuck Ribak, Dima Rinberg, Helen Scharfmann, Menahem Segal, Terry Sejnowski, László Seress, Alcino Silva, Bob Sloviter, David Smith, Larry Squire, Wendy Suzuki, Karel Svoboda, Gábor Tamás, David Tank, Jim Tepper, Alex Thomson, Susumu Tonegawa, Giulio Tononi, Matt Wilson, and Menno Witter. There are many more people who are important to me, and I wish I had the space to thank them all. Not all of these outstanding colleagues agree with my views, but we all agree that debate is the engine of progress. Novel truths are foes to old ones.

Without the freedom provided by my friend and colleague Dick Tsien, I would never have gotten started with this enterprise. Tamás Freund generously hosted me in his laboratory in Budapest, where I could peacefully focus on the problem of space and time as related to the brain. Going to scientific events, especially in far-away places, has a hidden advantage: long flights and waiting for connections in airports, with no worries about telephone calls, emails, review solicitations, or other distractions. Just read, think, and write. I wrote the bulk of this book on such trips. On two occasions, I missed my plane because I got deeply immersed in a difficult topic and neglected to respect the cruelty of time.

But my real thanks go to the generous souls who, at the expense of their own time, read and improved the various chapters, offered an invaluable mix of encouragement and criticism, saved me from error, or provided crucial insights and pointers to papers and books that I was not aware of. I am deeply appreciative of their support: László Acsády, László Barabási, Jimena Canales, George Dragoi, Janina Ferbinteanu, Tibor Koos, Dan Levenstein, Joe LeDoux, Andrew Maurer, Sam McKenzie, Lynn Nadel, Liset Menendez de la Prida, Adrien Peyrache, Marcus Raichle, Dima Rinberg, David Schneider, Jean-Jacque Slotine, Alcino Silva, and Ivan Soltész.

Sandra Aamodt was an extremely efficient and helpful supervisor of earlier versions of the text. Thanks to her experienced eyes and language skills, the readability of the manuscript improved tremendously. I would also like to acknowledge the able assistance of Elena Nikaronova for her artistic and expert help with figures and Aimee Chow for compiling the reference list.

Craig Panner, my editor at Oxford University Press, has been wonderful; I am grateful to him for shepherding me through this process. It was rewarding to have a pro like him on my side. I owe him a large debt of thanks.

Finally, and above all, I would like to express my eternal love and gratitude to my wife, Veronika Solt, for her constant support and encouragement and to my lovely daughters, Lili and Hanna, whose existence continues to make my life worthwhile.

1

The Problem

> The confusions which occupy us arise when language is like an engine idling, not when it is doing work.
> —Ludwig Wittgenstein[1]

> There is nothing so absurd that it cannot be believed as truth if repeated often enough.
> —William James[2]

> A dream we dream together is reality.
> —Burning Man 2017

The mystery is always in the middle. I learned this wisdom early as a course instructor in the Medical School of the University at Pécs, Hungary. In my neurophysiology seminars, I enthusiastically explained how the brain interacts with the body and the surrounding world. Sensory stimuli are transduced to electricity in the peripheral sensors, which then transmit impulses to the midbrain and primary sensory cortices and subsequently induce sensation. Conversely, on the motor arc, the direct cortical

---

1. "Philosophy is a battle against the bewitchment of our intelligence by means of our language" (Wittgenstein: Philosophical Investigations, 1973). See also Quine et al. (2013). "NeuroNotes" throughout this book aim to remind us how creativity and mental disease are intertwined (Andreasen, 1987; Kéri, 2009; Power et al., 2015; Oscoz-Irurozqui and Ortuño, 2016). [NeuroNote: Wittgenstein, a son of one of Austria's wealthiest families, was severely depressed. Three of his four brothers committed suicide; Gottlieb, 2009].

2. http://libertytree.ca/quotes/William.James.Quote.7EE1.

pathway from the large pyramidal cells of the primary motor cortex and multiple indirect pathways converge on the anterior horn motor neurons of the spinal cord, whose firing induces muscular contraction. There was a long list of anatomical details and biophysical properties of neurons that the curriculum demanded the students to memorize and the instructors to explain them. I was good at entertaining my students with the details, preparing them to answer the exam questions, and engaging them in solving mini-problems. Yet a minority of them—I should say the clever ones—were rarely satisfied with my textbook stories. "Where in the brain does perception occur?" and "What initiates my finger movement, you know, before the large pyramidal cells get fired?"—were the typical questions. "In the prefrontal cortex" was my usual answer, before I skillfully changed the subject or used a few Latin terms that nobody really understood but that sounded scientific enough so that my authoritative-appearing explanations temporarily satisfied them. My inability to give mechanistic and logical answers to these legitimate questions has haunted me ever since—as it likely does every self-respecting neuroscientist.[3] How do I explain something that I do not understand? Over the years, I realized that the problem is not unique to me. Many of my colleagues—whether they admit it or not—feel the same way. One reason is that the brain is complicated stuff and our science is still in its young phase, facing many unknowns. And most of the unknowns, the true mysteries of the brain, are hidden in the middle, far from the brain's sensory analyzers and motor effectors. Historically, research on the brain has been working its way in from the outside, hoping that such systematic exploration will take us some day to the middle and on through the middle to the output. I often wondered whether this is the right or the only way to succeed, and I wrote this book to offer a complementary strategy.

In this introductory chapter, I attempt to explain where I see the stumbling blocks and briefly summarize my alternative views. As will be evident throughout the following chapters, I do believe in the framework I propose. Some of my colleagues will side with me; others won't. This is, of course, expected whenever the frontiers of science are discussed and challenged, and I want to state this clearly up front. This book is not to explain the understood but is instead an invitation to think about the most fascinating problems humankind can address. An adventure into the unknown: us.

---

3. The term "neuroscientist" was introduced in 1969 when the Society for Neuroscience was founded in the United States.

ORIGIN OF TODAY'S FRAMEWORK IN NEUROSCIENCE

Scientific interest in the brain began with the epistemological problem of how the mind learns the "truth" and understands the veridical, objective world. Historically, investigations of the brain have moved from introspection to experimentation, and, along this journey, investigators have created numerous terms to express individual views. Philosophers and psychologists started this detective investigation by asking how our sensors—eyes, ears, and nose—sense the world "out there," and how they convey its features to our minds. The crux of the problem lies right here. Early thinkers, such as Aristotle, unintentionally assumed a dual role for the mind: making up both the *explanandum* (the thing to-be-explained) and providing the *explanans* (the things that explain).[4] They imagined things, gave them names, and now, millennia later, we search for neural mechanisms that might relate to their dreamed-up ideas.[5]

As new ideas about the mind were conceived, the list of things to be explained kept increasing, resulting in a progressive redivision of the brain's real estate. As a first attempt, or misstep, Franz Joseph Gall and his nineteenth-century followers claimed that our various mental faculties are localized in distinct brain areas and that these areas could be identified by the bumps and uneven geography of the skull—a practice that became known as *phrenology* (Figure 1.1). Gall suggested that the brain can be divided into separate "organs," which we would call "regions" today. Nineteen of the arbitrarily divided regions were responsible for faculties shared with other animals, such as reproduction, memory of things, and time. The remaining eight regions were specific to humans, like the sense of metaphysics, poetry, satire, and religion.[6] Today, phrenology is ridiculed as pseudoscience because we know that bumps on the skull have very little to do with the shape and regionalization of the brain. Gall represented to neuroscience what Jean-Baptiste Lamarck represented to

---

4. Aristotle (1908). We often commit similar mistakes of duality in neuroscience. To explain our results, we build a "realistic" computational model to convince ourselves and others that the model represents closely and reliably the "thing-to-be-explained." At the same time, the model also serves to explain the biological problem.

5. A concise introduction to this topic is by Vanderwolf (2007). See also Bullock (1970).

6. Gall's attempt to find homes for alleged functions in the body was not the first one. In Buddhism, especially in Kundalini yoga, "psychological centers" have been distributed throughout the entire body, known as chakras or "wheels." These levels are the genitals (energy), the navel (fire, insatiable power), heart (imaginary of art, dreams), the larynx (purification), mystic inwardly looking eye (authority), and the crown of the head (thoughts and feelings). The different levels are coordinated by the spine, representing a coiled serpent, in a harmonious rhythmic fashion. See also Jones et al. (2018).

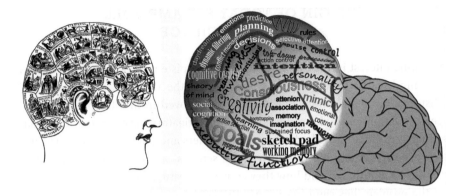

Figure 1.1. A: Franz Joseph Gall and his phrenologist followers believed that our various mental faculties are localized in distinct brain areas that could be identified by the bumps and uneven geography of the skull. Phrenology (or cranioscopy) is ridiculed as pseudoscience today. B: Imaging-based localization of our alleged cognitive faculties today. I found more than 100 cognitive terms associated with the prefrontal cortex alone, some of which are shown here.

evolution. A reminder that being wrong does not mean being useless in science. Surprisingly, very few have complained about the more serious nonsense, which is trying to find "boxes" in the brain for human-invented terms and concepts. This strategy itself is a bigger crime than its poor implementation—the failure to find the right regions.

There Are Too Many Notes

"There are simply too many notes. Just cut a few and it will be perfect," said the emperor to the young Mozart. While this was a ludicrous line in the movie *Amadeus*, it may be a useful message today in cognitive neuroscience jargon. There is simply not enough space in the brain for the many terms that have accumulated about the mind prior to brain research (Figure 1.1). Anyone versed in neuroanatomy can tell you that Korbinian Brodmann's monumental work on the cytoarchitectural organization of the cerebral cortex distinguished 52 areas in the human brain. Many investigators inferred that differences in intrinsic anatomical patterns amount to functional specialization. Contemporary methods using multimodal magnetic resonance imaging (MRI) have identified 180 cortical areas bounded by relatively sharp changes in cortical architecture, connectivity, and/or topography. Does this mean that now we have many more potential boxes in the brain to be equated with our preconceived ideas?[7] But even with this recent expansion of brain regions, there are still many more human-concocted terms than there are cortical areas. As I discuss later in the book, cognitive functions usually arise from a relationship between regions rather than from local activity in isolated areas. But even if we accept that interregional interactions are more important than computation in individual areas, searching for correspondences between a dreamed-up term and brain activity cannot be the right strategy for understanding the brain.

THE PHILOSOPHICAL ROOTS

If cognitive psychology has a birthdate, it would be 1890, when *The Principles of Psychology,* written by the American philosopher and psychologist William James, was published.[8] His treatment of the mind–world connection (the "stream of consciousness") had a huge impact on avant garde art and literature, and the influence of his *magnum opus* on cognitive science and today's neuroscience is hard to overestimate. While each of its chapters was an extraordinary contribution at the time, topics discussed therein are now familiar and acceptable to us. Just look at the table of contents of that 1890 work:

Volume 1
Chapter IV—Habit
Chapter VI —The mind-stuff theory
Chapter IX— The stream of thought
Chapter X—The consciousness of self
Chapter XI—Attention
Chapter XII—Conception
Chapter XIII—Discrimination and comparison
Chapter XIV—Association
Chapter XV—Perception of time
Chapter XVI—Memory

Volume 2
Chapter XVII—Sensation
Chapter XVIII—Imagination

---

7. Brodmann (1909); Glasser et al. (2016). For light reading about a recent quantitative cranioscopy analysis, see Parker Jones et al. (2018).

8. James (1890).

Chapter XIX—The perception of "things"
Chapter XX—The perception of space
Chapter XXI—The perception of reality
Chapter XXII—Reasoning
Chapter XXIII—The production of movement
Chapter XXIV—Instinct
Chapter XXV—The emotions
Chapter XXVI—Will

Over the years, these terms and concepts assumed their own lives, began to appear real, and became the de facto terminology of cognitive psychology and, subsequently, cognitive neuroscience.

Our Inherited Neuroscience Vocabulary

When neuroscience entered the scene in the twentieth century, it unconditionally adopted James's terms and formulated a program to find a home in the brain for each of them (e.g., in human imaging experiments) and to identify their underlying neuronal mechanisms (e.g., in neurophysiology). This strategy has continued to this day. The overwhelming majority of neuroscientists can pick one of the items from James's table of contents and declare, "this is the problem I am interested in, and I am trying to figure out its brain mechanisms." Notice, though, that this research program—finding correspondences between assumed mental constructs and the physical brain areas that are "responsible" for them—is not fundamentally different from phrenology. The difference is that instead of searching for correlations between mind-related terms and the skull landscape, today we deploy high-tech methods to collect information on the firing patterns of single neurons, neuronal connections, population interactions, gene expression, changes in functional MRI (fMRI), and other sophisticated variables. Yet the basic philosophy has remained the same: to explain how human-constructed ideas relate to brain activity.

Let's pause a bit here. How on earth do we expect that the terms in our dreamed-up vocabulary, concocted by enthusiastic predecessors hundreds (or even thousands) of years before we knew anything about the brain, map onto brain mechanisms with similar boundaries? Neuroscience, especially its cognitive branch, has become a victim of this inherited framework, a captive of its ancient nomenclature. We continue to refer to made-up words and concepts and look for their places in the brain by lesion studies, imaging, and other methods. The alleged boundaries between predetermined concepts guide the search, rather than the relationships between interactive brain processes. We identify hotspots of activity in the brain by some criteria, correlate them with James's and others' codified categories, and mark those places in the brain as their residences. To drive home my point with some sarcasm, I call this approach "neo-phrenology."[9] In both phrenology and its contemporary version, we take ideas and look for brain mechanisms that can explain those ideas. Throughout this book, I refer to this practice as the "outside-in" strategy, as opposed to my alternative "inside-out" framework (Chapters 3, 5 and 13).

That brings me to the second needed correction in the direction of modern neuroscience: the thing-to-be-explained should be the activities of the brain, not the invented terms. After all, the brain gives rise to behavior and cognition. The brain should be treated as an independent variable because behavior and cognition depend on brain activity, not the other way around. Yet, in our current research, we take terms like emotions, memory, and planning as the to-be-explained independent categories. We will address these issues head on when we discuss the role of correlations and causation in neuroscience in Chapter 2.

The Outside-In Program

How has the outside-in program come to dominate modern neuroscientific thought? We can find clues in James's Table of Contents. Most of the titles have something to do with sensation and perception, which we can call the inputs to the brain coming from the outside. This is not by chance; William James and most early psychologists were strongly influenced by British empiricism, which in turn is based on Christian philosophies. Under the empiricist framework, knowledge derives from the (human) brain's ability to perceive and interpret the objective world. According to the empiricist philosopher David Hume,[10] all our knowledge arises from perceptual associations and inductive reasoning to reveal cause-and-effect relationships.[11] Nothing can exist in the mind without first being detected by the senses because the mind is built out of sensations of the physical world. He listed three principles of association: resemblance, contiguity in time and place, and causation. For example, reading a poem can make you think of a related poem. Recalling one thing (cause) leads another thing to

---

9. Poldrack (2010) also sees a parallel between phrenology and the new fMRI mapping approach. However, his suggested program for a "search for selective associations" still remains within the outside-in framework.

10. [NeuroNote: Hume had several "nervous breakdown" episodes in his life.]

11. Associational theories assume that sequential events, such as our memories and action plans, are built by chaining associated pairs of events (*a* with *b, b* with *c*, etc). However, neuronal sequences (Chapter 7) contain important higher order links as well.

be remembered (effect). Western science and its methods are based largely on the inductive methodology that descended from Hume's perception-centered philosophy.

In the empiricism-inspired model, signals enter the brain from the outside, neuronal circuits process and perceive them, and some part of the brain decides whether or not to generate a motor response. The main emphasis is on perceptual processing and association, which are believed to be the main drivers of the brain's ability to comprehend and represent the outside world. In John Locke's words: "Ideas are clear, when they are such as the Objects themselves from whence they were taken."[12] Locke and his followers assumed that our sensors provide veridical descriptions of the objective and, I would add, observer-independent, world. An inevitable consequence of this perception-centered view is the assumption of an intelligent *homunculus* (i.e., a conscious selector, or, in modern-day neurospeak, "decision-maker").[13] Indirectly, it also leads to the thorny problem of a free will that links perception to action. When "consciousness" is mentioned in experimental neuroscience circles, eyebrows are still raised today. "Decision-making," on the other hand, has become a ubiquitous, colloquial term. This is interesting since decision-making is defined as a process of collecting pertinent information, identifying alternatives, estimating the consequences, weighing the evidence, making choices, and taking actions.[14] This definition is virtually identical to Thomas Aquinas's philosophical formulation of free will, except that his freedom of the self is granted to humans only, and they must make good choices (i.e., those that God likes).[15] In contrast, making decisions is the one forbidden thing in Indian Buddhism.[16]

Because of the empiricist emphasis on sensory inputs and stimulus associations as the sole source of knowledge, it is not surprising that James dedicated only a few pages to action in his book. In this "outside-in" framework, the

---

12. Locke (1690); as quoted in Wootton (2015).

13. Several other related terms, such as "supervisor" and "executive" or "top-down functions," refer to the same fundamental idea.

14. Gold and Shadlen (2007); Kable and Glimcher (2009); Shadlen and Kiani (2013).

15. MacDonald (1998). More recently, decision-making has been popularized as a computer science metaphor, which regards past, present, and the future as separate entities. In this framework, memories belong to the past and decisions are made in the "now," with the expectation of benefits in the future (see Chapter 10).

16. In far Eastern philosophies, the concept of "freedom" (Chinese *tzu-yu*; Japanese *jiyu*) refers to liberation from the human nexus or being away from others. More recently, decision-making has been popularized as a computer science metaphor, one that regards past, present, and the future as separate entities. In this framework, memories belong to the past and decisions are made in the "now" with the expectation of benefits in the future (see Chapter 10).

brain is fundamentally a passive device whose main job is to perceive, evaluate the sensory inputs, weight their importance, and then—and only then—decide whether and how to act. As you will see, I disagree with this perception-action "outside-in" model (Chapters 3 and 5).[17]

Following the empiricists' influence forward in time, we find that other important streams of thought in the nineteenth and twentieth centuries also adopted the outside-in approach. German Gestalt psychology, which has its own epistemological roots, also focused mainly on perceptual issues. *Cognitivism*, an offshoot of Gestalt psychology, was primarily interested in mental representation. It emphasized the input side of the brain and downplayed the importance of studying the output, although reaction time was often used to probe cognition.

The influence of the outside-in empiricist framework is also evident in Russian psychology, led by Ivan Petrovich Pavlov, even though such a connection has never been acknowledged in official Soviet scripts.[18] Pavlov regarded the brain as an associative device. Learning in his dogs required only pairing of the conditioned and unconditioned signals, such as sound and food. As a result of repeated associations between them, the previously ineffective sound of the metronome comes to generate salivation (i.e., the conditioned reflex), which previously only the food was able to elicit (the unconditioned reflex). According to Pavlov, these associations are made automatically, which is why he called them *reflexes*. His dogs were constrained by a harness for practical reasons. Their task was to attend and associate, but no behavior was needed except to salivate and eat the food. Pavlov's brain theory of classical conditioning is known as *stimulus substitution*; in essence, it assumes that, after conditioning the animal, the brain treats the conditioned signal in the same way as the unconditioned signal. Really? Is a dog not smart enough to understand that the metronome is not palatable?[19]

---

17. The current debate on consciousness also originates from the "brain as an input-processor" model (Dennett, 1991). According to Joaquín Fuster, "All goal directed behavior is performed within the broad context of the perception-action cycle" (Fuster, 2004). The explicit formulation of the perception–action cycle can be traced back to Jakob von Uexküll ("sensory-motor function circle"), who also coined the term *umwelt* (i.e., perceptual world). However, in contrast to the empiricist associationism, von Uexküll emphasized that the animal modifies its *Umwelt* (i.e., the surrounding world in which an organism exists and acts). von Uexküll introduced 75 new terms to explain his views (von Uexküll, 1934/2011). Situated, embodied cognition (Beer, 1990; Brooks, 1991; Noë, 2004), as an alternative for the associational models, grew out from this framework (Clark and Chalmers, 1998).

18. Pavlovian conditioning and the related dogma of imposed effects on the passively associating brain were a perfect fit for the ideology of the totalitarian Soviet system.

19. This short paragraph is an unfair treatment of the enormous value of the Pavlovian conditioning paradigm for the selfish aim of getting to my point quickly. Classical conditioning

Most contemporary computational models of brain function also fall under the outside-in framework (Chapter 13). Of these, perhaps the most influential is the model worked out by the British neuroscientist and thinker David Marr. He suggested that investigation of brain function should follow a three-level sequence of stages. First is a computational stage that defines the problem from an information processing perspective. The second, algorithmic stage should ask what representations the brain uses and how it manipulates those representations. Only after the first two stages are established can we move to the implementational stage; that is, how the algorithmic-mathematical model is solved at the level of synapses, neurons, and neural circuits.[20] I respectfully disagree with Marr's strategy. My first problem is the definition of "information processing" (see later discussion). Second, ample evidence is available to demonstrate that numerous algorithmic models are compatible with a given computational problem, but only one or a few are compatible with those implemented by neuronal circuits.[21] Third, learning, the most fundamental aspect of the brain, is largely missing from his outside-in analysis.[22] In contrast to the computational-algorithmic-implementation strategy, I argue throughout this book that understanding brain function should begin with brain mechanisms and explore how those mechanisms give rise to the performance we refer to as perception, action, emotion, and cognitive function.

Stimulus, Signal, and Reinforcement

A transient challenge to the outside-in framework was *behaviorism*. John B. Watson, the founder of behaviorism, attempted to position psychology as "a purely objective experimental branch of natural science" whose goal is the "prediction and control of behavior." His follower, B. F. Skinner, developed the

---

produced a plethora of interesting and important observations, including higher order conditioning, extinction, stimulus generalization and discrimination, associative strength, conditioned suppression, blocking, masking, latent learning, etc. The mathematical formulation of Rescorla and Wagner (1972) can account for many of the observations and yielded quantitative fits to a variety of experimental data. Yet stimulus substitution as an explanation for conditioning has persisted to date. Even Watson agreed: "The [conditioned stimulus] now becomes a substitute stimulus—it will call out the [response] whenever it stimulates the subject" (Watson, 1930; p. 21).

20. Marr (1982).

21. Prinz et al. (2004). In fact, it is hard to find an impactful neuroscience experiment which is based on Marr's suggested three-stage strategy.

22. As pointed out by Tomaso Poggio in his *Afterword* for the 2010 addition of Marr's *Vision* book.

theory of operant conditioning and suggested that behavior is solely determined by its consequences, depending on whether behavior leads to reward or punishment. Like Pavlov, the utopian views of Skinner advocated the *tabula rasa* or blank slate view of the brain, which can be shaped in any direction with appropriately dosed reinforcement. Although behaviorism departed from the assumption of passive associations and emphasized the importance of quantifiable motor behavior, it treated the brain as a "black box," which may explain why cognitive neuroscience did not embrace its philosophy and practice until recently.[23]

To explain why certain objects are more attractive than others, behaviorists introduced a hypothetical concept they called a *reinforcer* as an attempt to replace the homunculus decision-maker with something more tangible. Instead of the brain-resident homunculus or the unconditioned stimulus in Pavlovian conditioning, the selection is now made by the reinforcing positive and negative values of the reward and punishment, respectively. This is also an outside-in framework in the sense that explanation of behavior depends on outside factors.

Stimulus, signal, and reinforcer typically mean very different things to an experimenter. But do brains make such distinctions as well? Inputs from the environment and body arrive to the brain through the sensory interfaces. We often call these inputs "stimuli" or "signals." Both terms imply a significant relationship between events outside the brain and the brain's response to them. The term "stimulus" means that it exerts some change in brain activity because it must stimulate it. The term "signal" implies that the brain regards its change in response to the stimulus as important because it signals something useful to it. Thus, both the stimulus and signal refer to a relationship between the outside world and brain activity. Conversely, the absence of such relationship implies that the input does not exist from the point of view of the brain. Myriads of things in the surrounding world and our body remain unnoticed by our brains either because we do not have sensors for them (e.g., radio waves) or simply because they have no momentary significance (e.g., the T-shirt I am wearing while writing these sentences). There is nothing inherent in the physical things or events that would qualify them a stimulus or signal. Only when they are noticed by the brain do they become signals.

Now let's turn to reinforcement. In the parlance of behaviorist literature, the reinforcer refers to a special quality in a stimulus that brings about a change in the animal's behavior. Relatedly, reinforcement is the mechanism that increases or decreases the probability of recurrence of actions that lead to the positive

---

23. Watson (1930); Skinner (1938).

(pleasant) or negative (unpleasant) reinforcers, respectively. To register the occurrence of the reinforcer, the brain must perceive it. Thus, the reinforcing aspect of an input can be communicated only through sensory channels. Again, whether something is a reinforcer or not reflects a relationship between "things out there" and brain activity. The reinforcing quality is not inherent in any object or event. A chocolate may be a positive reinforcer, but if someone is forced to eat many pieces, it soon becomes aversive. If something is not noticed by the brain, it is irrelevant from the brain's point of view; it does not bring about a change in neuronal activity.

The reader may have noticed that the content of the preceding two paragraphs is identical, even though they seemingly refer to different things: signal and reinforcer. Perhaps they mean the same thing. Some inputs induce large changes in behavior while most others bring about small or unnoticeable effects. The reinforcer or reward is just another regular stimulus that acts through the exact same sensory channels (smell, taste, or pain) as the stimuli, but it may exert a stronger impact on brain activity. Viewed in this way, signals and reinforcers represent a continuous distribution on a wide scale in which the two ends of the distribution are designated as positive or negative and those close to the middle as neutral signals. I devote two chapters to the discussion of broad distributions of many things in the brain and discuss that often the discrete nature of words simply reflects large quantitative differences (Chapters 12 and 13).[24]

SHORTCOMINGS OF THE OUTSIDE-IN APPROACH

Ever since John von Neumann's comparison between the computer and the brain, computer metaphors have dominated the language of neuroscience. A metaphor is a powerful tool to convey an idea because it relates a mysterious phenomenon to an understood one. However, metaphors can also be misleading because they may give a false sense of understanding a novel phenomenon before it is actually known how the thing works. In the brain–computer metaphor, neuronal operations are likened to the machines' ability to transduce a

---

24. What determines the magnitude of the brain's reaction to a stimulus? Over the past decades, the term "reinforcer" got separated from the signal and acquired a new meaning. In contemporary neuroscience, it is identified with *dopamine*, a neuromodulator produced in the ventral tegmental area and substantia nigra. An entire new field with strong links to computational neuroscience has emerged, known as "reinforcement learning," with dopamine as the reinforcer or supervisor that assigns value to external stimuli. How stimuli can mobilize the dopaminergic neurons to alter and magnify the value of stimuli post hoc, the so-called "credit assignment problem," is currently strongly debated. Reinforcement-learning algorithms are inspired by behaviorism (Sutton and Barto, 1998; Schultz, 1998, 2015).

real-world thing into symbols, called *representations*.[25] By this logic, symbols in the brain—the firing patterns of neurons, for example—should correspond to some real-world stimuli in the same way that computer algorithms correspond to, or represent, inputs from the objective world. However, unlike in the computer, the same stimuli often induce highly variable responses in the brain depending on brain states and testing conditions[26] (Chapter 6).

We often pretend that we know which features interest the brain, and correlate them with neuronal activity. Such arbitrariness is appropriate and, in fact, essential for symbolic representations by machines because they are trained to produce a "best match" between the experimenter-chosen features and symbols.[27] But brains are different. Representation of the objective world and finding truth are not their main jobs. The brain evolved not to represent anything but to help its host body to survive and reproduce. In addition, brains produce many effects that do not correspond to anything tangible in the physical world. Pleasure and fear, desire, irrational numbers or the entire discipline of math, and even Santa Claus feel real without corresponding to something veridical outside the brain.[28]

According to the outside-in framework, the best strategy to understand mechanisms of perception and cognition is to present various stimuli (e.g., an object or aspects of an object) and examine the spatiotemporal distribution of the evoked neuronal responses in the brain (Figure 1.2).[29] To increase the rigor of the discovery process, initially "simple" stimuli are presented, such as vertical and horizontal gratings for vision or beeps of particular frequencies for hearing research. After the data are collected and interpreted, researchers can search for

---

25. *Representation* is a Platonic concept, referring to the idea that we humans cannot comprehend the real world (Kant's *noumenon*) but only a representative of it (*phenomenon*). Both in classical Cartesian dualism and Lockean empiricism, the mind relies on the representation of *noumena* (Skarda and Freeman, 1987; Berrios, 2018).

26. A similar distinction was made between representation and meaning by Walter Freeman in his book *How Brains Make Up Their Minds* (1999). See also Eliasmith and Anderson (2003).

27. A digital camera can reliably "represent" what it "sees" in a different format (bits), but such information has no meaning to the camera. The many smart programs associated with picture compression, face recognition, and segmentation can reveal many important and useful statistical relationships about the inputs. In fact, by knowing the inputs and the computed results, one can effectively figure out the algorithms and even the electronic circuits involved. But this strategy often fails in the investigation of the brain (Chapter 3).

28. James (1890); Milner (1996); Damasio and Damasio (1994); LeDoux (2014, 2015).

29. Engel et al. (2001); Hebb (1949); James (1890); Milner (1996); von der Malsburg (1994); Hubel and Wiesel (1962,1974); Rieke et al. (1997).

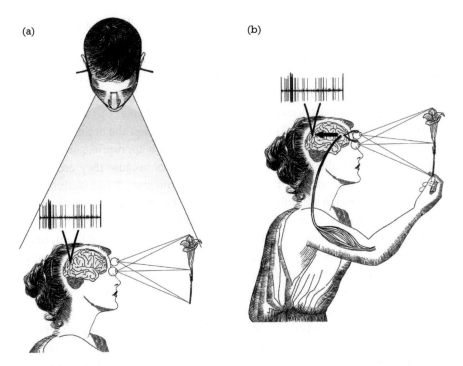

Figure 1.2. Outside-in and inside-out strategies. A: Outside-in. A stimulus is presented (flower) and responses (e.g., action potentials of a single neurons) are recorded from the sensory cortex. The experimenter has access to both the stimulus and the neuronal responses and can establish a reliable relationship between the input stimulus and neuronal activity. However, neurons in the sensory cortex do not "see" the stimulus. All they can register is action potential–induced inputs from other upstream neurons. B: Inside-out. The meaning of the stimulus can be grounded by comparing it something else. In addition to inputs from the retina, sensory neurons also receive input from movement-producing neurons in the motor area (called a "reafference" signal; discussed in detail in Chapter 3). Whenever a motor command is sent to the eye muscles and the arm, sensory neurons are informed about such outputs (*arrow*). By moving the eyes or moving the flower by hand, the visual neurons can get a "second opinion" about the flower. These action-induced signals are critical for providing meaning to the stimulus.

neuronal responses to increasingly complex stimuli. This strategy anticipates that (eventually) one should be able to explain how an object is represented in the brain from the observed relationship between the stimuli and the neuronal responses.

This research program was pioneered by the legendary scientific duo, David Hubel and Thorsten Wiesel, the Nobel Prize winners in Physiology and Medicine in 1981, who introduced single-neuron recordings to study the visual system. They made a long series of extraordinary discoveries that continue to influence sensory research today. The theoretical framework of a passively associating brain was a welcome strategy to experimenters because neuronal responses to sensory stimuli could be registered also in anesthetized or paralyzed animals. When experiments were subsequently performed in waking animals, their heads and even the eyes were fixed to eliminate inconvenient behavior-induced variability in the recorded brain signals. However, building a complex percept in the brain from its assumed elementary components turned out to be a formidable hurdle given the astronomical realm of possible combinations.[30] Luckily, it turned out that natural visual scenes and sounds represent only a small subset of all possible combinations, so sensory researchers recently began to use natural images and sounds. They discovered that brains are "fine-tuned" for naturally occurring stimuli, and thus many aspects of neural activity were missed with the previous artificial stimuli. For example, even in a small brain, like that of a frog, transmission of acoustic information is most efficient when the spectrum of sounds matches the spectrum of natural frog calls.[31] But the nature of the stimulus is only a technical problem. The bigger issue is that the brain is not a passive device that unconditionally associates signals.

A tacit assumption of the outside-in[32] framework is that attributes or features of objects are bound together by some brain mechanism. What makes an object an entity is the coherent fusion of its components, which separates it from its surroundings. For example, a car has a particular shape with a body and wheels, color, sound, and a characteristic movement pattern. But there is a problem with this philosophy, which is that the "attributes" of an object are not a fixed set that reside in the object but are constructed in the brain of the observer. Brain responses do not intrinsically "represent" stimuli until the experimenter interprets the data.[33] Judgments of "similar" or "different" are subjective (i.e., observer-dependent). Yes, an experimenter can quantitatively define the relationship between stimuli and the spatiotemporal firing patterns of recorded neurons (Figure 1.2). However, this relationship does not mean that neurons in

---

30. An illuminating review on how methods influence scientific thinking is that by Evarts (1973). I often wondered how many more recordings, monkeys, and cats would be needed to explain vision with the outside-in strategy, even if all 30,000 neuroscientists were working on the visual system.

31. Attneave (1954); Singh and Theunissen (2003); Wang et al. (1995); Rieke et al. (1995).

32. Related terms to the "outside-in" framework are "empiricist," "associational," and "representational" strategies, and their various combinations. *Principles of Neural Science*, a 1,700-page "bible" of neuroscience and the most widely used textbook for educating neuroscientists, also follows the outside-in framework (Kandel et al., 2012).

33. Skarda and Freeman (1987); Werner (1988).

the brain of the experimental subject views the stimuli in the same way as the experimenter, nor does it mean that the brain uses these recorded signals in the same way or at all. This is the crux of the problem of the outside-in correlational approach. It cannot yield adequate understanding of perception or neuronal computation. I briefly summarize the outline of this argument here and return to it in more detail in Chapter 3.

ONLY ACTION CAN PROVIDE MEANING TO BRAIN ACTIVITY

Like the philosophical issues that plagued physics through the twentieth century, the issue in neuroscience is the "observer." There is a fundamental difference between an observer outside the brain, such as the experimenter, and an observer inside the brain, such as Descartes's hypothetical homunculus residing in the pineal gland.

First, let's take the perspective of an experimenter studying the visual system. Upon presentation of a visual stimulus, an image is converted to action potentials by retinal ganglion cells and conveyed to the primary visual cortex via neurons in the thalamic lateral geniculate body. The experimenter can place a recording device, such as an electrode array, inside the brain and monitor neural responses to different images presented to the eye. In this situation, the experimenter has a privileged point of view: he has access to both the images (external world) and the neuronal responses in the brain (Figure 1.2). From this perspective, he can find a reliable correlation between specific features of the presented images (input; the rose in Figure 1.2) and neuronal responses (output for present purposes), which allows him to make a guess about the possible input–output transformation rules. After establishing a correlation (often mistakenly called a *neuronal code*), he can recover stimulus properties from the neuronal responses and even predict the properties of untested stimulus patterns (e.g., another flower). In short, a particular constellation of neuronal spike patterns *becomes* information. But to whom is that information useful? The recorded signal in the brain, combined with a priori knowledge about the stimulus, can generate meaning only to the experimenter. The observed correlation between the image and neuronal response only means that something outside the brain was transformed into firing pattern changes.

What about the internal observers, the brain circuits? The neurons that "respond" to the image do not see the image, just patterns of neural spiking arriving from their peer neurons. Neurons in the visual cortex are blind to events that happen in the world and so is the rest of the brain or even a magical homunculus (i.e., the little human) watching the action of all visual cortical neurons

(Figure 1.2). Imagine that the little human is the experimenter who, in a magic experiment, can register the activity of all neurons in the brain. Without access to information outside the skull, such knowledge cannot reveal what the brain is sensing; it cannot see the rose. In the language of the neural code, the outside-in framework has shown only that decoding of stimulus properties from neural activity is possible *in principle* and only *if an observer has the code book* (i.e., the set of stimulus–response correlations). However, the brain only has its own neural "responses," and the outside-in framework does not mention how such a code would be generated or read by neural responses alone.

All any neuron in the brain ever "sees" is that some change occurred in the firing patterns of its upstream peers. It cannot sense whether such change is caused by an external perturbation or by the brain's perpetual self-organized activity (Chapters 7 and 8). Thus, neurons embedded in networks of other neurons do not "know" what the brain's sensors are sensing[34]; they simply respond to their upstream inputs. In other words, unlike our ideal experimenter, the neurons have no way of relating or *comparing* their spikes to anything else because they only receive retinal correspondences or processed "representations" of the sensory input. But establishing correspondences without knowing the rules by which those correspondences are constructed is like comparing Mansi words with Khanty words when we understand neither language.[35] Only after we have defined the vocabulary of one language can we understand the corresponding meaning of words in the other. Similarly, without further information, sensory neurons can attach no meaning whatsoever to their spikes. Put simply, the mind's eye is blind.[36]

To explain my point differently, let me invite you to do a thought experiment. Let's connect the output of a video camera (or the optic nerve of the retina) to cultured neurons in a dish, derived from the visual cortex of the smartest person in the world, so that a fraction of the neurons is stimulated by the

---

34. A classic paper on this topic is by Lettvin et al. (1959). See also Eggermont (2007).

35. This is not to say that everything is equal to a newborn brain. Phylogenetic experience can provide the necessary cipher for certain patterns, such as human faces, for babies. But, for most of the things in the world, we have to create a cipher through action-based grounding.

36. This "blindness," in my view, applies also to the Global Workspace version (Baars, 1988; Dehaene and Changeux, 2011) of the Cartesian theater metaphor of consciousness, since the Global Workspace, even it if involves distributed processing in the entire brain, faces the exact same problem: having no access to the external information. Realizing the shortcomings of the Cartesian theater idea, Dennett and Kinsbourne (1992) suggested an alternative solution, what they call the "Multiple Drafts" model, reminiscent of E. Roy John's distributed consciousness model (1976). However, this model is also a "passive" interpreter and has no access to outside world information and, therefore, faces the same "grounding problem" (Harnard, 1990).

camera output. The part of the tissue culture that is stimulated by the camera output can be called the *sensory region*. We may discover that repeated visual patterns, such as the picture of the rose, will induce somewhat similar neuronal responses. We may even detect some changes in synaptic connectivity among neurons as a result of repetition. It is also possible, though, that some members of the cultured neuronal network will be preoccupied with each other and ignore the stimulation altogether.[37] The experimenter can establish a relationship between the applied stimuli and the corresponding neuronal responses and declare that some coding was identified. But the tissue culture does not "see" the rose. So far, this is analogous to the passive, representational brain model and the experimental approach used in many laboratories. Since the neurons in the culture have no direct access to the outside world, they have no way of verifying or "grounding" their patterns to the events outside. The term "grounding" refers to the ability of the brain's circuits to connect it to something meaningful for the owner of the brain. It's kind of a second opinion from a more reliable source.[38] Using our dictionary metaphor, if you know the meaning of words in an English dictionary, you can ground the meaning of corresponding words in a dictionary of any other language.

The situation in the dish may change dramatically if the spiking output of some neurons is connected to a robot that can move the camera based on the spiking pattern of these experimenter-designated "motor" neurons in the culture. Now neurons in the dish have two functions. One function is generating an output (move the camera) and the second is responding to the camera-supplied signals (sense the inputs). Because the tissue culture circuit is supposedly well interconnected, neurons in the "sensory" region will receive two types of inputs: input from the camera signals and an additional input from their motor neuron partners, which inform them that it was their action that induced movement of the camera and, therefore, changes in the sensory input. Now, we have a closed-loop

---

37. It helps if the tissue culture is "awakened" by adding a cocktail of subcortical neurotransmitters, mimicking the aroused state of the brain (Hinard et al., 2012).

38. Steve Harnad (1990) explains the need of "symbol grounding" in understanding. An effective computer algorithm can discover correspondences between two sets of symbols without understanding the meaning of any of them. Handheld translators have enormous vocabularies and can convert commonly spoken words and phrases to dozens of other languages at an amazing speed and even pronounce the words in the target language—all done by a few chips and a microprocessor. However, the semantic "meaning" of symbols of one set (e.g., a new language) can be constructed only if the translator's knowledge is grounded to an already known set of symbols (e.g., your mother tongue). Without grounding, a theorem remains disconnected from empirical phenomena. See also Fields (2014). Grounding is also related to the "credit assignment problem"; that is, the determination of how collective action of neurons relates to behavior.

system in a dish, an output-driven sensory feedback device. The addition of the robot provided the cultured neurons with the same advantage as the experimenter: now the neurons can sense both the outside word and the internal computation and, therefore, can compare their joint impact.[39]

Now let's assign a goal to our closed-loop system—for example, to find the rose in a picture. Whenever the camera is focused on the rose by the chance behavior of the cultured neurons, a magic potion (let's call it dopamine) is spritzed onto the culture to strengthen the connections between those neurons that successfully moved the camera to the desired spot in the visual scene and the neurons that were activated by the rose. This simple modification ("plasticity") may increase the probability that such neuronal activity will happen again and lead to pointing the camera at the rose more often.[40] Thus, out of the large realm of possible neuronal network patterns, at least one unique event acquired a meaning. By connecting a bunch of otherwise useless tissue culture neurons with the world through an action-perception arc, we have just built a machine with a purpose: a brain-like device.[41] In fact, we may not need a magic potion if we assume that the brain is already prewired to some extent so that its connectivity is not random but guided by some evolution-shaped statistical rule (Chapter 11). The internal feedback from the action-inducing neurons to sensory neurons may be regarded as a second opinion, the needed grounding signal that can attach meaning to input-induced neuronal responses (Chapter 3).

---

39. For an attempt to achieve such comparison by neurons, see Demarse et al. (2001). A computational model using such building blocks is by Weng (2004); Choe et al. (2007).

40. Artificial intelligence (AI) devices using cameras give equal priority to all image pixels at the earliest stage of processing. In contrast, the visual system prioritizes and isolates the most relevant information based on previous knowledge and moves the eyes to attend those aspects (Olshausen et al., 1993). Inspired by the brain's solution, recent AI architectures serially take glimpses of the input image and update their internal state representations before sampling the next location. The goal of AI agents is to select the best outcome that results in maximal "reward" (Mnih et al., 2014). The main success of AI agents, in, for example, defeating human players in Atari games, is that those games are also designed by humans using artificial algorithms. The current instantiations of AI agents are still pretty helpless when it comes to developing and solving temporally extended planning strategies.

41. William Grey Walter at the Burden Neurological Institute in Bristol, England built an electronic "tortoise," among the first autonomous robots, which he called *Machina speculatrix*. His *Machina* had only two "neurons" (vacuum tubes) but could be assigned a goal, which was to find its way to a recharging station when it ran low on battery power. Walter (1950) argued that goal-directed action is a main characteristic of animals and that animated gadgets with a goal, such as his robots, can emulate such behaviors. A similar argument was made subsequently by Brooks (1991): brains have no representations. Instead, "it is a collection of competing behaviors. Out of the local chaos of their interactions there emerges, in the eyes of an observer, a coherent pattern of behavior."

In our tissue culture example, we arbitrarily designated two sets of neurons and called them *sensory input* to where the stimuli from the camera were delivered and *motor output* because these neurons were connected to the robot moving the camera. The tissue culture has no a priori bias toward these designations, and the two areas are reciprocally interconnected to each other. In real brains, the sensory and motor areas have somewhat different internal connectivity but the similarities are more striking than the differences. The differences are even more blurred when we compare their partner structures, often referred to as *higher order sensory* and *supplementary motor areas*.[42] These areas are reciprocally connected, and the traffic moves not only from the sensory to the motor circuits but equally so from the motor to the sensory areas. This latter motor-to-sensory projection has a fundamental impact on many levels of brain operations, as we will see in the subsequent chapters.

The picture I just painted is unfortunately an oversimplification. I did not mean to create the impression that our tissue culture thought experiment can explain the mechanisms of a thinking, feeling brain. It was only meant to illustrate the minimum necessary requirements of a brain-like system. Random connections and plasticity are not enough (Chapter 12). A large part of brain structure is genetically determined, and self-organized activity is as important for the development of brain dynamics as is sensory stimulation even in primary sensory areas of the brain[43] (Chapter 5).

THE INSIDE-OUT, READER-CENTRIC FRAMEWORK

Can James's Table of Contents and the outside-in strategy be renegotiated? I do not think it would get us far. My point is not that measuring brain responses

---

42. Mrganka Sur and colleagues (Sharma et al., 2000) "rewired" the brain of ferrets so that the visual information was connected to the auditory system. Neurons in the auditory cortex of these animals responded in many ways like the visual cortex of intact animals, including visual orientation modules. Impressively, the ferrets used their newly wired brain to avoid visual objects.

43. Brain circuits can assemble themselves under genetic guidance to provide "proto-maps." When the thalamic input to the neocortex is disrupted in Gbx-2 mutant mice, neocortical region-specific gene expression and the main cortical layers and divisions develop normally (Miyashita-Lin et al., 1999). Deletion of a single protein, Munc18-1, in mice abolishes neurotransmitter secretion from synaptic vesicles throughout development. However, this does not prevent an apparently normal initial assembly of the brain. However, persistent absence of synaptic activity leads to synapse degeneration and the mice die (Verhage et al., 2000). The constellation of ion channels in individual neurons is also driven by their activity.

to stimuli yields no valuable data or insight. In fact, a primary slogan in my laboratory is that "it is impossible to find nothing" in the brain. An enormous amount of precious knowledge has been obtained by identifying neuronal "correspondences" and "representations" of the sensory world over the past decades. Until recently, the empiricist outside-in approach has generated most of the available knowledge about sensory coding, and it continues to be an important strategy. However, when studying the more central parts of the brain involved in cognition, the limitations of a strict stimulus–response strategy become apparent.[44] The ultimate goal of brain research cannot just be to uncover correspondences between external signals and neuronal responses. Building a dictionary of representations from scratch would be an enormous effort, but it is easy to see why such a program has limited success (Chapter 4). The outside-in approach interrogates brain circuits by asking what those circuits can *potentially* do, rather than what they actually do, which produces an illusory understanding of brain computation. Overall, I suggest that searching for alternative approaches is warranted.

The essence of the complementary strategy I suggest is simple: understand the brain from within. In this inside-out framework, the key issue is the mechanisms by which stimuli and situations become meaningful percepts and experiences for the brain.[45] The main emphasis is on how the brain's outputs, reflected by the animal's actions, influence incoming signals. By linking an otherwise meaningless brain pattern to action, that pattern can gain meaning and significance to the organism. Within the brain, the emphasis is on how downstream networks make use of the messages obtained from their upstream partners. In this framework, the goal of the brain is to explore the world and register the consequences of successful exploratory actions to improve the efficacy of future actions. Thus, an action–perception loop learns to make sense of sensory inputs. Perception is what we do (Chapter 3).[46]

44. Rieke et al. (1997); Friston (2010, 2012).

45. The outside-in versus inside-out dichotomy should not be confused with the top-down and bottom-up distinctions. The former refers to the relationship between the world and the brain, whereas the latter refers mainly to the spread of activity in anatomical space. It is also distinct from Paul MacLean's (1970) concept of the *triune brain*, a division into reptilian, paleomammalian, and neo-mammalian brain. The relationship of the inside-out framework to embodied cognition and predictive coding will be addressed in Chapters 3 and 5.

46. In my younger days, I thought these ideas were solely mine. But, with time, I discovered that many people, often independently from each other, have come to a similar conclusion (e.g., Merleau-Ponty, 1945/2005; Mackay, 1967; Bach-y-Rita, 1983; Held and Hein, 1983; Paillard, 1991; Varela et al., 1991; Berthoz, 1997; Bialek et al., 1999; Järvilehto, 1999; O'Regan and Noë, 2001; Llinás, 2002; Noë, 2004, 2009; Choe et al., 2007; Chemero, 2009; Scharnowski et al.,

Nearly all chapters of this book revolve around an action-centered brain because I want to make it as clear as possible that no meaning or advantage emerges for the brain without the ability to calibrate neural patterns by behavior-induced consequences. Without generating outputs to move and optimize the sensors, and calibrate brain circuits at some point in our lives, I claim, no perception or cognition exists.[47]

A major advantage of inside-out approach is that it is free of philosophical connotations. It takes brain mechanisms as independent variables, as opposed to attempting to find correspondences between subjectively derived categories and brain responses. Brains largely organize themselves, rather than being enslaved to input signals. I am interested in how internally generated, self-organized patterns in the brain acquire "meaning" through action, which becomes what we call "experience." Instead of looking for correlations between experimenter-selected signals and activity patterns of neurons, we should ask a more important question: How do those firing patterns affect downstream "reader" neurons? We can call this strategy the *neural observer*, or *reader-centric*, view. Before you accuse me of simply replacing the homunculus with the "reader," please consult Chapters 4 and 5. The term "reader/observer" is meant to be an engineering term, referring to an actuator (such as a robot, muscle, or simply a downstream neuron) that responds to a particular pattern of upstream neuronal activity but not to others; only a particular key of a bunch on a keychain opens a particular lock. In the subsequent chapters, I will give you a taste of how this is done in the brain by discussing a set of recent findings that illustrate progress toward an inside-out research program.

2013). Interaction between the body and mind was natural to the Greek philosophers. Aristotle might have said that we learn about the world by our actions on it. Only after the concept of the soul fused with the mind came the full separation, especially with Descartes's declaration of the independence of *res extensa* (corporeal substance or the external stuff) and the God-given *res cogitans* (the thinking thing or soul). Perhaps Henri Poincaré was the first thinker who speculated that the only way sensations become relevant and turn into an experience is by relating them to the body and its movements. "To localize an object simply means to represent oneself the movements that would be necessary to reach it" (Poincaré, 1905; p. 47). The volume by Prinz et al. (2013) is an excellent update on the renewed interest in action science. Tolman's cognitive map theory (1948) is another important and major departure from the outside-in framework but it is not action-based.

47. Once a brain pattern acquires meaning (i.e., it has been calibrated through action), it remains meaningful as long as memory is intact, without further action (Chapters 5 and 8). Learning through action also applies to species knowledge acquired through evolutionary selection (Watson and Szathmáry, 2016).

BLANK PAPER OR PREEXISTING CONSTRAINTS?

The choice of outside-in and inside-out, action-centric frameworks also shapes our ideas about fundamental brain operations. One of the oldest outside-in views was formulated by Aristotle, who suggested that we are born with a blank slate (or *tabula rasa*) on which experiences are written.[48] The *tabula rasa* view is an almost inevitable assumption of the outside-in framework because, under the empiricist view, the goal of the brain is to learn and represent the truth with its veridical details. This view has influenced thinking in Christian and Persian philosophies, British empiricism, and the Marxist doctrine, and it has become the leading school of thought in cognitive and social sciences.[49] Similar to the free will idea that humans can do anything, *tabula rasa* thinking implies that anything can be written on a blank brain. Although I have never met a neuroscientist colleague who would openly subscribe to the *tabula rasa* view, many experiments and modeling studies today are still performed according to this associationist philosophy-based framework, or its modern-day version: "connectionism." The main idea of connectionism, especially in cognitive psychology and artificial intelligence research,[50] is that perception and cognition can be transcribed into networks of simple and uniform neurons, which are interconnected relatively randomly with largely similar synaptic connections. Nothing is further from the facts, as I explain in Chapter 12 and briefly summarize here.

Early behavioral observations already argued against the idea of the brain as a blank slate. Researchers demonstrated repeatedly that animals do not associate everything equally and cannot be trained to do all tricks the experimenter expects them to do. Behaviors that relate to the animal's ecological niche can be trained easily because the brain is predisposed or "prepared' to do things that have survival and reproductive advantage. For example, "spontaneous alternation," the tendency in rodents to choose different paths during foraging, is an instance of biological preparedness for the rapid acquisition

---

48. The *tabula rasa* concept was more explicitly developed by the Persian Muslim Avicenna (Abu Ali al-Hussain Ibn Sina). Avicenna was an important precursor of British empiricists. His syllogistic, deductive method of reasoning influenced Francis Bacon, John Locke, and John Stuart Mill and continues to influence scientific reasoning even today. Steven Pinker's *Blank Slate* (2003) is an excellent overview on this old philosophical problem.

49. Read, for example, Popper (1959).

50. The classic works on connectionist networks have been summarized in Rumelhart et al. (1986) and McClelland et al. (1986). The much more sophisticated recent neuronal networks also fall under the *tabula rasa* framework. The leading author on connectionism in cognitive psychology is Donald Hebb (1949).

of species-specific learning. Returning to the same location for food within a limited time window is not an efficient strategy because choosing an alternate route will more likely lead to reward. In contrast, associations that would be detrimental to survival are called "contraprepared." For example, it is virtually impossible to train a rat to rear on its hindlimbs to avoid an unpleasant electric shock to the feet since rearing is an exploratory action and incompatible with the hiding and freezing behaviors deployed in case of danger.[51] A practical message here is that if training takes weeks or longer, the experimenter should seriously consider that the brain signals associated with such extended shaping of behavior may not reveal much about the intended question because each animal may use different tricks to solve the task, and these may remain hidden for the experimenter.

Neuroscience has also accumulated much experimental evidence against the *tabula rasa* model. The most important is the recognition that a great deal of activity in the brain is self-organized instead of being driven by outside signals. This self-generated persistent activity is supported by the numerous local and brain-wide neuronal rhythms, which we will discuss in Chapter 6. For now, it is sufficient to note that these oscillations are not only part of and help to stabilize neuronal dynamics but also offer a substrate for syntactical organization of neuronal messages. Each period of an oscillation can be conceived as a frame that contains a particular constellation of spiking neurons. Metaphorically, we can call it a "neuronal letter." In turn, the cycles of the many simultaneously acting oscillations can concatenate neuronal letters to compose neural words and sentences in a virtually infinite number of ways. This brain syntax is the main focus of the discussion in Chapters 6 and 7.

---

51. Each of these telegraphic statements represents large chapters of cognitive and behavioral psychology. Skinner believed that any behavior could be shaped by his method of "successive approximations" and was able to teach rodents and other animals to do some extraordinary things. However, his students (Breland and Breland, 1961) countered that animals often misbehave and develop "autoshaped" or "superstitious rituals" (e.g., Catania and Cutts, 1963). The literature on autoshaping is a rich source of information for the expected surprises that today's users of the increasingly popular "virtual reality" tasks will have to face (Buzsáki, 1982). The prepared and contraprepared spectrum of behaviors, developed by Martin Seligman, had a great influence on human phobia research (Seligman, 1971, 1975). Phobias (e.g., to spiders, snakes, and heights) can be viewed as a prominent prepared category because they are commonly associated with objects and situations that have threatened us throughout our evolutionary history. Joe LeDoux, the leading authority on threat behavior, amplifies this view further by explaining that brain circuits that detect and respond to threats in rodents are not necessarily the same as those that produce conscious fear in humans (LeDoux, 2015).

Wide-Range Distributions: Good Enough and Precise Solutions

As an alternative to the empiricist outside-in *tabula rasa* view, I raise the possibility that the brain already starts out as a nonsensical dictionary. It comes with evolutionarily preserved, preconfigured internal syntactical rules that can generate a huge repertoire of neuronal patterns. These patterns are regarded as initially nonsense neuronal words which can acquire meaning through experience. Under this hypothesis, learning does not create brand new brain activity patterns from scratch but is instead a "fitting process" of the experience onto a preexisting neuronal pattern.

The preconfigured brain is supported by a mechanism I refer to as "high-diversity" organization (Chapters 12 and 13). Diversity and large variations are the norm in the brain, whether we are analyzing its microscopic or macroscopic connectivity or dealing with its dynamics at a small or a large scale. Neuronal firing rates, connection strength between neurons, and the magnitude of their concerted action can vary by three to four orders of magnitude. I speculate that this diversity, which respects a logarithmic rule (Chapter 12), is the essential backbone that provides stability, resilience, and robustness to brain networks against competing requirements for wide dynamic range, plasticity, and redundancy. There are many advantages of a preconfigured brain[52] over the blank slate model, the most important of which is stability of brain dynamics because adding new experience after each learning event does not perturb much the overall state of neuronal networks. I will discuss that even the inexperienced brain has a huge reservoir of unique neuronal patterns (Chapter 13), each of which has the potential to acquire significance through the experience of unique events or situations.

The tails of these wide-range distributions, such as neurons with high and slow firing rates and strong and weak synaptic connections, can support seemingly opposing functions, such as judging a situation as familiar or novel. Neural circuits involving a minority of highly interactive and well-connected neurons may appear as a constraint. However, the benefit of such apparent constraint is that the collective action of such oligarchic circuits with skewed distribution of resources allows the brain to generalize across situations and offer its "best guess" under any circumstances, so that no situation is regarded

---

52. "Preconfigured" usually means experience-independent. The backbone of brain connectivity and its emerging dynamics are genetically defined (Chapter 12). In a broader sense, the term "preconfigured" or "preexisting" is also often used to refer to a brain with an existing knowledge base, such as mine at this instant of time after numerous years of action-based calibration.

as completely unknown. I call this aspect of neuronal organization the "good-enough" brain. The good-enough brain uses only a small fraction of its resources, but its member neurons are always on the alert. On the other hand, the remaining majority of slow-firing member neurons with weak synaptic connections have complementary properties, such as high plasticity and slow but precise and reliable processing. This other extreme of network organization comprises the "precision" brain, whose main job is to redefine a situation or an event and determine its distinctness in high detail. The "good enough" and "precision" aspects of brain organization complement, and in fact, assume, each other (Chapters 12 and 13).

In the preconfigured brain model, learning is a *matching* process, in which preexisting neuronal patterns, initially nonsensical to the organism, acquire meaning with the help of experience. As discussed earlier, attaching meaning or significance to any neuronal pattern requires grounding, which can be provided only by action, the ultimate source of knowledge (Chapter 3). In contrast to the *tabula rasa* model, in which the complexity of neuronal dynamics scales with the amount of experience, in the preconfigured brain network homeostasis is only minimally and transiently affected by new learning because nothing is added to the circuitry; only rearrangements take place. The total magnitude of plasticity in brain circuits and the overall spiking activity remain constant over time, at least in the healthy adult brain (Chapter 12). I make no claim that my strategy is definitely right, but the alternative formulation that I am suggesting might provide different insights than the current representational framework.

In contrast to the preconfigured brain, the outside-in framework assumes that there are obvious things in the outside world, which we often call information, and the brain's job is to incorporate or absorb that information. But how?

THE BRAIN "CODES INFORMATION": OR DOES IT?

This often-heard statement should communicate a lot of content to everyone. After all, we live in an Information Age in which everyone is familiar with such terms as "information" and "coding." But when we attempt to explain and define information coding in the brain, we pause because we do not have a widely accepted, disciplined definition. When terms are not defined precisely, they often acquire many disguised interpretations and, therefore, convey different meanings to different individuals. "Is my grant proposal rejected or approved?" From the perspective of information theory, the answer represents a single bit of information (yes or no).⁵³ But for the receiver of such information (me), it induces profound and long-lasting changes in my brain.

While the mathematical formalism of information theory is neutral to information "content," the same cannot be said for the brain. For a behaving organism, "decoding" information cannot be divorced from "meaning": the experience-based prediction of potential consequences of that information. There is no such thing as invariant information content in a stimulus.⁵⁴ "Features" or attributes of a stimulus are not objective physical characteristics of the stimulus. Instead, the relevance of the stimulus depends on changing internal states and the brain's history with similar situations. In the following chapters, I hope to show you that this situational relevance is not a small matter but has a profound and observable effect on the nature of neural activity and its relationship to the sensory environment. The same visual stimulus (e.g., the rejection letter) evokes very different neuronal patterns in a frog's brain, my brain, or the brains of other humans. The induced pattern even in my brain depends on its current state. If the letter arrived the same week when I received the news about other rejections, it would induce a different effect compared to an alternative situation when the letter arrived a day after I won the lottery jackpot.

---

53. Once agreed by the sender and receiver agents, everything can be represented by symbols; for example, the head or tail outcome of coin flipping by 1 and 0. Each flip produces one bit of information, which can be concatenated by some agreed rules. In the American Standard Code for Information Interchange (ASCII), seven bits represents a character; for example, 1100101 is code for the letter "e." Each block representing a single letter is separated by an agreed SPACE code (0100000). As in Morse code or any other coding scheme, spaces are the essential part of packaging messages (Chapter 6). Without space codes, the information may become completely undecodable. Physics uses information as a conceptual tool to understand interactions and relationships. Information can be mathematically defined as the negative of the logarithm of the probability distribution ($I = \log_2 N$). When $n = 2$, information I is 1, thus the unit of information is a choice between two possibilities, measured in "bits" (Shannon, 1948). The *meaning* of the observed events does not matter in this definition of information.

54. I accept that one solution would be to stick to the framework of Shannonian theory or entropy (Shannon, 1948) and use it to define relationships between world events and brain events. A connection to information theory is *free energy* that quantifies the amount of prediction error or surprise. *Prediction error* is the difference between the representations encoded at one level and the top-down predictions generated by the brain's internal model: a comparison. More formally, it is a variational approximation to negative log likelihood of some data given an internal model of those data (Friston, 2010). However, in this definition, we should abandon the wider use of information as it is typically applied in neuroscience.

Computation performed by computer programs or machines is referred to, in general, as "information processing."[55] In reality, the information is not in the processing. It *becomes* information only when interpreted by an observer that recognizes its meaning and significance, be it a human interpreter or a mechanical actuator. A computer program can effectively control a robot or other man-made machines, thus giving rise to the illusion that information resides in the program. However, the process is based on a human-designed solution even if it involves a complex trial-and-error learning process, including "deep learning" in artificial intelligence programs.[56] In short, information is not inherent in the computation (in machines or brains) but becomes such when it is interpreted.[57]

In neuroscience, information is often understood as acquired knowledge, a broader concept than in information theory. Claude Shannon warned us that information theory is "aimed in a very specific direction, a direction that is not necessarily relevant to such fields as psychology."[58] In the forthcoming chapters, I discuss why the brain is not an information-absorbing, perpetual coding device, as it is often portrayed, but a venture-seeking explorer, an action-obsessed agent constantly controlling the body's actuators and sensors to test its hypotheses. The brain ceaselessly interacts with, rather than just detects, the external world in order to remain itself. It is through such exploration that correlations and interactions acquire meaning and become information (Chapter 5). Brains do not process information: they create it.

Now let me try to give a broad definition of coding as it applies to the brain. In essence, coding is an agreement for communication between the sender and receiver, making the content of the encoded information a secret to outsiders.

---

55. The term "processing" has an interpretative connotation. It refers to a procedure that takes something and turns it into something else, a sort of format transformation. This is what is usually referred to as "representation." The often used "information-processing" expression relies on this view and tacitly implies that representation is the final product. However, representation is not a thing; it is a process or an event (Brette, 2015, 2017). Therefore, representation (and perception) cannot be described as a transformation of one thing into another thing. See also Jeannerod (1994).

56. Deep learning (or hierarchical deep machine learning or deep structured learning) is part of a broader family of machine-learning algorithms that attempts to model high-level abstractions in data by using a "deep graph" with multiple processing layers, akin to the associational networks sandwiched between sensory and motor areas in the brain. For the origin of this highly successful and rapidly expanding field, see Hinton et al. (1995) and Sejnowski (2018).

57. The thoughtful paper by Chiel and Beer (1997) outlines a broad program for brain-inspired robot design. See also König and Luksch (1998).

58. Shannon (1956).

If the code is not known to the receiver, the packets carry no inherent information (Chapter 7). There are many forms of coding (e.g., Morse code or the genetic code), but the fundamental features remain the same. The encoding–decoding process is done via a cipher,[59] a transforming tool that encrypts messages by mutually agreed syntactical rules, so the information looks like nonsense to the uninitiated. Encryption is not always intentional, of course; sometimes it just happens. Evolution has no aim of secrecy, yet biological codes are often mysterious to us—that is, until we uncover the cipher. In the outside-in philosophy, the words in James's list are the senders and the brain is the receiver. However, terms such as "emotion," "will," "attention," and others are not senders of anything and possess no inherent information. Space and time are not sensed directly by the brain either (Chapter 10), and no messages are being sent by them. If we want to understand information, we should start with its creator: the brain.

CODE-BREAKING EXERCISES

The American writer, Edgar Allen Poe, whose hobby was cryptography, believed that "human ingenuity cannot concoct a cipher which human ingenuity cannot resolve." But modern science claims that it is possible to encrypt messages that no human or machine can break.[60] Does the cipher of brain dynamics fall into this highly secured category, or can it be outsmarted?

As we search for ciphers in the brain, we may learn from previous deciphering successes. The Englishman Thomas Young and the Frenchman Jean-François Champollion are credited with deciphering Egyptian hieroglyphics by using a written cipher: the Rosetta Stone. This most famous artifact of the British Museum contains the same script in three languages (Greek, Demotic, and hieroglyphics). As a polymath, Young's strategy was to find correspondences of well-known names of gods and royalties, such as Ptolemy and Cleopatra, between the words in the Greek text and the glyphs. This approach provided fast initial progress but could not uncover the generative rules of hieroglyphic language. That task needed Champollion's genius and his understanding of syntactic

---

59. From the Arabic *sifr*, meaning "zip" or "blank." Deciphering is the process of breaking the secret (i.e., decoding). After reading extensively about ciphers, I realized that brain research badly needs such concepts. Yet very little specific research has tried to uncover and understand such mechanisms, even though we pay extensive lip service to "breaking the brain's code."

60. Poe (1841). An accessible review on the current state of ciphers is by Eckert and Renner (2014). The key is to make choices that are independent of everything preexisting and are hence unpredictable: a device-independent cryptography.

rules in Hebrew, Syriac, Greek, Arabic, Aramaic, Persian, and, critically, Coptic and Demotic. He hypothesized that phonetic signs were not just used for the names of royals but also represented a general rule. This insight suggested that Egyptian hieroglyphs could be read and understood by transforming the glyphs to sounds and concatenating them.[61] Thus, the nonsensical messages of a forgotten language became information. However, without the Rosetta stone, the secrets of the Egyptian hieroglyphs might well have remained forever hidden.

My second example of a cipher is connected to the most celebrated (posthumously) code breaker, Alan Turing, a genius by many accounts. He and his team at Bletchley Park, north of London, are credited with breaking the code of the secret messages of the German Navy in World War II.[62] With all my admiration and respect for Turing's contributions, I must make a correction: neither he nor his team "broke the code." They received the cipher, the key syntactical tool of encryption, in the form of Enigma machines,[63] from the Polish Cipher Bureau early during the war.[64] Without this access to the syntactical rules of the cipher, the secret messages sent to the German U-boats by the Kriegsmarine Headquarters would likely have remained gobbledygook to the allies. Alan Turing's Rosetta Stone was the Enigma's hardware.

Is deciphering a secret code possible without (some) a priori knowledge of the syntax? Our third example, discovery of the genetic code and the translation of DNA to RNA to protein, was such a case. The cipher in this case was the double spiral. The two strands of the helix consist of polynucleotides made up from simpler units of nucleotides. Each nucleotide is composed of a nitrogen-containing nucleobase—cytosine (C), guanine (G), adenine (A),

---

61. Silvestre deSacy, Champollion's master, had already found a relationship between pronounced sounds of Demotic symbols and their meaning (Allen, 1960). Champollion is often criticized for not giving enough credit to deSacy and Young. Eventually, 24 glyphs, representing single consonants, were identified, similar to an alphabet.

62. Turing's short and complex life continues to inspire generations of creative people. The movie *The Imitation Game* (2014) is perhaps the best known, albeit heavily romanticized, account of his code-breaking years and personal life.

63. If the letters are shuffled properly, as can be achieved today with fast computers, the cipher is practically unbreakable.

64. In a book about the history of British Intelligence in World War II, mathematician Marian Rejewski noted, "we quickly found the [wirings] within the [new rotors], but [their] introduction . . . raised the number of possible sequences of [rotors] from 6 to 60 . . . and hence also raised tenfold the work of finding the keys. Thus, the change was not qualitative but quantitative" (as quoted in the appendix of Kozaczuk, 1984) Another important predecessor of Turing's decoding venture was the work of another British cryptographer, Dillwyn Knox, who worked on deciphering Enigma-encrypted messages of the Nationalists in the Spanish Civil War (Batey, 2009).

or thymine (T)—as well as a monosaccharide sugar called deoxyribose and a phosphate group. The syntax is elegantly simple. A is always paired with T, and C paired with G, providing the rule for making the double-helix DNA: the hydrogen bonds between these matched pairs of nucleotides bind the nitrogenous bases of the two separate polynucleotide strands. This DNA sequence is ultimately translated into a string of amino acids, the building blocks that make up proteins. This is done via RNA, which transcribes a particular set of three nucleotides into a particular amino acid.[65] Using this messenger RNA as a template, the protein molecule is then built by linking neighboring amino acids.

These examples of code-breaking practices and their players offer an important lesson for brain scientists in our search for a neuronal "code." When the syntax is known, breaking a code is a hopeful exercise. Uncovering the meanings of the hieroglyphs took a few thinkers and a few decades of hard work. A few years of nerve-wracking search for correspondences, with several hundred people working together, were required to decipher the messages encrypted by the Enigma machines, even though the cipher was available. In the case of the genetic code, with well-understood goals (a combination of four elements to code in triplets for 20 known amino acids), the efforts of several hundreds of scholars over the span of a few decades provided a mechanistic explanation for a "miracle" of nature, the hypothetical force that produces complex organisms from simple seeds. So what should be our strategy for identifying the brain's cipher? Collecting correspondences and correlations is not enough; we also need to understand the neuronal syntax (Chapters 6 to 8) and to "ground" our correlational observations with a firmer "second opinion" (Chapters 3 and 5). Only with such independent knowledge can we hope to figure out the secret meanings of the brain's vocabulary.

At this point, the reader may rightly ask: Why does it matter to logically proceed from inside out rather than the other way around? Can we not understand relationships either way? Answering this will require discussion of the complex problem of causality in neuroscience, the topic we are going to explore in the next chapter.

---

65. James Watson, Francis Crick, and Maurice Wilkins were acknowledged for their discovery of the DNA code with the Nobel Prize in 1962. The Russian physicist Georgiy Gamow had already calculated, using mathematical modeling, that a three-letter nucleotide code could define all 20 amino acids. He formed the "RNA Tie Club" with 20 regular members (one for each amino acid), and four honorary members (one for each nucleotide; Gamow, Watson, Crick, and Sydney Brenner) with the goal of wide cooperation in understanding how RNAs can build proteins. Yet the breakthrough came from outsiders, Marshall W. Nirenberg, Har Gobind Khorana, and Robert W. Holley, who were awarded the Nobel Prize "for their interpretation of the genetic code and its function in protein synthesis" in 1968. See http://www.nobelprize.org/educational/medicine/gene-code/history.html.

SUMMARY

In this introductory chapter, I presented some contrasting views and research strategies that can provide complementary knowledge about brain operations. The first comparison was between "outside-in" and "inside-out" strategies. The outside-in or perception-action framework has deep philosophical roots, starting with Aristotle and most explicitly elaborated by the empiricist philosopher David Hume. According to this framework, all our knowledge arises from perceptual associations and inductive reasoning of cause-and-effect relationships. This philosophy is the foundation of Western science and continues to influence thinking in contemporary neuroscience. An inevitable consequence of this perceptual representation-centered view is the assumption of a hidden homunculus that decides whether to respond or not. In contrast, my recommended inside-out framework takes action as the primary source of knowledge. Action validates the meaning and significance of sensory signals by providing a second opinion. Without such "grounding," information cannot emerge.

Next, we discussed the relationship between the *tabula rasa* (blank slate) and preconfigured brain models. In the empiricist outside-in model, the brain starts out as blank paper onto which new information is cumulatively written. Modification of brain circuits scales with the amount of newly learned knowledge by juxtaposition and superposition. A contrasting view is that the brain is a dictionary with preexisting internal dynamics and syntactical rules but filled with initially nonsense neuronal words. A large reservoir of unique neuronal patterns has the potential to acquire significance for the animal through exploratory action and represents a distinct event or situation. In this alternative model, the diversity of brain components, such as firing rates, synaptic connection strengths, and the magnitude of collective behavior of neurons, leads to wide distributions. The two tails of this distribution offer complementary advantages: the "good-enough" brain can generalize and act fast; the "precision" brain is slow but careful and offers needed details in many situations.

2

Causation and Logic in Neuroscience

Nihil in terra sine causa fit.

—The Bible

Everything connects to everything else.

—Leonardo da Vinci[1]

The law of causation ... is a relic of a bygone age, surviving, like the monarchy, only because it is erroneously supposed to do no harm.

—Bertrand Russell (1992)

After coming to America as a postdoctoral fellow, I gradually learned how strongly our upbringing affects our interpretation of the world. Different cultures shape the brain differently, making it difficult to empathize with a new society that holds viewpoints that are distinct from the social milieu where you spent your formative years. Shaking hands while looking at another's eyes induces different emotions than bowing and expressing submission. Most people agree that our upbringing is linked to our prejudices. I often wondered whether cultural differences can appear not only in moral and emotional issues but also in scientific thinking. Even practices that appear trivial, such as reading and writing from left to right or vice versa, may influence the workings of the brain.[2]

---

1. Brulent and Wamsley (2008).

2. Scholz et al. (2009). Learning a second language can induce structural changes in white matter (Schlegel et al., 2012). The brains of musicians are favorite targets of researchers examining practice-induced structural changes in the brain. Use-dependent brain plasticity has been consistently demonstrated in the brains of professional musicians and in children receiving instrumental musical training for several months (Hyde et al., 2009; Lutz, 2009).

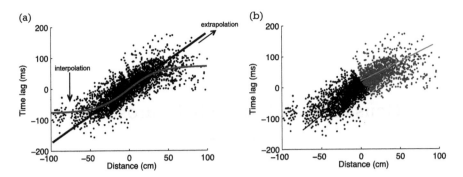

Figure 2.1. Interpreting graphs. A: Each dot corresponds to the measured distance the animal walked in an experimental apparatus plotted against time difference between the spikes of a neuron pair. The black line is a linear regression of all values. Taking into consideration the tails of the distribution, the sigmoid (*gray line*) is a better fit. B: An alternative interpretation is that there are actually two clouds (*black and gray*), which have different relationships between distance representation and the time lag of spikes.

Can we somehow put ourselves outside our subjective world and use objective methods to make valid judgments without being influenced by our pasts? If my reasoning has some merit, then examining the same problems from multiple angles may be useful in understanding the brain. Let's start by evaluating our everyday tools for reasoning in the Western world: logic and correlations and the necessary and sufficient conditions of causation.

OBSERVATIONS, CO-OCCURRENCES, AND CORRELATIONS

Hypotheses in the neurosciences are most often tested by models of associations between observed properties of the world. Figure 2.1 shows a generic example of a relationship between two observed variables: in this illustration, between the distances walked by an animal and the temporal relationships between spikes of neuron pairs corresponding to those distances.[3] Inspection of the cloud of dots

---

3. These data are modified from Diba and Buzsáki (2008), showing the distance measure between place field peaks of a pair of hippocampal place neurons (*x* coordinate) while the rat ran on a linear track and the time difference between the spikes of the same neuron pair. The physiological explanation of the flat ends of the sigmoid curve is that the distance-to-duration conversion is brought about by theta oscillation dynamics. The duration of the theta cycle is 140 ms (±70 ms in the figure), which limits the available distance representation to approximately 1 meter in the dorsal hippocampus. To resolve longer distances, the animal must move in the environment (see further discussion in Chapter 7).

suggests a reliable relationship: longer distances correspond to larger intervals between spikes. A simple way to quantify this relationship is to fit a "regression line" over the values (the solid black line). The line is a great simplification, an assumption that all measurements are "represented" by the regression line. The slope of the line and the magnitude of the variance of the values on both sides of the line indicate the reliability and strength of the relationship. This regularity is a "co-relationship" or correlation, in which approximately the same number of values lie above and below the regression line.[4]

While a plot such as that in Figure 2.1 shows only the values of observations, we often wish to read much more from a static cloud of data points. For example, there are many empty areas without measured values on the graph. We can simply assume that those values exist so that, if we had more time and more funding to generate more data, those new points would fill in the blanks between the already-measured ones. This operation, performed intuitively or with mathematical rigor, is called *interpolation* (the arrow in Figure 2.1), but we can also call it *explanation* or *deduction*. *Postdiction* would be another possible term because interpolation is done based on data collected in the past. If we feel comfortable filling the graph with imagined data between the dots, perhaps it is not much of a jump to imagine more data that *extend* the cloud, again following the trend in the observations. This operation is called *extrapolation* or *induction*. Alternatively, we can call it *prediction* or *guessing* from the currently available data, an act of "logical inference."[5] It is an abstraction from the present, which amounts to predicting the future. How far can we go in our extrapolation of the graph? Just because two sets of variables correlate now, there is no guarantee that this correlation will continue into the future.

Interpolation and extrapolation, as mathematical abstractions, are independent of the objects or observations being represented. The graph itself is an abstract relationship. The dots have lost their physical correspondences and are now embedded in a mathematical graph. In fact, the same dots could represent

---

4. The coordinates are both rotationally and translationally symmetric. There is no special need for the *y* and *x* axis to point north and east. The graph can be oriented in any direction as long as the relationships between *x* and *y* values are unchanged. This symmetry applies to Euclidian space. In Minkowski spacetime, the symmetry applies to time as well (Chapter 10).

5. Conflation among expressions such as "explanation," "prediction," "postdiction," and "cause" is common in interpreting statistical data, but distinguishing them is important for advancing scientific knowledge. The statistical literature often lacks a thorough discussion of the many differences that arise in the process of modeling explanatory versus predictive processes. Explaining (the past) and predicting (the future) are logically different (Shmueli, 2010). To "predict" (from the Latin *pre-* in front of, and *dicere* -to speak) means to verbally declare what is going to happen in the future. Postdiction is an inference. Thus, statistics is not free of subjective interpretation.

any relationship and are thus substrate-free. For example, they could represent an observed relationship between emotional magnitude and neuronal firing. In contrast, the interpretation of the relationship always depends on their substrate. Real things have a physical substance. Because the substrates of the variables are missing from the graph, the experimenter adds them in his or her interpretation to deduce a substrate-dependent mechanism that supposedly explains the observed relationships. This interpretational desire is the reason we give various names, such as *contingency, dependence, interdependence, indication, anticipation, prediction,* or *forecasting* (I might add *fortune-telling* or *crystal gazing*), to such simple x-y relationships.[6]

In our example in Figure 2.1, we took distance as the independent variable in the tradition of the outside-in framework. The distance between the two place fields in the maze can be measured objectively by a ruler, and the goal is to learn how distance is "represented" in the brain (e.g., by time offsets). In a sense, we are asking how distance predicts duration. However, as you will see in Chapter 7, the brain internally generates time offsets between neuronal spikes, and, with the help of the locomotion speed of the animal, it calculates the distances. In the inside-out framework, the cause is a brain mechanism and the consequence is distance estimation. Distances are meaningless to a naïve brain without calibrating the visual cues with action (Chapter 3). Thus, depending on the framework, the cause-and-effect relationships can change.

However, another experimenter may disagree with our interpretation of the x-y relationship. At the ends of the clouds, there are fewer data points, but those seem to lose the linear relationship characteristic of the other data points. Perhaps a sigmoid function (S-shape curve, in gray) is a better fit. To a third observer, the cloud of dots may appear to be split (Figure 2.1B). When all the dots are lumped together in our mind, the single-cloud interpretation justifies the use of the term "integration" or "pattern completion," by analogy to completing a picture or a sentence with missing details. But the more experienced eye of the third observer suggests that there are actually two clouds, and, therefore, two relationships exist, a case for segregation or pattern separation. Do the points in the graph reflect a single integrated relationship or two segregated ones? The regression problem then becomes a classification problem. Our scientific vocabulary and views are often formed by data plots even if they are more sophisticated than those shown in this simple figure. Without an arbitrator or supervisor, we cannot tell who is right and who is wrong.

---

6. Francis Bacon declared that experiments and observations are sufficient to obtain reliable ideas about the world. In contrast, René Descartes, although he had faith in mathematics, argued that bare facts are never self-explanatory: they require interpreting. A graph will never tell how thing happened, only how they *could* have happened.

CAUSALITY AND SELF-CAUSATION IN THE BRAIN

In Figure 2.1, we identified some regularity between two variables: one is the thing-to-be-explained (*explanandum*), called the dependent (y) variable, and the other is the thing-that-explains (*explanans*), called the independent (x) variable.[7] In more general terms, we can also call them the "defined concept" and the "defining concept." Of course, our strongest desire is to conclude that one variable predicts or causes the other. Ideally, the defining concept is more general than the defined concept, with some asymmetry between them. "The brain is a machine"; however, the other way around—"the machine is a brain"—does not apply. Accordingly, our choice to call one variable independent and the other dependent is often arbitrary and may reflect our preconceived bias. In some cases, we are able to manipulate the independent variable experimentally and observe its effect on the dependent variable. In neuroscience we often observe quantities which we cannot manipulate directly or for which there is no way to "independently" manipulate a single variable without indirectly manipulating many more properties of the system. In this case, the choice of calling one variable independent and the other dependent is often arbitrary and reflects a bias. In fact, the plot itself is symmetric and its statistical description would be the same if we reversed the x and y coordinates: the ratios would not change. Yet, in neuroscience, we place behavioral observations or stimulus features on the x axis (in this instance, the distance walked) and brain measures on the y axis. This tradition takes its origin from the outside-in framework as discussed in Chapter 1. Putting aside this historical issue, may we be able to make a causal inference from such data by plotting and evaluating the correlational measurements in some other way?[8] Equally important, we should ask about our confidence in the prediction: the probability of being right or wrong. Are these observational data enough, or do we need to do additional experiments?[9] To show convincingly that one variable causes the other, we must

---

7. Hempel and Oppenheim (1948).

8. Our expectations often influence our interpretation. From the outside-in framework, one can assume that the brain "measures" distances in the environment and represents them by temporal offsets between the representing neurons. However, experiments designed to challenge such correlations showed that the temporal offsets between hippocampal neurons are determined by internal rules of brain dynamics that remain the same across increasing size environments. As a result, the spatial resolution of the neurons decreases in larger environments. Distance representations in the hippocampus scale with the size of the surrounding world (Diba and Buzsáki, 2008). Thus, brain network dynamics is the independent variable that explains distance, not the other way around.

9. What if we remove Jupiter from the solar system? One astronomer would say that we can calculate the consequences from the available data (no experiment is needed). But another

vary the independent values experimentally and examine how the dependent variable follows such changes.[10] Direct manipulation of physical objects is the most basic way for arguing the persuasive rule of cause and effect. When a child discovers that flipping a light switch can make the room appear and disappear, he repeatedly practices it to learn that he is the sole causative agent of the consequences. Similarly, when a rat presses a lever that delivers food, it repeatedly tests the consequences of its action. It seems that functioning in the world requires animals to develop the concept of cause–effect relationships.

Both mathematics and scientific thinking require logic. All mathematical theories can be reduced to some collection of axioms or set of rules on which an internally consistent system is built. These axioms should not be violated. But science is supposedly different. It has no a priori rules. We simply observe regularities in nature and call them "laws."[11] When we interfere with those regularities, we become part of the relationship and therefore may alter it. As Heisenberg, the father of uncertainty in quantum mechanics, famously stated: "What we observe is not nature, but nature exposed to our method of questioning." This healthy skepticism is important because our measuring instruments may affect the cause–effect relationship.

Every action we take has an effect on the system we wish to observe. However, this cause-and-effect situation runs deeper than just a difficulty in experimental science—it forms the root of scientific inference in the Western world: once the cause is revealed, the problem is "explained," the truth is found. In the traditional wisdom, there are no effects without causes, and no logical argument can be delivered without causation. Western philosophy is built on a foundation of causality. Yet much of the world has moved forward and onward

---

astronomer would disagree and claim that the impact is unpredictable given the complex relationships among the celestial bodies (and many more such perturbations, such as removal of other planets or perhaps galaxies, are needed to get a firm answer). These are the types of questions that neuroscience constantly faces.

10. Without manipulating the system experimentally, one can add or leave out variables to "regress out" noncausal agents. It is never foolproof, but often this is the best one can do without interfering with the system. Descriptive statistics that reveal correlations are a bit like passive associations. In neither case can one make causal grounding. The source of knowledge is in the manipulation.

11. Laws, as organization tools of scientific thinking, have specific theological associations. Physicists, including Newton, who introduced these laws have assumed that things are "ruled" or "governed" by God, who was regarded as changeless and eternal. In Christian and many other religions everything must have a cause and laws and Order of Nature are assumed to be fixed in their forms (Toulmin and Goodfield, 1965). However, laws are substrate-free. They are not physical things, so they cannot exert an effect on anything else, short of divine interventions. They simply reflect regularities—the earth orbits around the sun.

without accepting such causality rules. The idea of causation is quite different in Asian cultures than in European cultures. For example, in Buddhist philosophy, nothing exists as a singular and independent entity because things and events depend on multiple, co-arising conditions.[12] A Buddhist monk or teacher lights three incense sticks leaning against each other, as a demonstration that if any of the sticks falls, the others will fall as well. Similarly, Confucians approached causation in an all-encompassing manner, searching for a network of connections. Historical changes are explained by coincidences of events instead of a particular cause. In Muslim philosophy, causal relations have no inherent necessity at all. Instead, they are just expressions of what God does.[13] Therefore, it should not surprise us that people from different cultures—yet with the same kind of brains—judge the affairs of the world surrounding us differently.[14] So why do we assume that the European philosophy of science is the superior or only way to view the world and ourselves, especially given that many other cultures have achieved great scientific discoveries? Is it really true that when the cause is identified, the truth is found? Even Descartes hesitated to accept such a link. First, he states: "If a cause allows all the phenomena to be clearly deduced from it, then it is virtually impossible that it should not be true," but then he immediately backtracks and adds: "Nevertheless, I want causes that I shall set out here to be regarded simply as hypotheses" because we often make "assumptions which are agreed to be false."[15] The modern meaning of the term "hypothesis" refers to a conjecture which can be rejected by observations and experiments,

---

12. "Origination" (or *pratityasamutpada* in Sanskrit) refers to the principle of interdependent causation and explains the twelvefold chain that describes the endless rebirth of life (Dalai Lama and Chodron, 2017). In metaphysics, *accidentalism* denies the doctrine that everything occurs or results from a definite cause.

13. Leaman (1985).

14. Carl Jung also rebelled against the Kantian axiom that "nothing can happen without being caused." Instead, he argued for "an acausal connecting principle," or the principle of "meaningful coincidences," based on Eastern philosophies. He and the physicist Wolfgang Pauli coined the term "synchronicity" to replace causality as a principle of explanation. Explanation "cannot be a question of cause and effect, but of a falling together in time, a kind of simultaneity" (Jung, 1973). Jung explained everything with synchronicity, including telepathy. For a statistician, coincidence is not a surprising, meaningful event because "with a large enough sample, any outrageous thing is likely to happen with no apparent causal connection" (Diaconis and Mostelle, 1989). These issues are eloquently exposed in Arthur Koestler's book *The Roots of Coincidence* (1973). [NeuroNote: Koestler attempted suicide in Lisbon in 1940 after he failed to return to England, but he survived. After he developed Parkinson's disease in his seventies, he and his wife committed double-suicide.]

15. Descartes (1984).

thus eliminating potential causes rather than proving the truth. Scientific knowledge is tentative and not definitive; it is progressive but not absolute.

Cause and Effect

Causality is a subtle metaphysical notion, an abstraction that indicates how events in the world progress. The common notion of causation can be traced back to Aristotle and is a statement about how objects influence each other across space and time. From the correlation graph (Figure 2.1), we cannot make the claim that x is the cause of y without extra manipulations.[16] Causation is defined as a chain of dependency, where the cause always precedes the effect: the cause is in the past of a present event. An event, A, is a cause of event B if the presence of A is necessary for B to happen. For A to have any effect on B, event A has to happen first.[17] David Hume remarked that "if the first object had not been, the second never had existed." In a deterministic framework, the presence of the cause is a necessary and sufficient condition for an effect to happen. Causality and determinism are often intertwined. Of course, an effect can have many co-occurring causes, but we are compelled to believe that we should be able to identify at least one. For example, when an individual gene alone cannot account for a mental disease, it can still be viewed as a conditional cause when co-occurring with other causes. When enough necessary components occur together, they may be sufficient to induce the disease. Under this framework, the triad of correlations and necessary and sufficient conditions complement each other. In another formulation, one can distinguish an *immediate* or *proximal cause* (also called *precipitating cause*: a stroke in the motor area caused paralysis), *predisposing cause* (she got upset), and *sustaining cause* (she had high blood pressure). Neither the predisposing nor the sustaining course alone could be designated as necessary and sufficient condition. Yet the proximal cause may not have happened without the sustaining and predisposing causes.

---

16. One step forward in interpreting a graph is to vary the independent variable. In this case, the designation of the independent and dependent variables is critical because regression of x on y is not the same as regression of y on x.

17. This logic stems from a Newtonian framework, which has a "time arrow." When this fundamental requirement of causation (i.e., the notion of time) is removed from its definition, things become rather complicated. In quantum-mechanical examples—such as entanglement—two particles can have interdependent or "entangled" properties. The usefulness of "classical" causation and the role of time as part of causality are intensely debated in both general relativity and quantum mechanics (Hardy, 2007; Brukner, 2014; Ball, 2017). As logic is timeless, it is not clear why time should be part of causation. We visit the nature of time and its alleged role in brain operations in Chapter 10.

Although Aristotle's formulation of the cause-and-effect relationship has undergone extensive revisions over the past millennia, it has remained the main tool of reasoning in the laboratory, likely due to its simplicity. However, even in everyday life, one can often see how such logic fails. My shadow follows me, and without me it never occurs. My shadow and I (often) correlate. I am the necessary (but not sufficient) condition because my shadow is not always detectable. In darkness, the suspected causal rule is violated because the occurrence of my shadow is conditional upon the presence of light. However, light is not a cause and not even a co-contributing component but a condition.[18]

Similar violations are found in neuroscience. A weak stimulus, such as an ant climbing on my leg, can sometimes induce reliable firing of neurons in multiple brain circuits and is prominently noticed (an effect). At other times, the same cause (the ant) may not induce a noticeable effect at all (for example, when I am intensely preoccupied with writing this book chapter). Again, the change in brain state is not considered a co-cause, but it is a necessary condition for the detectable cause–effect relationship. Later, I highlight some of the pitfalls of our everyday reliance on causation as a scientific rule. My goal, of course, is not to disqualify formalized reasoning but to call attention to why the simple recipe of logic fails in neuroscience, especially when things become really interesting.

Because of the omnipresence of logical inference and causation in science, both philosophers and scientists have undertaken multiple rounds of explanations and attempts to refine them. Each time, these concepts were approached with a particular point of view and purpose, and, not unexpectedly, the resulting explanations vary. Philosophers are primarily interested in the ontological question of whether causation is an inherent property of the physical world independent of the human mind or whether it has epistemic limits. As we learned in Chapter 1, David Hume's subjectivism separated the objective world from our first-person, subjective experiences and introduced the idea that our mind transforms and "represents" the objective reality. He criticized rationalists who assumed that we can acquire knowledge of the world by trusting our minds. For instance, we observe a regular relationship between lightning and thunder, but the inference that lightning causes thunder is only a subjective mental operation, so there may be no objective causality between

---

18. In Boolean logic, it could be called a "gate." In algebra, Boolean logic follows Aristotelian logic in a formalized manner, where all values are reduced to a binary true (1) or false (0). Everything in the middle is excluded or forced to belong to the true or false category (you either perceive something or not). The binary nature of Boolean algebra is what makes it so suitable for computer science, using the true and false statements with the fundamental operators And, Or, Xor and Not. Boole's symbolic system makes computation and logic inseparable. Early brain models assumed that brain operations follow a Boolean logic (McCulloch and Pitts, 1943).

the two. Hume thought that we do not experience the cause itself; we simply infer it from the regular appearance of the cause and effect. Immanuel Kant countered Hume's interpretation by arguing that there is a universal principle of causality that cannot be based on experience. Kant promoted the idea that the realm of things in themselves (*das Ding an sich*) is the cause of the realm of appearances (phenomena).[19] Our experience is shaped by a priori causality because the mind interprets the sensory inputs arriving from an objective, observer-independent world.

Hume's idea of representation has shaped the thinking of cognitive psychologists and neuroscientists alike and became the de facto scientific explanation of how "attributes" of objects become neuronal "representations." His subjectivism has been modernized to a subjective realism in which the mind represents, but does not faithfully mirror, the outside reality. Therefore, my representation of truths and judgments can only become similar to your representations through a mutually agreed consensus. Each brain in isolation would generate a somewhat different representation of the same object, color, or sound. If objective and subjective reality are separate, then our mental influence of causation does not necessarily apply to the relationships present in the physical world.

In a deterministic system with one-way interactions among the constituents, cause-and-effect relationships can be reliably described by Aristotelian–Humean arguments because the cause regularly precedes and induces the effect.[20] Determinism heavily rests on causality. In the Newtonian world, cause-and-effect relationships seem to work well since nothing moves unless it is being moved.[21] If the initial conditions and all variations are known in a system, in principle, all consequences can be accounted for. However, in practice, our

---

19. Interestingly, Muslim theology makes a similar distinction between *batin* (things-in-themselves or *noumena*) and *zahir* (phenomena). Muslim scholars believe that categories of causality and existence apply only to the noumenal realm, but not to *zahir*. Muhammad exists, but we do not know how (El-Bizri, 2000); we just have to believe it. For a concise exposure to fresh thinking about causation, see Schaffer (2016).

20. The concept of cause, as usually presupposed in standard causal modeling, is a "mover" and cannot be used to reduce probability. Can the absence or reduction of something be a cause? Inhibition in the brain is such a case. The primary role of a special, dedicated group of neurons in the brain is to inhibit excitatory neurons and each other. Thus, elevation of firing rates of excitatory cells can be brought about by excitation via other excitatory neurons or by reduction of inhibition ("disinhibition cause?").

21. Nearly every tool and machine is based on the logic of cause-and-effect relationship. Machines may represent the exteriorized reflection of this logic (Chapter 9), or, conversely, the functioning of the tool and machine components could have led to internalization of the observed logic of things (Chapter 5).

knowledge of the real world is limited by inevitably noisy measurements with limited precision, and such wisdom does not offer much solace. Even if we did have complete knowledge of the universe's past, the time needed to compute the future might be as long as the elapsed past.

Troubles with Causation

Problems with causality surfaced early in contemporary physics with the birth of quantum physics, nonlinear dynamics, chaos theory, and quantum entanglement. Quantum physics underwent a multitude of interpretations, including deterministic causality. On the other hand, in the framework of general relativity, time became symmetric and the foundations of cause–effect logic were shaken.

To understand complex economic data, Clive Granger reintroduced the time-asymmetry axiom (Chapter 10) and a novel statistical method for testing causality. "The past and present may cause the future, but the future cannot cause the past." His method provides predictability, as opposed to correlation, to identify the cause–consequence relationship with the key requirement of separability; that is, that information about a causative factor is independently unique from the variable. Strictly speaking, Granger's method does not present a definition of causality; it only defines a metric by which we can measure how much knowledge of one signal can enable you to predict another signal at a later time. It is ambiguous to the underlying source of predictability.

Yet the absence of the cause and effect between activities in brain regions does not prevent neuroscientists from talking comfortably about causal inferences in correlational data from the brain-derived data all the time. Granger's method has become popular in neuroscience in attempts to interpret cause-and-effect relationships, especially in complex functional magnetic resonance imaging (fMRI), electroencephalograms (EEG), and magnetoencephalographic (MEG) data.[22]

---

22. Statistical models used for testing causality are almost always association-based models applied to observational data. The theory itself is expected to provide the proof of causality, but, in practice, the ground truth is often absent (see Granger, 1969; Wiener, 1956; Mannino and Bressler, 2015). Causality as a tool of inference has grown from a nebulous concept into a mathematical theory with significant applications in many fields of sciences (Pearl, 1995). In addition to Granger's causality measures, other methods, such as directed coherence, partial coherence, transfer entropy, and dynamic causal modeling, have been used to assess the direction of activity flow in the brain (Barnett et al., 2009; Seth 2005). "Proper" statistical methodology for testing causality has been advanced, such as specially designed experiments or association-based, specialized causal inference methods for observational data, including causal diagrams (Pearl, 1995), discovery algorithms (Spirtes at al., 2000), probability trees

But many situations common in neuroscience will result in Granger's formalism giving the incorrect answer. Consider the following real-life observation. In recording from groups of neurons in area A and area B in the brain, we find a reliable correlation between the peaks of spiking activity of the populations, with neuronal spikes in area A, on average, consistently and reliably preceding the peak spiking activity in area B. Logic and Granger's formula dictate that it is likely that activity of neurons in area A causes the increased activity in area B. But we may be surprised to learn that axons of neurons in area B innervate neurons in area A unidirectionally; in short, area B is upstream to area A. To explain the contradiction between the anatomical ground truth and the physiological observations, let's assume that neurons in area A respond to inputs from B with a low threshold (i.e., before neurons in area B reach their population peak activity) and stop firing after their first spike. This results in peak spiking of the population in area A *earlier* than the peak in area B! Neuronal patterns often have such properties. In fact, it is a basic feature of a common feedforward inhibition motif found throughout the brain. Neurons can stop firing under continuous and even increasing drive because intracellular mechanisms may prevent sustained firing in neurons of area A (such response is known as *spike frequency adaptation*) or because strong feedback from inhibitory neurons prevents the occurrence of further spiking. Without knowing the anatomical connections and biophysical/circuit properties of neurons, Granger's formula gives the false conclusion that area A causes activity in area B. This example illustrates just one of the many pitfalls that have to be considered in any "causal" analysis.[23] Of course, this example does not disqualify the existence and importance of cause-and-effect relationships; it only reveals that the method is not always appropriate.

Hidden common causes are the most frequent sources of misinterpretations based on correlational arguments. Take as an example the observation that

---

(Shafer, 1996), propensity scores (Rosenbaum and Rubin, 1983) and convergent cross mapping (Sugihara et al., 2012). Applying these methods to the brain often requires the assumption that the brain is a deterministic, nonlinear dynamic system, although this assumption is questionable (Friston et al., 2012).

23. Another striking illusion of causality has been described in the entorhinal-hippocampal circuits. While superficial layer neurons of the entorhinal cortex are known to innervate hippocampal neurons but not vice versa, population activity in the downstream hippocampal neurons within theta oscillations cycles peak *earlier* than peak activity in the upstream entorhinal neurons (Mizuseki et al., 2009). The explanation in this case is that it is the fast spiking minority of neurons that fire early in the theta phase, and they have a much stronger impact on their target interneurons than the majority of slow firing cells. We discuss in Chapter 12 that skewed distributions are the main reason that mean population-based calculations are often misleading.

springtime brings flowers and higher birthrates. Nobody argues against this reliable relationship. However, even if this relationship is acknowledged, the issue of causality remains ambiguous because the likely cause is not that spring induces labor, but that couples spend more time together during the long dark nights of winter. Similarly, there is a reliable correlation between the blooming of flowers and birthrate increases. If we are not aware of the nonrecorded "third-party" variable (spring), we may suspect a causal relationship between these two variables as well. This example may seem silly for an easily relatable relationship such as flowers and birth rate, but when we are observing variables for which we do not have the luxury of everyday intuition, the reader can see how easy it would be to mistake correlation for causation. Hidden or permissive common causes are the most frequent sources of misinterpretations based on correlational arguments.[24] For these reasons, deterministic causality has been challenged in several other areas of science as well.

Probabilistic Causation

Events can follow each other with high probability, very low probability, or anything in between, so probabilistic influences fluctuate over a wide range from 0, meaning chance, to 1, meaning certainty. However, it is not obvious which aspects of the distribution are "causal" and which are not. Probability is an abstract mathematical concept and not a "mover" needed for causation in Newtonian physics. Determining probabilistic causation is a two-step process: the first step is the description of patterns and probabilistic facts (e.g., the substrate-free illustration of Figure 2.1) and the second is to draw conclusion from the measured probabilities. This second step is not inherent in the observed facts, however, but is an interpretation of the experimenter. When we observe a regular succession between things, we feel compelled to assume a cause. But how likely does succession have to be to reach a threshold to be perceived as "causal"? Might your threshold be different from mine? We single out high- and low-probability events from a continuous distribution and call them cause and chance. Confusion about chance, low probability, and high probability is a major contributor to wishfully interpreted "causal" relationships between gender and performance, genes and intelligence quotient (IQ), and religion and moral values. Conflation between explanation (postdiction) and

---

24. Calcott (2017) discusses the distinction between permissive and instructive factors. These factors are not really causal, yet their presence may often be critical. *Permissiveness* can be viewed as a gating function that allows the cause to exert its effect. Instructive factors may be a prerequisite but can induce a final effect only in conjunction with other factors.

prediction is also common in neuroscience data. Yet correlating, explaining, predicting, and causing are different.

Like causality, the notion of probability is often entangled in metaphysical disputes, and its interpretation appeals to mysterious properties.[25] Such considerations led Bertrand Russell to suggest that we remove causation from scientific thinking: "In the motions of mutually gravitating bodies, there is nothing that can be called a cause, and nothing that can be called an effect; there is merely a formula."[26] In the mathematical equations of physics, there is no room or need for causes. Do we need them for neuroscience? Paradoxically, Russell's quantitative probabilistic formulation managed to save rather than eliminate causation as the tool of scientific explanation. The concept of probabilistic causation replaced deterministic substrate-dependent causation.[27] Spurious regularities, a headache for deterministic causality, had become quantifiable and therefore could be contrasted with "true" regularities. However, making causal inferences from correlational data is harder than is usually thought, mainly because probabilistic causation is a statistical inference, whereas causation should involve physical substrates ("enhanced neuronal synchrony is the cause of epilepsy"). The correlation between various genes and schizophrenia leaves no doubt that aberrations in such genes increase the probability of the disease. On the other hand, claiming that my friend's schizophrenia was caused by particular genes is hard to support. Correlations need to be "challenged" to improve their explanatory power. Adding true meaning to correlations requires a second validation step, a hypothesis-testing practice based on some perturbation of the explanatory independent variable. For this reason, we need to do many independent experiments and worry more about replicability than causal explanation.

---

25. Despite the solid mathematical bases of probability and statistical inference, the connection between probability relations and correlational dependencies is often vague.

26. Russell et al. (1992). The cause (C) has a causal influence on the effect (E) if conditional probability ($E \mid C$) for E reliably changes under free variation of C. In an open, linear system, this requirement is not an issue, but in complex interconnected systems, such as the brain, the assumption of free variation of C is most often not tenable.

27. Moyal (1949); Churchland (2002); Mannino and Bressler (2015). While the term "probabilistic" is a useful theoretical construct, its use in real life may not be always practical. Consider Richard Feynman's explanation of the increase of entropy as defined by the second law of thermodynamics, which says that entropy is always increasing. "Things are irreversible only in a sense that going one way is likely, but going the other way, although it is possible and is according to the laws of physics, would not happen in a million years" (Feynman, 1965, p. 112). Similarly, in probabilistic causation, everything can be explained by singular causes in principle, but most of them may be statistically and practically improbable.

MUTUAL AND SELF-CAUSATION

Probabilistic causation appears more attractive and perhaps more appropriate for describing interactions in complex systems such as the brain. After all, it is only in interactions that nature affects the world. One may wonder whether transforming the philosophical problem to a precise mathematical description makes the word "causation" more valid. This issue is especially relevant in mutually interconnected and strongly recurrent complex systems with amplifying-damping feedback loops and emerging properties. Brain networks are robust and balanced, so sometimes even very strong stimuli may fail to affect the brain. Other times, minute perturbations can have a large impact on neuronal activity depending on the state of these networks. The most important feature of complex systems is perpetual activity supported by self-organized dynamics (Chapter 5). Ever since an EEG signal was first detected from the surface of the brain and scalp, its ever-changing electric landscape has been called "spontaneous." Related terms, such as *endogenous, autogenous, autochthonous, autopoietic, autocatakinetic,* or *self-assembled* are more popular in other complex systems, but they refer to the same idea.[28] These concepts have proved to be difficult to tackle because, by definition, the generation of spontaneous activity occurs independently of outside influences. Thus it must be induced by some self-cause, not by external causes, although the latter are much easier to study.

How can we separate causes from effects in self-organized brain networks with multiple parallel and interacting feedback loops? In many cases, brain patterns, such as neuronal oscillations, are not caused but are released or permitted. For example, sleep spindles in the thalamocortical system can be induced occasionally by some sensory stimuli but most often they just occur because the decreased activity of subcortical neuromodulators during sleep do not prevent their occurrence. Spontaneous patterns emerge and evolve without obvious trigger events. A pertinent analogy is the emergence of a tune from the interaction of jazz musicians. When they find a coherent melody, it decreases the degree of freedom of each player and forces them to time their actions to the emergent tune. But—and this is where it gets interesting—emergence is not the sum of multiple decomposable physical causes but a result of a multitude of *interactions*. Emergence is not a "thing" that affects neurons or humans.[29]

---

28. The term "self-organization" was introduced by the British psychiatrist W. Ross Ashby (Ashby, 1947). Increasing order in self-organized systems is discussed in Kampis (1991), an accessible text on self-organization. For a more complex treatment of the subject, see Maturana and Varela (1980).

29. Hermann Haken (1984) introduced the term "synergetics" to explain emergence and self-organized patterns in open systems far from thermodynamic equilibrium. His "order

Just because two sets of data move together over time, that does not necessarily mean they are connected by a causal link. Correlations are not causation, as conventional wisdom rightly says. Therefore, cause does not reduce to correlation. To understand a machine, it is not enough to look at it carefully. We must take it apart and put it together or remove parts from it and see what happens. Of course, in a closed system, such as a car, if all parts and all their linear interactions are known, in principle, the state of the system can be described at any time. However, in open systems, such as the brain with its complex dynamics, such complete description is rarely possible, mainly because the interactions are nonlinear and hard to predict.

Challenging Correlations

Perturbation, a targeted challenge to the system, is a powerful method for investigating a system provided that perturbation is done using a correct design and the resultant changes are interpreted properly.[30] In neurology and brain sciences, the perturbation method is one of the oldest. Damage or degeneration

parameter" is the synergy or simultaneous action of emergence. In synergetic systems, upward (local-to-global) and downward (global-to-local) causations are simultaneous. Thus, the "cause" is not one or the other, but it is the configuration of relations. One can call such causation reciprocal or circular, terms that are forbidden by Aristotelian and Humean logic (*circulus vitiosus*). However, in complex systems, circular causation is neither vicious nor virtuous but a norm. Reciprocal interactions weave the correlations. An excellent discussion of synergetics in neuroscience is by Kelso (1995), with an abbreviated version in Bressler and Kelso (2001).

30. Controlling neuronal activity in real time was only a dream for neuroscientists not so long ago. Since the introduction of optogenetics into mainstream neuroscience by Karl Deisseroth and Ed Boyden (Boyden et al., 2005), we now have the ability to turn on or off genetically identified neurons at will. This revolutionary tool allows us to challenge long-held views and generate new advances in our understanding of brain function in both health and disease. Unfortunately, when new techniques are invented, optimism and hype often trump humility. Optogenetic techniques are too often advertised as causal tools to discover the relationship between activity patterns in specific neuronal circuits and behavior. However, because neurons are both embedded in circuits and contribute to circuit function, their perturbation can bring about secondary and higher order changes, which need to be separated from the primary action of optical stimulation (Miesenbock, 2009). This separation is rarely straightforward because the network may change in an unpredictable manner. For this reason, it is not easy to tease out whether the observed variable is caused by silencing of a specific group of neurons or is the unseen consequences of novel patterns in their targets. Without knowledge of the precise effect, the assumed cause ceases to be a true cause. A potential improvement is the combination of correlation and perturbation methods. To fully exploit the advantages of a perturbation technique for brain circuit analysis, one should try to limit the perturbation to a small number of neurons that are continuously monitored together with their neighbors and upstream targets (Buzsáki et al., 2015; Wolff and Ölveczky, 2018).

of certain brain parts as a result of injury or disease has offered numerous and powerful insights into the contribution of various brain structures to normal function, suggesting that different parts of the brain are specialized for different functions. Complex functions, such as memory, emotions, and planning, have found their homes in the neural substrates of the hippocampus, amygdala, and prefrontal cortex, respectively, with the often explicit conclusion that neuronal activity in these structures is causally related to the respective behaviors. In animal experiments, structures or their parts can be deliberately damaged, "turned on," or "turned off" as a "causal" manipulation. The ensuing behavioral, cognitive deficit can then be interpreted to indicate that, in the intact brain, the perturbed structure is responsible for that behavior.

But slam on the brakes. Such a conclusion is not always warranted. For example, massive degeneration of substantia nigra neurons (loss of function of dopamine-producing neurons) is followed by muscular rigidity, slowed movement (bradykinesia), and tremor. However, decreased dopamine is a necessary but not always sufficient condition for the symptoms to occur, and similar symptoms may also occur in the absence of dopamine dysfunction. Similarly, subcortical denervation of the hippocampus in rodents results in more severe cognitive deficits than damage to the hippocampus itself. This happens not only because the hippocampal circuits can no longer perform their physiological jobs but also because the denervated hippocampus induces abnormal neuronal activity, adding insult to injury. Conversely, if behavior is unaffected long after lesioning a particular brain structure or circuit, one cannot rule out its important contribution to that behavior because the brain circuits have redundancies, degeneracies, and plasticity that can often compensate for an acute impairment. These examples illustrate why, in a densely interconnected dynamical system like the brain, even minor and local perturbations may induce unexpected activity in downstream and distant structures. Furthermore, functional recovery may occur via alternative behavioral strategies and adaptive repurposing of structures that are targets of the damaged area. The emerging nonphysiological activity in these downstream structures can be as damaging as the absence of their upstream partners. These considerations point out the caveats and limitations of using causation in our search for physiological function in the complex labyrinth of brain dynamics.

In conclusion, statistical correlations are similar to associations. In neither case can one rely on causal grounding, which requires active manipulations. In the strongly interconnected complex networks of the brain, perturbation may bring about secondary and higher order changes, which need to be separated from the primary effects. This is rarely straightforward as complex networks often respond to challenges in an unpredictable manner. Since there is no simple recipe for how to deal with emergent systems, correlational and perturbation methods should be combined whenever possible because they represent complementary ways of analyzing the same phenomena. This is the best we can do.

The difficulty in applying causation as the sole tool for inference will often reappear in this book. In the next chapter, we will examine the relationship between perception and action. The two are often correlated. As we learned in this chapter, correlation is symmetric—in the mathematical sense. However, in searching for a cause with an assumed direction, we designate one set of variables independent and the other dependent and make the relationship asymmetric in our minds. To verify our intuition, we must vary one of the variables and quantify the consequence on the other variable. From this perspective, there should be a distinction in whether we talk about a perception–action cycle (where perception is the critical variable for action) or an action–perceptions cycle (where action is critical for perception). While in everyday parlance such distinction may sound prosaic and hair-splitting, I submit that the failure to recognize the importance of this difference is a main contributor to why the outside-in framework continues to dominate cognitive neuroscience. In an uncalibrated (naïve) brain, changing the sensory inputs may not impact the motor response much (save phylogenetically learned fixed action patterns), whereas moving the sensors by actions brings about large changes in the activity of sensory structures. The next chapter discusses how confusion between cause and effect has impacted thinking in perception research.

SUMMARY

In neuroscience, we often compare our observational data against association-based models, such as correlations, and plot our data as a relationship between presumed independent and dependent variables. When we identify some regularity between two variables (find a correlation), we are tempted to regard one set of variables as the thing-to-be-explained and the other set as the thing-that-explains. This is the basic logic of cause-and-effect relationships. Interpreting correlations is a two-step process. The first step is the description of facts (the correlation), followed by a second, biased step—an assumption of a cause-and-effect relationship from a symmetric statistical relationship.

Causality is the most critical pillar of scientific inference in the Western world. Revealing a cause amounts to an explanation. However, other cultures that do not rely on cause-and-effect arguments can also arrive at valid scientific conclusions. The concept of causation is especially problematic in self-organized systems with amplifying-damping feedback loops, such as the brain. Causes in such systems are often circular or multidirectional; events are not caused but emerge from the *interaction* of multiple elements.

3

Perception from Action

Our sensory systems are organized to give us detailed and accurate view of reality, exactly as we would expect if truth about the outside world helps us navigate it more effectively.

—Robert Trivers[1]

The hardest thing is to do something which is close to nothing.

—Marina Abramovic[2]

The eye obeys exactly the action of the mind. In enumeration the names of persons or of countries, as France, Germany, Spain, Turkey, the eyes wink at each new name.

—Ralph Waldo Emerson (1899)[3]

---

1. Robert Trivers, one of the most prominent evolutionary biologists and sociobiologists, made this statement as recently as 2011 (Trivers, 2011). Many other prominent scientists share the view that our perception is veridical and is a window on truth, "veridicality is an essential characteristic of perception and cognition" (Pizlo et al., 2014). David Marr (1982) also expressed the view that humans "definitely do compute explicit properties of the real surfaces out there" but he denied this ability in non-human species. Other theorists make such dichotomy even among senses; for example, that vision is veridical but taste is not (Pizlo et al., 2014).

2. Abramovic (2016).

3. Chapter 5 on Behavior, in Emerson's essay collection, *The Conduct of Life* (1899), is full of outstanding observations on how human beings and their thoughts are exteriorized by eye movements. "The eyes of man converse as much as the tongues, with the advantage that their ocular dialect needs no dictionary, but is understood all the world over" (p. 173).

In the outside-in framework, the separation of perception from action is an important distinction. Perception-action separation is both intuitive and reasonable in the context of early anatomical studies of the nervous system. In the nineteenth century, François Magendie discovered that the motor neuron fibers that innervate skeletal muscles exit from the anterior root of the spinal cord, while the sensory neuron fibers that carry information about touch from the skin and the contraction states of the muscles from their own stretch-detecting receptors (i.e., proprioceptors) enter through the dorsal root.[4] That is, distinct sets of neurons deal with body sensation and muscle control. A few decades later, the Ukrainian neuroanatomist Vladimir A. Betz noted that the brain maintains the anterior-posterior segregation of the spinal cord: structures dedicated to vision, hearing, taste, and somatosensation reside largely in the posterior half of the human brain, whereas areas behind the eye, called *frontal structures*, are mainly devoted to motor functions,[5] in keeping with the anterior motor and posterior sensory organization of the spinal cord. However, while anatomy supports the separation of sensation and motor commands, it has no bearing on the temporal ordering of events, the connections between "sensory" and "motor" areas in the brain, or the postulated causal relationship between perception and action.

Furthermore, while sensation and perception are obviously related, they are different. Some authors use the terms interchangeably; others refer to sensation in lower animals and perception in higher animals. A more rigorous distinction is that sensation is the instantaneous feeling that receptors are being stimulated, whereas perception compares sensation with memories of similar experience to identify the evoking stimulus. Similarly, although "action" and "motor output" belong to the same general category, "action" is a more general term than "motor." Outputs from the brain can not only affect skeletal muscles

---

4. Jørgensen (2003); Bell (1811); Magendie (1822).

5. Betz noticed that the anterior (motor) areas in humans contain large pyramidal cells in deep layers, whereas in the posterior part the cells were smaller. This observation led to his best-known contribution, deducing the function of the giant pyramidal neurons of the primary motor cortex (known today as "Betz cells"), which are the origin of the corticospinal tract. "The sulcus of Rolando divides the cerebral surface into two parts; an anterior in which the large pyramidal nerve cells predominate. . . . Undoubtedly these cells have all the attributes of so-called 'motor cells' and definitely continue as cerebral nerve fibres" (Betz, 1874, pp. 578–580). In addition to the human precentral gyrus, Betz found these cells in the same location in dogs, chimpanzees, baboons, and other primates, and, based on the "brilliant physiological results" of Fritsch and Hitzig, he concluded that "these cells have all the attributes of the so-called 'motor cells' and very definitely continue as 'cerebral nerve fibres' (for further details, see Kushchayev et al., 2012). The anterior cingulate cortex of hominids also contains such large neurons, and much speculation has been devoted to the significance of these "spindle cells," also known as *von Economo neurons*, in intelligence (Allman et al., 2001).

but also exert autonomic effects (e.g., change heart rate) and control endocrine function (e.g., milk production). Furthermore, thought and imagination can also be conceived as actions (Chapters 5 and 9).[6] Thus, the interaction of the perceptual and action systems is a complex problem.

Most details of the world around us pass without our notice. Stimuli become salient or meaningful through learning and become percepts. But then if perceptual awareness is understood as an interpretation of sensation by the brain, it is an active process. Physiologically, awareness requires that the signal's trace be distributed in many brain structures and linger in neuronal networks for some time.[7] This fact is interesting because initiating movement also has a volitional component; we are aware of our voluntary actions, as opposed to reflex movements and automatic, well-learned actions, such as walking. Therefore, there is terrain between sensory inputs and motor outputs, a hypothetical central (or "top level") processor. This poorly understood but often speculated about ground is usually referred to by the name of *homunculus*, "decision-maker" or volition.[8]

In the outside-in framework, sensory information is distributed to higher order cortical areas and, consecutively, funneled to the motor areas. The alleged brain route from sensation to motor response is often referred to as the "perception–action loop." Decisions are made somewhere in the unexplained territory between sensory inputs and motor outputs. There is no need for anatomical connections going in the opposite direction.

---

6. That thought is an action (see also Llinás, 2002) is, of course, a very different idea from the outside-in view that our thoughts are the result of the synthesis of sensory inputs.

7. Many experiments suggest that conscious recognition requires the recruitment of a large number of neurons in an extended, distributed complex brain circuit. According to Libet (2005), this is longer than 0.5 second, which he calls "mind time," supporting a sensory input–conscious decision–action arc model. See criticism of this requirement in Goodale et al. (1986).

8. The little man (or rather a "little rat" in most experiments) is of course only a metaphor for an assumed brain mechanism or, alternatively, the soul. The origin of this homunculus thinking is religion and folk psychology. The soul is often interpreted as an alternative version of the self. Sometime, such as during dreams, the soul is believed to leave the body and even empower it with supernatural abilities. After death, soul and body divorce, and the soul leaves behind the helpless dead body for good. Because the soul has superb interpreting powers, if anything similar exists in the brain, it should have similar interpretive powers. The yin-yang dichotomy of Confucianism reflects a similar yet very different general idea: the yin derives from heaven, whereas yang is from the Earth. Everything is a product of yin and yang. They cannot be divorced from each other. Many, like, Daniel Dennett, liken the homunculus literally to a person watching a movie projected from the retina and other modalities, in what he calls a "Cartesian Theater" where it all comes together; this is in reference to Descartes's dualist view of the mind (Dennett, 1991). The homunculus is often used as a fallacy argument for the infinite regress of logic.

I have already suggested that the brain's primary goal is to generate actions and evaluate their consequences with the help of its sensors.[9] Combining the historical perspective of cognitive neuroscience (Chapter 1) and the experiments presented later, I will conclude that the correct order is "action–perception" cycle. First, let's examine experiments that lead to my conclusion.

PHI PHENOMENA AND PASSIVE OBSERVERS

Occasionally, our eyes (actually our brains) play tricks on us. Among the best-known illusions is apparent motion or the *phi phenomenon*. In its simplest version, two small balls at two discrete locations on the screen alternate over a short interval (say 60 ms). All humans, and most likely other mammals, see a single ball moving back and forth between the two positions, even though there is no movement whatsoever on the screen.[10] If the two balls have different colors in addition to motion, we also experience an abrupt color change in the middle of the ball's illusory passage between the two locations. This is very interesting because the color change is felt 25 ms or so *before* the second color is actually flashed.[11] How can the perceptual effect occur before its physical cause? Phi phenomena have kept philosophers and psychologists busy over the past century offering numerous explanations. They disagree on details, such as whether the mind generates the movement by filling in the missing pieces or the intervening apparent motion is produced retrospectively and somehow

---

9. Hoffman's interface model of perception (Hoffman, 1998; Hoffman et al., 2005) also emphasizes that we do not simply passively view the world, but also act on it. We can interact with the environment not because we perceive its objective reality but because the world has sufficient regularity that allows for the prediction of our actions on it. However, Hoffman states that "having a perceptual experience does not require motor movements" (Hoffman et al., 2005). I agree that this may be the case in an already movement-calibrated brain, yet I maintain that without action calibration at some stage of life perception does not arise.

10. The phi illusion was discovered by the German psychologist Max Wertheimer (1912). For an extended discussion, see Kolers and von Grünau (1976) or Dennett and Kinsbourne (1992). Many contemporary neuroscience researchers use illusions and subjective feedback about them as tools to probe neuronal responses (Dehaene and Changeux, 2011; Koch, 2004). The title of Giulio Tononi's excellent and entertaining book on consciousness-related topics is *Phi* (2012).

11. There are many interesting practical and theoretical implications of the phi phenomena. In a movie theater, we do not perceive a rapid flipping of still frames at 24 frames per second. Instead, we have the illusion of continuous movement without interruptions. Neon light advertisements also use the same principle to create motion perception.

projected backward in time, but nearly all of the theories call upon consciousness as *explanans*, the thing that explains, rather than the phenomenon to be explained.

Phi phenomena, as well as ambiguous pictures, Rorschach patches, and other illusions, are favorite territory of consciousness researchers. This research program often focuses on perception, asking, for example, how the brain can extract and bind sensory features to produce a unified object. But this program stops short of explaining the perception–action cycle[12] because it says nothing about action. This neglect creates a huge gap between perception and action that requires filling it with mental constructs such as volition and decision-making. This separation largely explains why research on perception and action have taken independent paths in neuroscience.

If action is a prerequisite of perception, as I claim, then there should be no perception without action. You can show that this is indeed the case by fixing your gaze steadily on a stationary target. No matter what surrounds the gaze target, even if pink elephants are running around, it will all disappear after just a few seconds. This never-failing demonstration is known as the *Troxler effect*.[13]

Another way to probe the primacy of action is to get rid of its consequences altogether. One such manipulation is called *retinal stabilization*. In an early, ingenious instantiation, a suction cup with a mirror on it was attached to the cornea. (Today, we would call it a contact lens.) When a picture or movie is projected onto the mirror, you can no longer investigate the scene with your eye movements because the image on the retina moves with your eyes, in the same direction and at the same speed and amplitude. As you can guess by now, without the brain's ability to obtain a "second opinion" by eye movement investigation, vision ceases.[14]

---

12. For a discussion of the involvement of higher order cortical areas, see Quintana and Fuster (1999).

13. Named after Ignaz Troxler, a Swiss physician who first demonstrated the effect (Troxler et al., 1804). We must fix our gaze to inspect the details of a scene, but if we fixate a while, it fades from our brain. See it yourself: https://en.wikipedia.org/wiki/Troxler%27s_fading. One explanation of the Troxler effect and related illusions is neuronal "adaptation" in the retina or further up in the visual system, without the actual identification of the cellular and circuit mechanisms of the adaptation (Martinez-Conde et al., 2004). The moment you move your eyes, sight returns. Zen Buddhism effectively exploits this feature of the visual system for meditation. In the famous zen Rock Garden (Karesansui) in the Ryoanji Temple in Kyoto, the meditator can focus on one of the fifteen rocks arising from flat white sand and make the rest of the (visual) world disappear. It works well; I tried it.

14. The most popular citation for the stabilization experiments is Yarbus's book (1967). Alfred Yarbus was a Soviet psychologist who recorded eye movements in humans while they viewed objects and scenes. He showed that the eye movements reliably track the observer's interest in

ACTION SPEED LIMITS PERCEPTION SPEED

Instead of asking why the brain can be fooled by artificial or rare patterns, perhaps it is more productive to ask why the brain reliably produces illusions.[15] As autoshaped behaviors are not mistakes (Chapter 1), illusions are not mistakes either. They are due to the brain's ability to efficiently extract important features from garbage. Some distinctions are of paramount importance, whereas we fail to observe most of the enormous variability in the environment.[16] Perception is not the veridical representation of the objective world even in case of vision, to which sense the human brain devoted a very large part of its real estate. Even the most sophisticated observer, without the knowledge that the color balls in the phi illusion do not move, cannot tell whether the brain is faithfully following the motion of a ball or whether two stationary balls are being flashed in a rapid sequence. There is no way to distinguish the two possibilities because brain mechanisms to track such fast changes consciously do not exist.

So why not have a faster perceptual system? If the primary goal of the brain were to perceive and process, it would be odd for evolution to compromise speed. The answer may lie in the mechanics of our body. Conditionally accepting my conjecture that the brain's fundamental goal is to produce action, there is no reason to invest in speed if the muscle system that it controls is slow. The contraction speed of vertebrate muscles is determined by the properties of *myosin*, a contractile protein that is largely preserved across all mammals independent of body size. Therefore, small and large brains must deal with problems at the same temporal scale. In the phi illusion, jerky or saccadic movements of the eye are needed to refocus the color vision region (fovea) of the retina from

---

the details of the objects. In the case of the human face, the investigation of eyes, mouth, cheek, or nose regions depends on what emotions are being identified. See also Riggs and Ratliff (1952) and Ditchburn and Ginsborg (1952). There is a resurgence of interest in eye-tracking techniques today, perhaps due to the recognition that they provide useful information about the "mind's opinion and intentions." Wearable systems can identify one's interest in particular details of scenes and advertisements with high precision. Every major technology-oriented company, from Sony to Google, offers such devices.

15. Visual areas V1 and MT may be involved in phi phenomena as they respond to real movement and apparent movement (e.g., induced by alternating flashing of stationary stimuli). Neuronal spiking in these cortical regions varies systematically with spatial and temporal properties of the stimuli that closely parallel the psychophysically defined limits of apparent motion in human participants (Newsome et al., 1986).

16. Visual artists exploit illusory perceptual experiences for aesthetic goals. Victor Vasarely, who was born in my hometown of Pécs, Hungary, and other painters create impossible three-dimensional objects on planar surfaces. The impossibility becomes apparent only after carefully scanning the continuity of the lines. Or think of another painter: Maurits Escher.

one ball to the other. These ballistic eye movements occur three to four times a second on average and are the fastest movements in the human body (100–300 degrees/s or 20–30 ms during reading).[17] The skeletal muscles are considerably slower than those that move the eyeballs. Nobody can move his fingers faster than 20 taps per second, not even Franz Liszt.[18] Even if an imagined super brain could perceive, decide, and send a movement command in less than a millisecond, it would have little benefit for the muscles, which must temporally coordinate their actions over a sluggish tens to hundreds of milliseconds. From this evolutionary perspective, we have one more reason to believe that perception is secondary to action.

Perception–Action or Action–Perception?

The clever skeptic, of course, can argue that there is no reason to prioritize the "chicken and egg" two-way problem of action–perception because it is a cycle anyway. An often-used argument in favor of the perception–action cause–effect direction is the existence of simple reflexes, such as the *patellar reflex*. When a neurologist strikes the patellar tendon with a reflex hammer, it stretches the muscle spindle receptors in the quadriceps muscle. The muscle receptors are peripheral axon processes of the neurons in the dorsal root ganglion of the spinal cord, and they convey the stretch-induced change to the ventral horn neurons of the cord, which in turn send their axons to innervate the same quadriceps muscle. Every time the neurologist strikes the tendon, the monosynaptic response is activated and a kick is observed. At the same time, another axon branch of the neuron in the dorsal root ganglion sends the same spike message to another neuron in the spinal cord which, in turn, conveys the information to the thalamus. In line with the Bell-Magendie rule, so the argument goes, the sensorimotor reflex must be the evolutionary precursor of perception. But who is doing the sensing? The neuron in the ganglion, the ventral horn motor

---

17. Myosin is a contractile protein found in skeletal muscles. The contraction of human myosin is only twofold slower than the contraction of myosin in the rat, which is more than 100-fold smaller (Szent-Györgyi, 1951). For eye-movement speed in humans, see Fischer and Ramsperger (1984). Muscles are the second cell type in the body that can rapidly change their membrane potentials (and generate action potentials). Several key molecules in muscles have similar functions in neurons. Both muscles and neurons can generate energy by both aerobic and anaerobic metabolism (e.g., Buzsáki et al., 2007).

18. In skeletal muscles with sensory feedback, fast afferent processing is performed by two large side appendages of the thalamocortical systems: the olivocerebellar system and the basal ganglia. In his best-selling book (*I of the Vortex*; 2002), Rodolfo Llinás eloquently describes the evolutionary reasons that the brain should lag behind the body's speed.

neuron, or the thalamus? What if some disease destroys the motor neurons but leaves the sensory pathways intact from the muscles and skin to the spinal cord, thalamus, and the rest of the brain. Of course, no stretch reflex will occur. But the issue is whether the patient will sense the strike of the hammer. I vote for a no answer because the strike-induced neuronal activation has no meaning to the brain since it has not been grounded.

Let's put my argument into a broader evolutionary perspective: an animal that can only act has a higher probability of survival than an animal that can only perceive. Sensing, or being aware of sensing, has no utility whatsoever unless the organism can act. On the other hand, movement can be useful even without any sensory information. In seawater, where our ancient animal ancestors lived, with abundant food around, rhythmic movement is sufficient to feed a simple creature. Once a movement control mechanism is in place—and only after this step—it makes sense to develop sensors that more efficiently guide movements toward food or shelter to improve the chances of survival. This argument is supported by another observation that the nervous system and muscles emerged together in the jellyfish.[19] Only through doing something do animals have some purpose of being.

From the perspective of evolution, therefore, there is a fundamental distinction between "perception–action" and "action–perception." If perception were the primary objective of brain design, many of our perceptual systems would appear to be built poorly. Movement, on the other hand, has to have useful consequences to call it an "action." Disconnecting its consequences on sensing the surrounding world breaks the action–perception loop and renders the system useless.

Such interruption of the cycle can occur naturally. During rapid eye movement (REM) sleep, we temporally lose our body, as our muscles cease to be controlled by the nervous system. When we wake, we immediately get our body back. Occasionally, these brain functions get de-coordinated in time, and we wake up a few seconds before muscle control is reinstated. This terrifying feeling is known as *sleep paralysis*. The brain is confused because stimuli impinge on our sensors, but they make no sense and induce no meaning since we cannot act on them. Demonic voices are heard, grotesque incubi dance on our chest, and we feel threatened. We are not dreaming but fully alert. These hallucinations, like the illusions we discussed earlier, are an inevitable

---

19. Llinás (2002) points out an even more striking example of the primacy of action in the tunicate, a marine invertebrate animal. In their larval stage, tunicates resemble a tadpole and move around happily looking for a home base. As soon as they find a suitable place to settle, they metamorphose into a barrel-like sedentary adult form. In this sessile existence, there is no need for movement, so tunicates digest most of their brains.

consequence of brain-body-environment interactions.[20] We connect to the world not through our sensors (although they are essential) but through our actions. This is the only way that sensation/perception can become "grounded" to the real world as experience (Chapter 1). The distance between two trees and two mountain peaks may appear identical on the retina. It is only through walking and moving one's eyes that such distinctions can be learned by the brain. Again, I reiterate that the brain's main function is not veridical perception and representation the objective world with its mostly meaningless details, but to learn from the consequences of the brain's actions about those aspects of the environment that matter for particular goals, such as reduction of hunger.

Recent anatomical and physiological experiments provide abundant support in favor of the critical role of action in perceptual computation. Primary sensory areas, and especially areas designated as "higher order sensory," receive numerous inputs from brain regions defined as "motor." In fact, beyond cortical areas which receive direct, thalamus-relayed inputs from the senses and which send axons to the spinal cord, all other areas show very strong and widespread interconnections.[21] Neurophysiological experiments in rodents document that more than half of the variability of neuronal activity across the entire cortex, including those we refer to as primary sensory, can be explained by self-generated action parameters such as body motion, whisker movements, orofacial movements, and pupil diameter changes. Such strong action-modulation of brain activity belittles responses to sensory stimuli. The involvement of wide brain areas in action suggests that nearly every cortical neuron receives two types of afferents, conveying both action and sensory inputs, although in different proportions.[22]

---

20. I was scared to death during my first sleep-paralysis experience. But safe recoveries from the momentary early morning attacks allowed me to internalize the experience (as discussed in Chapter 5), and they no longer bother me. In other words, I "grounded" these experiences by assigning a neutral meaning to them. You may have had similar adventures; nightmares occur in about one-third of the adult population (Ohayon et al., 1999) though usually during non-REM sleep, which is different from sleep paralysis. Horrific experiences in rare cases of anesthesia arise from the same mechanism: the brain can (mis)interpret sensory inputs without its ability to act. Anxiety and panic attacks may involve similar mechanisms due to our inability or perceived inability to change a situation and to be in charge.

21. Petreanu et al. (2012); Economo et al. (2016); Chen et al. (2018); Han et al. (2018); Harris et al. (2018).

22. Stringer et al. (2018) concluded that "an understanding of the function of sensory cortex will likely be impossible without measuring and understanding its relation to ongoing behavior, beyond easily-characterized measures such as running and pupil dilation." See also Musall et al. (2018).

SHORT-CIRCUITING
THE ACTION–PERCEPTION CYCLE

When we shout loudly near someone's ears, she may have difficulty hearing for a few seconds. In contrast, we can resume normal conversation right after our own shouting is heard by our ears. We are protected in that case by multiple mechanisms in our auditory system, as the Greeks speculated long ago. Today, we have a name for it: *corollary discharge*. As the name implies, this ancillary activity occurs simultaneously with the action output. The action-initiating circuits of the brain send action potentials not only to the downstream motor pathways but also synchronously to other areas in the brain. This secondary activity provides a feedback-reporting mechanism for self-organized action. "I am the agent that brought about change in the sensors."

A particular sensory modality may be activated when a stimulus unexpectedly occurs or when the animal is actively searching for the stimulus or expecting it. These two types of stimuli evoke different responses, like the feelings when you touch or are touched. An unexpected stimulus can induce either a generalized response, such as a startle reaction in response to a loud noise, or an orienting response.[23] The orienting response is the brain's active attempt to learn more about an unexpected event.

As discussed in Chapter 1, neurons cannot interpret the relevance of the signals conveyed by sensory inputs because they cannot ground their response without some independent verification. For neuronal networks to interpret the world, they need two types of information for comparison. The extra information can be supplied by the movement-induced changes of the brain's sensors. Only by comparing two signals, one of which is grounded by movement or previous knowledge (which was also generated by action at some point in life), can the brain figure out what happened out there. This distinction is easy to demonstrate by closing one eye and moving the open eyeball from the side

---

23. The term "orienting reflex" was coined by Ivan Pavlov and studied most extensively by Evgeny Sokolov (1960, 1963). Sokolov noted that if a novel signal is not followed by reward, the animal quickly learns to ignore it, a process he called *habituation*. Interestingly, Sokolov used the same comparator circuit as used in the corollary discharge model (Figure 3.1) to describe how the brain (particularly the hippocampus) can detect whether something is novel or not. This is perhaps not surprising. When an eye saccade occurs, the brain makes a prediction about the expected content of the scene after the saccade based on the current state of its sensory knowledge (a Bayesian prior; Chapter 13). If nothing new happens, no sensory change is registered. However, if a novel thing appears during the saccade, there is a mismatch between expected and actual inputs (i.e., if the prior model is erroneous, the model is updated). The process of updating can be called perception. Behaviorally, a mismatch between the predicted and detected states of the sensors will trigger an action: the orienting response.

with your finger. The entire visual field will appear to move at the pace of your finger's movement. In contrast, when you move your head back and forth or scan the world with your eyes, producing a similar pattern of stimulation on the retina, the world appears stationary even if you blink or make saccadic eye movements. Critically, if your eye muscles were paralyzed for some reason, and you intended to make an eye movement to the right, then you *would* observe a shift of the visual field in the same direction.[24] What would be a proactive brain mechanism to compensate for such confusing shifts?

The Proactive Corollary Discharge Mechanism

Distinguishing the real and perceived worlds has entertained both philosophers and artists prior to neuroscience. Perhaps the Italian painter Filippo Brunelleschi was the first to test it experimentally. On a wooden panel about twelve inches square, he painted the Baptistery of Florence in its surroundings. In the center, he made a hole and invited the viewer to look through the back of the painting while he held a mirror. Brunelleschi carefully positioned the painting (i.e., its reflection in the mirror) so that the real baptistery and the painting lined up perfectly. Through the mirror, the real world looked more two-dimensional and the painting more three-dimensional, thus the viewer could not distinguish easily between the two mirror reflections. His goal was to demonstrate that only by moving the head back and forth can reality and image be distinguished, making and unmaking the viewer's own world. It is the viewer's voluntary action of head movement in this artistic peep-show that can distinguish real change from apparent change.[25]

Erich von Holst and Horst Mittelstaedt, working on fish and invertebrates in Germany, and Roger Sperry, studying the optokinetic reflex in humans at the California Institute of Technology, went further than Brunelleschi. They wondered whether a motor command exits the brain unnoticed or leaves a trace behind.[26] They came to the conclusion that there must be brain mechanisms that separate self-generated stimulation from externally induced stimulation

---

24. This is exactly what patients with paralysis of the eye muscles see. Herr Professor Kornmüller (1931) famously anaesthetized his own eye muscles to investigate what happens when the volitional act is blocked. He reported an apparent shift of the visual field, which could not be distinguished from a similar magnitude of true shift of visual input.

25. As described in Wootton (2015, p. 165).

26. The paper by von Holst and Horst Mittelstaedt (1950) is an extraordinary read. In thirty pages, the entire history of motor control and sensory physiology is exposed, many experiments from flies to fish are presented, and the findings are summarized in quantitative models. The authors alert the reader that the brain of even the simplest creatures consists of more than "simply a set of connecting cables between receptors and muscles!" They not only demonstrated

of the brain's sensors. This hypothetical mechanism assumes that a copy of the motor command signal leaves an image of itself somewhere in the brain. Sperry called it "corollary discharge," while the German duo referred to it as "reafferenz," both of which likened it to the negative of a photograph. In turn, the stimulation from the sensor during the movement is the positive print that, when united with the negative, makes the image disappear. Experiments by Sperry, von Holst, and Mittelstaedt were among the first to demonstrate that action directly influences sensation. Neurons giving rise to an eventual motor command send a copy to neurons dealing with the sensor-supplied signals. These latter neurons (call them *comparators*) can compare what comes in and what goes out from the brain[27] (Figure 3.1).

The corollary discharge (i.e., the mechanisms that allows the sensory system to be informed about an action command) is an invention of the brain. No such internal reafferentation exists from the ventral horn motoneurons to the sensory side of the spinal cord. So Betz's intuition is only partially right. There is indeed a geometric separation of the posterior "sensory" and anterior "motor" areas in the brain. However, their way of communication is very different from that of the spinal cord. The spinal cord is a modular system, with each module utilizing pretty much the same algorithm and intersegmental coordination. It is mostly about reflexes and locomotion supported by a large number of neurons.

---

that an insect can discriminate between its own movements and movements in its environment but were confident enough to generalize their findings: "the reafference principle applies throughout the CNS [central nervous system] from the lowest phenomena (internal and external control of the limbs, relations of different parts of the body to each other) to the highest (orientation in space, perception, illusions)." Compare this with Sperry's description "[The] kinetic component may arise centrally as part of the excitation pattern of the overt movement. Thus, any excitation pattern that normally results in a movement that will cause a displacement of the visual image in the retina may have a corollary discharge into the visual centers to compensate for the retinal displacement. This implies an anticipatory adjustment in the visual centers specific for each movement with regard to its direction and speed . . . with the retinal field rotated 180 degrees, any such anticipatory adjustment would be in diametric disharmony with the retinal input, and would therefore cause accentuation rather than cancellation of the illusory movement" (Sperry, 1950, p. 488). The essential features of corollary discharge are also the foundation of predictive coding (Rao and Ballard, 1999; Kilner et al., 2007) and motor control theory (Shumway-Cook and Woollacott, 1995; Wolpert et al., 1995; Kawato, 1999; Grush, 2004).

27. The corollary discharge mechanism represents a basic pattern of loop organization of the brain. Output patterns inform input analyzer circuits. We will discuss extensions of this fundamental mechanism in multiple chapters. The comparator circuit has been long viewed as a fundamental mechanism of many brain functions (Sokolov, 1960, 1963). MacKay's (1956) epistemological automaton compares actual and expected sensory inputs. Excitatory recurrent feedback circuits might be viewed as an internalized version of the corollary discharge mechanism.

Chapter 3. Perception from Action

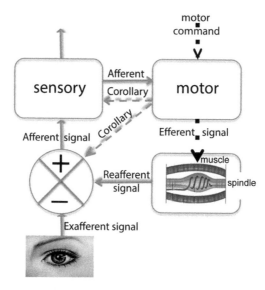

Figure 3.1. Schematics of the corollary discharge mechanism. Motor command signal is sent from the motor areas to eye muscles (efferent signal). At the same time, corollary discharge (*dashed arrows*) is also sent to comparator mechanisms in the sensory system. The comparator performs a subtraction or divisive normalization on the external (exafferent) signal determined by the corollary discharge. In addition, the magnitude of the reafferent signal from the tension sensor in the muscle can exert a delayed affect on the sensory signal. Projections from motor cortex to sensory cortical areas are a common architectural feature in all mammals.

In fact, the volume of the spinal cord in the elephant is larger than the brain of a macaque monkey. So the "smartness" of the brain is due not only to large numbers of neurons and synaptic connections. What also matters is how those connections are organized. The corollary recurrent loop from the output to the input is such a key motif of the cortex. I will return to recurrency, positive feedback, and corollary discharge in several chapters to argue that these mechanisms are the most important grounding processes of cognitive operations.

Two Types of Corollary Discharge Mechanisms

There are at least two ways to distinguish the sensory stimulation produced by one's own movement from real motion in the surrounding world. The first one can be called *adaptive filtering*, an example of which can be found in crickets. Have you ever tried to catch a chirping cricket? Crickets are wildly loud, but when you move toward them, they detect your motion and become quiet. For this reason, farmers in rural Japan used crickets as sentinels to protect their households. They put them in small cages where they sang during the dark hours of the night, except when they sensed something moving. The ensuing silence then served to alert the householder.[28] Crickets do not have ears, but they are equipped with a pair of tympanal organs on their legs. This super-sensitive device can detect vibrating air molecules (sound), which provides their primary alarm mechanism to avoid predators. In addition to listening, male crickets also "sing"; that is, they have a file and scraper body part on their wings that makes a loud *chirping* noise. Their 100-decibel chirps are several-fold louder than the sound of your footsteps, yet the cricket can be fully alert while chirping. Its brain counteracts the massive influx of auditory, proprioceptive, and mechanoreceptive information during sound production by inhibiting its central auditory neurons with a corollary discharge synchronously with sound production. Remarkably, in the cricket's brain, the mechanisms responsible for this forward signaling are accomplished by a single inhibitory neuron. When this corollary discharge neuron is active, it selectively inhibits the response of auditory neurons associated with self-generated chirps. A subtraction or division/normalization process (Chapter 11) in these auditory neurons removes the loud self-generated vibration signal yet maintains the cricket's ability to detect much quieter external stimuli.[29]

Distinguishing real changes in the environment from apparent sensory changes due to one's own movement can be also accomplished through a second mechanism that could be called "time division" because corollary discharge and signal detection occur at different times. For example, when you blink or produce saccadic eye movements, the input from the world to the retina disappears or gets blurred, but we do not notice that. An ideal observer could "see" everything around itself all the time, but in the real world the "seeing" is a compromise between spatial resolution and speed. A fly's eye and a digital camera both have a certain number of pixels to work with. The camera typically scans the target pixel by pixel, and its effective resolution is determined by the number

---

28. As told by Dethier (1987). His essay is worth reading for both scientific insight and the aesthetic beauty of the narrative.

29. In a remarkable series of studies, Poulet and Hedwig (2006) used simultaneous intracellular recordings of auditory neurons in the prothoracic ganglion, where primary auditory signals are processed, while searching for the corollary discharge in the mesothoracic ganglion, which houses part of the chirp pattern-generating network. The dendrites and cell body of the inhibitory corollary discharge neuron are surrounded by the chirp pattern generator neurons, which excite it, and its extremely large axon tree forms numerous synapses with the auditory neuropil in the prothoracic ganglion. The neuron generates bursts of spikes in synchrony with each chirp but does not respond to external sound. Its destruction has no impact on chirp production, so its sole job description is to provide corollary discharge to the auditory neurons. To date, this is the clearest mechanistic description of such a mechanism.

of pixels the processor handles in a given amount of time. The compound eye of the fly has approximately 3,000 pixels, and, by flying fast (perhaps fifty times faster than we walk, relative to body size), the fly generates an image shift on the retina, called *optic flow*, that scans a large part of its environment. The fly's erratic-looking zig-zag flights are actually stop-go saccadic patterns that actively shape its visual input by enhancing depth resolution and distinguishing figures from the background.

Other creatures scan the world differently. Frogs lack eye muscles, but respiration moves their eyes rhythmically. The preying mantis moves its entire head in saccadic scans. Spiders, with their body anchored to their net, protrude their eyes and rotate them like miniature periscopes to view the world. Vertebrates with superior brain processing power exploit both spatial resolution and scanning strategies. The eyes of primates, including humans, have many cones in the foveal region for accurate vision and many more rods surrounding the fovea to detect motion. Primate eyes can move either slowly (explorative) or quickly (saccadic). Like spiders, we can stay stationary yet monitor a very large part of our surroundings as long as our brain moves our eyes. During saccadic eye movements, the visual scene does not get blurred. Instead, it is suppressed momentarily by postulated corollary discharge. The visual field appears to be at one place or another, without experiencing the in-between scan. You can check this by looking into a mirror. You see yourself continuously and do not notice your saccadic eye movements. The suppression mechanism alerts visual processing areas in the brain that the forthcoming disruption is the result of your brain's command, not a change in the visual world.

In the primate brain, the neuronal hardware of this forward signaling is a bit more complicated than in the cricket, but the principle is the same. Neurons in the frontal eye field area of the neocortex are good candidates for the job because they receive both visual and corollary discharge information about eye saccades from the medial dorsal nucleus of the thalamus. After comparing the visual input with the grounding information about eye movement, frontal eye field neurons inform their targets whether the stimulus remained stable or moved, the strength and timing of their corollary discharge response, and the magnitude of stimulus translation if the scene moved during the saccade. Inactivation of the action path, for example by infusing local anesthetic in the thalamic medial dorsal nucleus to abolish spiking activity locally, will result in a failure to inform the visual neurons about the necessary grounding information. As a result, frontal eye field neurons supply nonsense information to their targets. Under such conditions, the experimental monkey reports that the visual scene jumps with each saccade just as humans feel after eye muscle paralysis. In further support for the role of the frontal eye field, experiments in monkeys show that microstimulation of this area enhances the responsiveness of visual cortical neurons with spatially overlapping receptive fields.[30] The just-discussed experiments and observations indicate that the self-motion–generating circuits provide a prediction signal for the sensory system so that the expected signal can be compared with the one conveyed to the brain by the sensors.

The Illusion of Continuous Vision

Sampling of the visual world is not continuous because it is interrupted by saccades, with each saccade causing a loss of up to 10% in sampling time. However, this loss has important advantages. First, blurred vision is prevented because during the saccade the visual input is suppressed. Second, corollary discharge is an important timing signal that helps coordinate neuronal activity beyond the visual system. Third, suppressed spiking in several types of visual neurons during the saccade allows them to replenish their resources, for example by restituting dendritic sodium and calcium ion channels inactivated during intense spiking. As a result, after the saccade, the visual system is transiently more sensitive to stimulation, a considerable gain.

Action in the form of muscle activity (again!) may be the evolutionary origin of intermittent sampling of sensory inputs. The muscle's fastest reaction is a twitch followed by a refractory period, during which it cannot be activated. This mechanical constraint may explain why sensors adapted to such a regime. Regardless of its origin, intermittent sampling is advantageous for transmitting information. First, intermittency introduces a way to chunk the information, similar to spaces between words. Thus, every neuronal message can have a beginning and stop code (Chapter 6). Second, intermittency simplifies how information processing can be coordinated across brain areas because it introduces the means of generating a clear time reference frame that can be shared by all mechanisms involved. Blind people discovered the supremacy of intermittent

---

30. Neurons in the parietal cortex (particularly in the lateral intraparietal area) can also function as corollary discharge comparators and drive neurons in the frontal eye field. Other brain regions, including the lateral geniculate body, superior colliculus, and even early visual areas V1 to V4 appear to be involved. In addition to suppression of retinal inputs, many interesting things happen in the brain around the time of saccadic eye movements, including compression of space and time and displacement of part of the world depending on the direction of the saccade (see Chapter 10). The magno- and parvocellular pathways in the early stages of the visual system are affected differentially. It is mainly the magnocellular pathway of the lateral geniculate body that is suppressed by saccades and informs the motion centers. In addition to corollary mechanisms, visual motion caused by the eye movement itself can contribute to masking vision. Duhamel et al. (1992); Umeno and Goldberg (1997); Sommer and Wurtz (2006); Crapse and Sommer (2012); Moore et al. (2003). For reviews, see Ross et al. (2001); Crapse and Sommer (2008).

sensory sampling by realizing that forms constructed from raised dots were more easily discriminated by the fingertip than continuous raised lines.[31] Sensing text with fingers also begins with movement.

PERCEPTION AND ACTIVE SENSING

Not everyone has the luxury of saccadic eye movements. A research participant named AI was unable to move her eyes since birth because of fibrosis in her eye muscles. Yet she had no complaints about her vision. She could read and write, although her reading speed was slow. She accomplished this feat, compensating for her lack of eye movements, by making head saccades. These saw-tooth movements to her right had an average duration of 200 ms. At the end of the line, her head jerked back to the left side of the page to read the next line. During picture viewing, AI's head movements scanned the important details of the picture, similarly to the way we deploy eye saccades to extract interesting features from a scene (Figure 3.2). By actively moving her head (and consequently her eyes), AI could enhance her visual sensing ability.[32]

Active sensing refers to a brain-initiated search as opposed to the response to an expected event.[33] Perhaps another word, "observation," equally well captures the same process. In the real world, stimuli are not given to the brain. It has to acquire them. The sensitivity of sensors depends, in part, on the effectors that can move them and maximize their efficacy. It is a bit like the relationship between a camera and its user. The camera takes the photo, but what is captured depends on the action of the user. Unexpected stimuli, when they occur independently of movement, trigger search behavior and optimize the sensors. The physiological foundation of such active sensing is also corollary discharge.

Active sensing occurs in two forms: the entire body can translocate (e.g., the fly's solution for vision), or sensors are moved locally (e.g., the spider's solution

---

31. Louis Braille, who was blind from childhood, went on to create the Braille tactile writing system, a sort of parallel Morse code.

32. Gilchrist et al. (1997).

33. Ehud Ahissar at the Weizmann Institute refers to active sensing as "closed-loop" sensing, in contrast to the perception–action semi-loop, in which the interaction between the sensory organ and the environment is the first step (Ahissar and Assa, 2016). James J. Gibson's affordance theory can be considered as a forerunner of active sensing. According to his theory, not only is the world perceived in terms of object shapes and spatial relationships, but these things also offer intuitive possibilities for action (affordances). The environment "offers the animal, what it provides or furnishes, either for good or ill" (Gibson, 1979). While Gibson tried to eliminate the decision-maker between perception and action, he still thought that perception drives action.

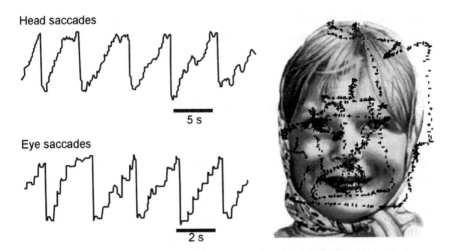

Figure 3.2. Head movements of patient AI with paralyzed eyes (head saccades) and eye movements from a control subject (eye saccades) during text reading recorded using a head-mounted search coil. Although AI is considerably slower overall, and her head stability is not as good as that of the control subject, her saccade strategy is the same as that of the normal subject. Right, AI's head movements while viewing a picture. Redrawn after Gilchrist et al. (1997).

for vision), but the purpose is the same. The two strategies can be exploited by the same animal. For example, when a hunting dog is taking a bird scent sample on the ground, it sniffs the grass rhythmically, but to chase the same scent in the air, it runs fast with its nose up to create a scent flow. In either case, the physiological foundation to make sense of such active sensing behavior is corollary discharge.

Mechanisms of Active Sensing

Researchers understand the neuronal mechanisms underlying certain forms of active sensing. Some animals can both sense and generate electric fields. The weakly electric fish (*Eigenmannia*) actively probes its environment with electricity. It has a specialized structure called the *electric organ* in its tail, which emits electric fields. It also has electroreceptors called *tuberous receptors* in its skin that sense electric fields for object detection *(electrolocation)* and communication with other electric fish *(electrocommunication)* in muddy waters.

In electrolocation, the fish's brain analyzes the frequency and phase difference of its emitted sinusoid electric field and the version returned from objects. When two fish are near each other, the frequency difference of the fields emitted by each is determined by the combination of the two fish's discharge patterns. Thus, the problem of sorting out self-generated electric field from fields produced by other fish is the same as that of touching something versus being touched—only the medium is electrical, not mechanical. The solution of the brain is also similar: a circuit analogous to the corollary discharge signal comparator. The firing pattern in the electric organ is controlled by motor neurons that are driven by the motor command nucleus. The axons of the neurons from the command nucleus bifurcate, with one pathway going to the medullary relay nucleus whose axons project along the spinal cord and excite electromotoneurons that generate the electric organ discharges. The other collateral pathway innervates the bulbar area, which contains the corollary discharge comparator circuit (Figure 3.1). This circuit is analogous, albeit different in its details, to the corollary circuit of the cricket.[34]

Echolocation, also called *bio-sonar*, is another well-known active sense that works by emitting sound or ultrasound calls into the environment and analyzing the returned echoes to locate objects and prey. Bats and marine mammals are the animals best known for their sonar sensing. Whales use a series of clicks during their dives. The precision of bat echolocation is good enough to identify mosquitoes even when both the insect and the bat are flying. The sound localization mechanisms in the brainstem act by detecting the time differences of sounds between the two ears. The self-vocalized sounds are strongly attenuated between the cochlear nerve and the inferior colliculus, early stations of the auditory system, as well as in the auditory cortex.[35]

The neuronal mechanism used in echolocation is similar to our ability to use self-generated sounds to tell whether we are in a small or large room. Normally, we do not use this option much because we have the luxury of vision. However, this ability can be enhanced if needed. Some blind individuals can replace their eyesight by making loud footsteps, tapping a cane, or producing mouth clicks and listening to the returning echoes. These active sensing methods are good enough for crossing roads, even riding a bicycle or rollerblading. Some exceptional blind individuals can identify an object's distance, size, and texture, such as the difference between a metal fence and a wooden fence.[36]

---

34. Walter Heiligenberg's book is a masterful combination of ethology, electrophysiology, and modeling on electric sensing in Eigenmannia used for jamming avoidance (Heiligenberg, 1991). For an update, see Chacron (2007).

35. Suga and Schlegel (1972); Hechavarría (2013).

36. Echolocation is a trainable skill that can provide blind people with self-reliance in their daily life. Ben Underwood and Daniel Kish, two Californians who were blind from childhood, are the best-known individuals to use these sonar techniques (Kremer, 2012; Kolarik et al., 2017).

In imaging studies, blind research participants show increased activity in visual cortical areas, and to some degree in the mid-temporal region, when they actively localize objects based on their acquired sonar methods. Such activation is strong in individuals who became blind early in childhood, weaker in people who lost their eyesight later in life, and absent in normal control participants. Surprisingly, no differences are detected in the auditory cortex. These observations imply that the visual cortex, deprived of its natural input, can be trained to adopt a novel function, such as echolocation.[37] To succeed, self-initiated action is a precondition.

Olfaction

Smelling provides an elegant example of active sensing in mammals. In olfactory research, every investigator knows that whatever questions they study, the phase of respiration should be taken into account because respiratory movement modulates perception. The goal of sniffing is to optimally sample the odorant and route it to the olfactory receptors. Sniffing is a rhythmic pattern (4–10 sniffs per second) that generates turbulence in the nose so that odorants, which would only slowly diffuse during regular breathing, come in direct contact with the receptor. As a result of sniffing, the sampling becomes intermittent: effective during inhalation and reduced or absent during exhalation. But that is not all. The entire olfactory system gets informed ahead of time about the expected arrival of the odorant by a corollary discharge signal, initiated by exhalation (or more likely due to relaxation of the inspiration motor action). While the exact paths and mechanisms of this signaling are not well understood, ample physiological evidence shows that the entire system, from the olfactory receptor neurons to the olfactory (piriform) cortex, generates rhythmic patterns in phase with the sniff cycles. From the olfactory receptor neurons to the mitral cells (the main output of the olfactory bulb), all neurons show large-amplitude membrane voltage oscillations and synchronous spiking to the sniff cycle. Pyramidal neurons in the olfactory cortex, where the significance and meaning of the odor information are likely extracted, also show strong inhalation-coupled spiking output.

One may argue that the mechanical effect of active sniffing, inducing turbulence in the nose, is sufficient to enhance odorant detection and identification without any need to inform neurons in the olfactory system about the phase of respiration. To counter that argument, researchers injected odorants into the

---

37. De Volder (1999); Thaler et al. (2011).

bloodstream at low concentrations in human volunteers, thus removing the air turbulence effect. Participants still detected the odorant more reliably when they sniffed, demonstrating that, even in the absence of the odorant in the nose, the motor action still provides a gain.[38] This finding supports the claim that feedback from action to the sensory system enhances perception.

Vision as a Searchlight

Corollary discharge from saccadic eye movements serves as a signal to inform the visual system whether the world or the eyeball is moving. The same circuit is involved in visual search, which is another name for active sensing. Vision or seeing is not a passive camera-like function but an active scanning of the environment.[39] High-acuity vision in humans (i.e., what allows you to read text in this book) is possible only with the cone receptors of the foveal region, corresponding to the central five degrees of the field of vision. One degree of visual angle is roughly the width of your index finger held at arm's length. If we fix our eyes on some part of a picture to keep the high-acuity receptors in place, we can see that the surrounding region becomes blurry. The effect is like pointing a flashlight toward part of the environment in darkness. To see what is out there, we must move the flashlight back and forth. During the day as well, we are taking in our surroundings in snippets, but because of the frequent scanning eye movements, our visual system's memory creates the illusion that we see the entire visual scene with high acuity and at once.

Even when we fix our eyes on a spot, eye movements assist vision via frequent "microsaccades," tiny jumps of the eye that move the scene across a few dozen photoreceptors. These movements ensure that our eyes do not fixate on a completely stationary picture. Microsaccades increase visual performance by enhancing the spiking activity of neurons in the visual cortex. When macaque

---

38. The authors (Bocca et al., 1965) interpreted their observations showing that the mechanical stimulation of the epithelium by sniffing lowered the threshold for detection of blood-borne molecules. A contemporary demonstration of bypassing the olfactory mucosa is by Dmitri Rinberg and colleagues (Smear et al., 2013). They optogenetically stimulated a single glomerulus, the functional unit of olfactory processing, and showed that this "implanted odorant" was sufficient for odor perception by simultaneously relaying identity, intensity, and timing signals.

39. I have often wondered where visual science would stand today if early visual researchers had been engineers or control theorists. Instead of taking the reductionist path of fixing heads and eyes, what if they had developed methods for detecting the position of the eyes in relationship to the environment in freely behaving animals and observed the mechanisms of seeing in action? It is notable that David Hubel focused his interest on the significance of eye movements in vision toward the end of his career (Martinez-Conde et al., 2004).

monkeys fixate on a target, and an optimally oriented line is centered over a cell's so-called receptive field[40] in the primary visual cortex, spiking activity increases after microsaccades. Importantly, such microsaccade-induced, transiently enhanced neuronal excitability is present in every visual area examined by different research groups and even in the lateral geniculate body, a thalamic structure that transmits the retinal information to the primary visual cortex. A great number of neurons are affected by eye movements, as might be expected because visual areas occupy a large part of the brain.[41] In other words, movements of the eye improve the visual performance of the sensory system and assist with the maintenance of perception during visual fixation. Saccades thus can be considered "visual sniffs," which provide a sensory gain (Chapter 11).

In terms of visual function, there is no real difference between regular and microsaccades. During eye fixation, the saccades are smaller ("micro"), while during "voluntary" visual search, they are larger. Both types of saccades are initiated in the superior colliculus in the midbrain. The widespread anatomical projections from this structure may explain why AI could effectively compensate for her loss of eye muscles by inducing saccade-like head movements (Figure 3.2). This observation emphasizes an important rule in neuronal organization: functions that are important for survival are widely and redundantly distributed in the brain—a robust safety mechanism.

Hearing

When our ears are plugged, we have trouble not only hearing others but also speaking or singing properly. Auditory feedback is very useful in speech.

---

40. Neurons respond to various constellation of inputs. The most robust response is believed to reflect the "relevant" input. Many visual neurons have preferred orientation selectivity, motion direction selectivity, color preference, etc., which we call the receptive field.

41. Many excellent original studies on eye saccades are compressed in these short paragraphs. Some of these studies are reviewed by Yarbus (1967) and Carpenter (1980) and Martinez-Conde et al. (2004). Martinez-Conde et al. emphasized the importance of eye movements in vision but considered them only to affect retinal processing: "microsaccades primarily modulate neural activity in early visual areas through retinal motion." Corollary discharge within the brain as a mechanism for visual gain is not discussed, even though saccades can affect the entire visual system. See also Otero-Millan (2008). Hofmann et al. (2013) discusses the many forms of active sensing in bats (echolocation) and invertebrates (e.g., electric communication). Wachowiak (2011) and Morillon et al. (2015) are excellent reviews on the role of motion in hearing and olfaction, respectively. Visual saccades affect even hippocampal neurons (Meister and Buffalo, 2016).

Distinguishing between our own speech and a recording of it is a piece of cake because the corollary discharge mechanism provides an intimate relationship between sound production and hearing. Each uttered sound leads to the next action, and every sound is preceded by action preparation. Sound utterances adjust the sound-controlling muscles and help to achieve the desired outcome.

As in other sensory systems, the auditory regions of the cortex also receive direct projections from motor areas, providing a substrate for conveying corollary discharge signals. A copy of the motor command from the premotor cortical region is sent to the auditory cortical areas, where it strongly suppresses both spontaneous and tone-evoked synaptic activity during a wide variety of natural movements. This is achieved by a so-called *feed-forward inhibitory mechanism*, which entails the activation of inhibitory interneurons in the auditory cortex and the consequent transient inhibition of the spiking activity of pyramidal neurons. As a result, sound-evoked cortical responses are selectively suppressed during movement.[42] In addition to cortical mechanisms, the muscles of the middle ear, which regulate the sound transmission from the eardrum to the cochlea, perform a similar preemptive action. Their activation by the movement command strongly attenuates the sound pressure before it can affect the cochlea. These corollary actions of the brain's output are responsible for making the sounds of the heavy timpani in Handel's *Messiah* pleasing, whereas the same sounds encountered unexpectedly would startle us.

Action-induced corollary signaling thus can transiently dampen the sensitivity of the auditory system to predictable sounds, while maintaining its responsiveness to unexpected stimuli. These same mechanisms, when derailed by disease, are implicated in tinnitus and auditory hallucinations.[43] Neuronal excitation sent from the motor areas to the auditory areas that does not occur during self-induced sounds are interpreted by the brain as real.

Body Sensation

The traditional separation of motor and sensory functions is even more prominent in the somatosensory system. This view is explicitly illustrated by the classical textbook figures of motor and sensory "homunculi" whose jobs are to perceive and act, respectively, with a one-way, sensory-to-motor

---

42. Paus et al. (1996); Houde and Jordan (1998); Zatorre et al. (2007); Eliades and Wang (2008); Nelson et al. (2013); Schneider et al. (2014).

43. Feinberg et al. (1978); Ford and Mathalon (2004); Langguth et al. (2005); Nelson et al. (2013).

communication between them.[44] But recent discoveries in rodents paint a different picture. Rodents have an elaborate whisker system adapted for tactile exploration, especially in the dark. Whiskers palpate objects just like fingers do. There are multiple loops between the input and output paths, which include short brainstem connections and longer thalamic, cortical, and cerebellar paths. Importantly, ample anatomical evidence shows that the traffic between sensory and motor areas is bidirectional even in the cortex. Signals of whisker motion reach the somatosensory area and can serve as corollary discharge. These signals enable neurons in this area to integrate the movements and touches of multiple whiskers over time, key components of object identification and navigation by active touch. In addition, the somatosensory area of whisker representation (called the *barrel cortex*) forms an equally direct and prominent motor control pathway. Activation of neurons in the "somatosensory" cortex is as effective in controlling whisker movements as stimulation of the anatomically designated motor cortex. Direct anatomical projections from the somatosensory cortex to the spinal cord are present not only in rodents but in monkeys as well, suggesting that similar mechanisms are at work in different species.[45]

A classical demonstration for the critical role of action in perception is the tactile-vision sensory substitution experiment by Paul Bach-y-Rita and colleagues at the University of Wisconsin, which allowed blind people to "see" using an array of electrodes placed on the tongue and stimulating the sensory terminals with weak electrical pulses. Blind participants were able to partially gain some visual function conveyed to the tongue by the output of a video camera. A crucial component of the successful outcome in each case was the participants' ability to control of the camera. Under passive tongue "viewing" conditions, only tickling the tongue was felt but vision-like sensation did not occur in the participants.[46]

Of course, you may counter my arguments by simple introspection and say that you can sit completely immobile and yet perfectly process the sensory flow. A touch on your hand or a bug flying in your visual field can be detected with no muscular effort whatsoever. Therefore, such examples can

---

44. Neurosurgeon Wilder G. Penfield, working with neurologist Herbert Jasper at the Montreal Neurological Institute, electrically stimulated the brain of awake patients with electrical probes and observed their motor responses or obtained their verbal feedback. The stimulation technique allowed them to create crude sensory and motor maps of the cortex (cartoon representations of the body surface, or *homunculi*). These somewhat misleading anatomically discrete maps are still used today, practically unaltered.

45. Ferezou et al. (2007); Mátyás et al. (2010); Hatsopoulos and Suminski (2011); Huber et al. (2012); O'Connor et al. (2013); McElvain et al. (2017).

46. Bach-y-Rita et al. (1969).

be taken as evidence in favor of the sensory-to-motor sequence of perception. However, such perception occurs in an already "calibrated" brain. Without actively sensing stimuli, without investigating them at some point in life, stimuli cannot not acquire a meaning to the brain. Once meaning (i.e., significance of a stimulus or event for the animal) emerges, it can be placed in a memory store. We can say it becomes *internalized*, and the internalized pattern can function as a grounding mechanism for further perceptual interpretations. I will expand on these issues in Chapter 5.

BODY TEACHING BRAIN

Although the brain has no a priori clues about what its sensors are sensing or what its effectors are effecting (Chapter 1), the developing brain does not start from scratch; it benefits enormously from inherited and early programs. But one-size-fits-all blueprints are not adequate to do the job because bodies come in different shapes and forms. The genetic blueprint is simply a "protomap." In the newborn rat, the spinothalamic tract and thalamocortical somatosensory pathways are already in place. This map thus has some correspondences with body parts. However, cross-talk among the body part representations is limited at this early stage because the corticocortical connections are just beginning to grow after birth. How do the growing axons "know" which neurons to innervate and which ones to avoid? The proximity relationships help but are not sufficient.

At the output end, the ventral horn motoneurons in the spinal cord, already wired to the muscles, begin to generate irregular, uncoordinated movements. These seemingly aimless movements in newborn rodents are the same as fetal movements or "baby kicks" observed in later stages of pregnancy in humans. Every expectant mother and physician knows that such kicks are important aspects of the normal development of the fetus. However, the biological utility of the kicks and their service to the brain have been clarified only recently. As you will see later, each kick helps the brain to learn about the physics of the body it controls.

Stretch sensors in muscles and tendons report the contractile state of the muscle to the spinal cord and eventually to the somatosensory cortex. In addition, twitches of skeletal muscles increase the probability that the skin over the muscle will touch another pup in the nest or touch the wall of the womb in case of human fetuses. Due to the physical constraints of the bones and joints, only a limited fraction of muscle movement combinations ever occurs, out of the potentially very large number of possibilities that could result from unrestrained combinations of muscle activity All these movement combinations are meaningful teaching patterns to the somatosensory thalamus and cortex because these are the combinations that will be used later in life. Random and independent induction of cortical activity by the 600 or so skeletal muscles that move the mammalian body would be of very little biological relevance. Furthermore, sending organized inputs to the brain cannot be achieved through merely a genetic blueprint because the metric relations among the body parts change rapidly as the body grows.

How can such dumb "training" from muscle twitch combinations contribute to the formation of the body map? In the newborn rat up, every twitch and limb jerk induces a "spindle-shaped" oscillatory pattern in the somatosensory cortex lasting for a few hundred milliseconds. In the adult animal, neurophysiologists would call such a pattern a thalamocortical "sleep spindle" because such patterns occur in adults only during non-REM sleep. Both in the pup and in a prematurely born human baby, these are the first organized cortical patterns.[47] When long-range corticocortical connections form after birth in the rat, the spindle oscillation can serve to bind together neuronal groups that are coactivated in the sensory cortical areas as a result of the simultaneous movement in neighboring agonistic muscles. Likewise, muscles with an antagonistic movement relationship in the body (e.g., the biceps and triceps of the upper arm) will induce consistent activity–silence relationships in their sensory cortex and create an inhibitory relationship between the respective neuronal groups. After the long-range corticocortical connections are in place a week or so after birth and the body map is formed, the spindles become confined to non-REM sleep. Then the pups can use their body map to respond to local touch and use the proprioceptive information from the muscles and muscle tendons for effortless ambulation.[48] Thus, the initially meaningless, action-induced feedback from sensors transduces the spatial layout of the body into temporal spiking relationships among neurons in the brain. This developmental process is how the brain acquires knowledge of the body it controls or, more appropriately, cooperates with. Thus, a dumb teacher (i.e., the stochastically occurring movement patterns) can increase the brain's smartness about its owner's body landscape. Once the body scheme is built, the relationship between spindles and muscle jerks disappears. The spindles become "internalized" (Chapter 5) and continue to occur during sleep as a self-organized pattern. Yet the relationship

---

47. Similar to early spindles, spontaneous retinal waves trigger activity in every part of the visual system during early developmental stages (Katz and Shatz, 1996).

48. Khazipov et al. (2004); Buzsáki (2006). The early spindles in premature human babies are called *delta-brush oscillations*, and the alternating pattern of silence and activity is known as *trace alternans*. As in rat pups, muscle twitches induce delta brushes at 29 to 31 weeks of postconceptional age in preterm neonates and in utero (Milh et al., 2007).

between movement and spindles does not completely vanish. The reader may have experienced occasional dramatic whole-body jerks just before sleep. Such movements trigger a sleep spindle, as in premature babies.

ACTION AFFECTS BODY MAPS

From my perspective, the term "somatosensory" cortex is a misnomer of the outside-in framework as it not only senses but also simulates the body.[49] This is most clearly demonstrated by the continued representation of missing body parts. The ownership of phantom limbs and the painful feelings that arises from "them" after their loss or amputation is a well-studied clinical problem. If the sensory cortex were truly sensory and fully dependent on external inputs, phantom limbs would not be constructed mentally.[50] Another striking example of body simulation by the somatosensory system is false body part illusions. In the laboratory, a research participant is seated with her left arm resting on a small table, hidden from her view by a screen, and a life-sized rubber model of a hand and arm is placed on the table directly in front of the participant. The participant is asked to fix her eyes on the artificial hand while the experimenter visibly touches the rubber hand with a paintbrush and, at the same time and unknown to the participant, also touches her hidden hand with another paintbrush. If the real and rubber arms are touched multiple times in synchrony, the average participant voluntarily accepts the rubber hand as her own. This simple procedure is sufficient to produce a feeling of ownership of a foreign body and incorporate it into the body scheme.[51] In another version of the experiment, the experimenter moves the participant's left index finger so that it touches the right rubber hand on the knuckle of the index finger, and, at the same time, the experimenter touches the participant's right index finger on the knuckle. If the

---

49. Michael Brecht expanded on these ideas and formulated a body model theory of the somatosensory cortex (Brecht, 2017). He suggests that layer 4 of somatosensory cortex mirrors the entire body and not just sensory afferents. This body model is continuously updated by layer 6 to layer 4 inputs and compares to an avatar rather than a mere sensory map.

50. Ramachandran et al. (1995). Similarly, blind people can imagine and dream about visual scenes. Conversely, patients with bilateral damage to the primary visual cortex can still respond to some visual stimuli and navigate around visual objects, even though they report seeing nothing and having no memory of the objects. This phenomenon is known as *blindsight* (Cowey, 2010; see it yourself here: https://www.youtube.com/watch?v=GwGmWqX0MnM. Monkeys with primary visual area lesions behave similarly; Cowey and Stoerig, 1995). These observations illustrate that "sensory" cortices do not simply serve to veridically "represent" the external world.

51. See a demonstration here: https://www.youtube.com/watch?v=sxwn1w7MJvk.

touches on the two hands are synchronized, the participant soon feels that she is touching her own hand. The illusory touch, in turn, will activate the ventral premotor cortex, intraparietal cortex, and cerebellum, indicating that the illusion reflects the detection of congruent multisensory signals from one's own body, as revealed in imaging experiments.[52]

These illusions derive from the lack of grounding by action. If the participant moves her hand and arm while the rubber hand is being touched, then no illusion arises. Furthermore, the participant loses the sense of ownership of the arm instantaneously when the experimenter moves it. An obvious direct test of the action–perception framework would be to deprive the brain entirely of its motor and autonomic outputs and examine whether sensation could be evoked in a brain that had never experienced touch in conjunction with action. Short of such an action-deprivation experiment, we could ask how the brain would handle an unusual body. For example, what if two brains shared one body? Nature has provided several such examples.

Two Brains Control One Body

Abby and Brittany are conjoined twins who live in Minnesota. They were charming children, and today they are self-confident adults. The twins have separate heads and spinal cords, a joint chest with two hearts, two hands, and two legs. Each brain and spinal cord innervates one hand and one leg. The sense of touch and limb control are restricted to the half of the body that each brain innervates. Yet Abby and Brittany can perform many movement patterns that require bilateral hand and leg coordination, such as walking, running, swimming, ball games, or driving a car. They have different tastes and even different career goals.[53] Because their brains have separate somatosensory innervation of parts of their shared body, many observers have wondered how they effectively coordinate their sensations to perform the synchronous or alternating movements needed for many simple or more complex activities. However, if action teaches perception, as I suggested in this chapter, then we can explain

---

52. Botvinick (2004); Ehrsson et al. (2005).

53. There are numerous YouTube sites on these twins; visit https://www.yahoo.com/tv/conjoined-twins--abby---brittany--get-their-own-reality-show--video-.html. Conjoined twins, when they are respectfully approached, could be a rich source of unique information on brain-body questions. Among these is the important question of whether a feeling of anxiety is a body-induced effect on the brain. For example, researchers could show anxiogenic pictures to one twin and can ask the other twin how she feels. Many other brain-body-brain questions could be explored in conjoined twins.

why movements of the "other body" are inevitably sensed by both brains. It is because the physical constraints of the skeletal system have enabled the incorporation of a body extension. The other body *becomes* part of the body scheme as a result of action-induced joint experience.

If action is so critical for perception, it should play a significant role in cognition as well. That this is indeed the case, I will discuss in Chapter 5. However, a prerequisite of that discussion is the understanding of the collective behavior of neurons, the topic of the next chapter.

SUMMARY

The short message of this long chapter is that perception is an action-based process, an exploration initiated by the brain. This message is fundamentally different from the representational view, fueled by the outside-in empiricist philosophy, which inevitably poses the question: What comes between perception and action? The homunculus with its decision-making power—and, I suspect, the problem of consciousness as well—are unavoidable logical consequences of the separation of perception from action.

I promote the alternative view: things and events in the world can acquire meaning only through brain-initiated actions. In this process, the brain does not represent the world in its numerous and largely irrelevant details but extracts those aspects that have become relevant to the organism by exploration. Thus the brain builds a simplified, customized model of the world by encoding the relationships of events to each other. These aspects of model building are uniquely different from brain to brain.

The critical physiological mechanism that grounds the sensory input to make it an experience is "corollary discharge": a reference copy of a motor command sent to a comparator circuit from the action-initiating brain areas. This comparator mechanism allows the brain to examine the relationship between a true change in the sensory input and a change due to self-initiated movement of the sensors. The same corollary discharge mechanism also serves active sensing, the process by which sensory receptors can be most efficiently utilized to sense the environment.

4

Neuronal Assembly

The Fundamental Unit of Communication

E pluribus unum. (Out of many, one.)
—Great Seal of the United States of America

United we stand, divided we fall.
—The Seal of Kentucky

Success is determined by the collective.
—László Barabási[1]

The outside-in framework offered an explicit recipe to neuroscience for how to explore the brain: present various stimuli and monitor brain responses. To study learning, we should register the relationship between the stimulus and neuronal response at various stages of learning. Similarly, to study motor behavior, we should establish a reliable relationship between neuronal activity and movement patterns. In principle, anything that the experimenter can conceive, including complex terms that describe various aspects of cognition (check out Figure 1.1 again), can be correlated with neuronal activity.

Recording neurons, one at a time, was one of the first popular techniques in the arsenal of neurophysiology.[2] The goal for the experimenter is to find the relevant stimulus that can induce reliable neuronal firing. For example, when recording from the ganglion cells of the retina, it would not make sense

---

1. Barabási (2018).
2. Hubel (1957); Evarts (1964).

to probe their responses with sounds and odorants. Instead, we use visual patterns to probe activity patterns in visual areas. However, as we move deeper into the brain, the idea of a relevant stimulus becomes increasingly more ambiguous. Furthermore, when we study more complex behaviors, such as emotions or memory, the correlational approach needs to be complemented by perturbational methods. Finding the best correlate of the activity of a single neuron and declaring that it responds exclusively or even primarily to, for example, the orientation or color of a visual stimulus, is a heroic venture because the same neurons should be tested over and over, while making sure that the brain state of the animal does not change in the meanwhile. Moreover, finding the physiologically relevant "tuning" response of the neuron may fail if we do not already know its job at least approximately. For example, a neuron may respond relatively selectively to a particular orientation of a black vertical bar even though its main contribution is to color identification. The outside-in approach can provide a tremendous amount of useful information, but it is inherently handicapped because a neuron can be probed only in a few conditions in a given experiment.

Although the spiking activity of a single neuron can offer only limited information, we could perhaps get an average picture of its physiological function by repeating the same experiment again and again, collecting more neurons under the same experimental conditions, and "pretending" that all neurons were recorded simultaneously. This idea is similar to asking each player of an orchestra to play his or her part separately and then combine the individual pieces into a whole. This is the rationale of the single-neuron doctrine.[3] A reasonable argument against the single-neuron framework is that, in a flexible brain circuit, neurons can be repurposed for multiple jobs, like a neuroscientist who can also function as a parent, a tennis partner, or a handyman, depending on the circumstances.[4] More importantly, with single neurons recorded one at a time, one cannot examine how neurons influence each other. A jazz song pieced together from the music played by isolated members will likely not sound right. Based on these considerations, theoreticians began to question that the single neuron is the fundamental unit of computation in the brain. Instead, the idea that a group of neurons that can align themselves for a particular purpose and

---

3. The classic paper in this field of research is by Barlow (1972).

4. For example, the same neurons in the entorhinal cortex can wear many hats. They may be classified as positional, speed, or head-directional neurons. Although a minority of neurons can appear as specialists, the majority have mixed properties. Instead of individual functional classes, all neurons may have multiplexed properties (Hardcastle et al., 2017). Particular behavioral correlates, designated by the experimenter, may in fact correspond to tails of wide distributions of response features (Chapter 12).

disband themselves when not needed emerged. This is the *cell assembly* or *neuronal ensemble* hypothesis.

CELL ASSEMBLY

The concept of neuronal assembly is one of those things that all neuroscientists talk about, weave into their important findings, and use to explain complex observations—even the workings of the mind—yet it lacks a rigorous definition. As a result, scientists define cell assembly differently. The concept is most often associated with Donald O. Hebb, who coined the term in his classic book *The Organization of Behavior*.[5] Hebb recognized that a single neuron cannot affect its targets reliably and suggested that a discrete, physically interconnected group of spiking neurons (the cell assembly) is the unit that can represent a distinct percept, cognitive entity, or concept. An assembly of neurons does not live in isolation but communicates effectively with other assemblies. Because of the assumed strong interconnectivity of the assembly members, the activation of a sufficient number of them can activate the entire assembly, a process described in early texts as "ignition" of the assembly. Hebb needed both a large group of interconnected neurons and a trigger because he was in search of "representations" of the outside world and wanted to understand how one signal leads to another.

The assembly idea likely grew out of the Berliner Gestalt psychology at the beginning of the past century, which defined the hypothetical brain substrates of perception on the principles of proximity, similarity, and good continuation.[6] Hebb's cell-assembly concept could provide brain-based explanations, at least in principle, for many experimental psychological observations. These

---

5. I am Donald Hebb's scientific "grandson." Cornelius (Case) Vanderwolf was one of his few students and also my postdoctoral advisor (http://neurotree.org/neurotree/). Despite our "family" relationship, it is fair to point out that Hebb was not the only person to conceive the assembly or ensemble idea. Similar suggestions were put forward by James (1890), Sherrington (1942), Nikolai Bernstein (1947 in Russian; English translation, 1967), and Konorski (1948). But perhaps the first credit should go to Yves Delage (1919): "every modification engraved in the neuron's vibratory mode as a result of its co-action with others leaves a trace that is more or less permanent in the vibratory mode resulting from its hereditary structures and from the effects of its previous coactions. Thus, its current vibratory mode reflects the entire history of its previous participations in diverse representations" (Translation by Frégnac et al., 2010). Delage thus had already combined the advantages of oscillations and cell groupings, and his description also relates to Hebb's second major concept (i.e., the spike timing-based plasticity rule). *Sub sole nihil novi est.*

6. The Gestalt idea of psychology has served as an important and recurring counterpoint to the single-neuron "doctrine" of physiology (Barlow, 1972).

hypothetical neuronal assemblies with assumed discrete boundaries could represent objects or even abstract entities of thought. Coupling two assemblies could serve as a neuronal substrate for associations, as activation of one assembly could then trigger another assembly. Flexibly linking several assemblies into a ring could underlie short-term memory, as reverberation of the activity could persist after the triggering stimulus had vanished. The chaining of such hypothetical cell assemblies, in turn, could support complex cognitive processes, such as memory recall, thinking, and planning, assuming that some internal process can indeed generate such sequences (Chapter 6).[7]

Hebb also realized that a second rule is necessary to flexibly connect assemblies and reinforce these connections so that they can support long-term associations. He suggested a general way that synaptic connections between pairs of neurons could be modified by appropriate timing of their action potentials. Modification of many such connections could form a new memory or an "engram."[8] A historical paradox is that his second rule, the Hebbian plasticity rule or simply *Hebb's rule*, an idea that he borrowed from others, is what earned him a prominent place in the neuroscience hall of fame. In his own words, "The general idea is an old one, that any two cells or systems of cells that are repeatedly active at the same time will tend to become 'associated,' so that activity in one facilitates activity in the other." Hebb's one-sentence rule became perhaps the most often cited quote in neuroscience, "When one cell repeatedly assists in firing another, the axon of the first cell develops synaptic knobs (or

---

7. Hebb's ideas shaped the thinking of numerous researchers who incorporated his ideas in various ways into their brain theories; for example, Miller (1956); Marr (1971); Braitenberg (1971); John (1972); Shaw et al. (1985); Damasio (1989); Abeles (1991); Churchland and Sejnowski (1992); Edelman (1987); Wickelgren (1999); Pulvermüller (2003); McGregor (1993); Miller (1996); Milner (1996); Kelso (1995); Mesulam (1998); Laurent (1999); Varela et al. (2001); Yuste et al. (2005); Harris (2005); Buzsáki (2010) and brain models, e.g., Willshaw et al. (1969); Palm and Aertsen (1986); Hopfield (1982); Amit (1988); Bienenstock (1994); Wennekers et al. (2003). The term "cell assembly" evolved over years, and one might argue that it no longer resembles what Hebb originally intended. In my discussion, I use Hebb's original definition.

8. The idea of the "engram," used to define a hypothetical group of neurons representing memories, is closely related to the concept of cell assembly. The term "engram" was introduced by German psychologist Richard Semon (1859–1918; see Schachter, 2001) and popularized by Karl Lashley (1930), who made a lifelong search to identify the engram. Lashley's "failure" to localize it could be explained by the lack of a good definition or its widely distributed nature. The Nobel Laurate Susumu Tonegawa at MIT has recently revived the term "engram" and performed a series of experiments with optogenetic methods to erase and implant memories into the hippocampus of mice (Tonegawa et al., 2015). For a review of the history and present status of the engram, see Schacter (2001), Josselyn et al. (2015).

enlarges them if they already exist) in contact with the soma of the second cell." Or, put more simply, "neurons that fire together wire together."[9]

Hebb's definition of a cell assembly relies on the structural and physiological connectivity of neurons. Specifically, its members are connected by excitatory synapses created or strengthened by experience. Once a cell assembly is formed, activation of a small group of members can reactivate its entire spatiotemporal signature. The assembly and plasticity concepts of Hebb's hypothesis defined a program in theoretical neuroscience that has prevailed to the current day. Based on this hypothesis, cognitive psychology established a grand and comprehensive research project to link psychological and physiological processes.[10] It opened the path to tracking down the physiological mechanisms of classification and categorization.[11]

While the concept of the cell assembly has proved useful, Hebb's definition made its alleged substrate virtually unidentifiable. In addition, tracking down the signatures of cell assemblies in physiological experiments with the available single-neuron method turned out to be a formidable task. Investigators noted early on the large trial-to-trial variability of neuronal firing patterns. Even the "best" neuron, which robustly responded to a stimulus on some trials, fired one lonely spike or remained completely silent on other trials. Under the cell

---

9. The original sentence is "neurons wire together if they fire together" (Löwel and Singer, 1992), although the phrase has been most widely popularized Carla Schatz. For demonstration of Hebb's rule, known also as "spike timing-dependent plasticity" (or STDP) see Magee and Johnston (1997); Markram et al. (1997); Bi and Poo (1998). Prior to Hebb, the timing rule was already well-known from Pavlovian conditioning. The conditional signal (CS) always must precede the unconditional signal (US), and both the CS and US should occur with similar probability to produce successful associations. Reversing the contingency actually makes the CS a predictor of US at a lower than chance probability. Before the in vitro experiments on STDP, Levy and Steward (1983) had already demonstrated the paramount importance of timing in anesthetized rats by associating the contralateral (weak) and ipsilateral (strong, teacher) entorhinal input-evoked local field potential (LFP) responses in the dentate gyrus.

10. The concepts of the cell assembly and spike timing-dependent plasticity have become the guiding rules in many forms of artificial neural networks. Due to connectedness, activity in a few neurons tends to activate all members of an assembly. As a result, the pattern as a whole becomes "auto-associated" and fixed to represent a particular item. The most popular model based on these principles is the *Hopfield attractor network* (Hopfield, 1982). Activity in the Hopfield network varies with time and can jump or move slowly from one stable state (called the "attractor") to the next. The jumps of the activity bump can be considered to be Hebb's phase sequence. The number of items (memories) that can be stored with tolerable interference scales with the number of neurons.

11. According to Gerald Edelman, the most fundamental brain operations are integration and segregation (Edelman, 1987; Tononi et al., 1994). Other antonym pairs, such as parsing versus grouping, differentiation versus generalization, and pattern separation versus pattern completion, are also frequently used.

assembly hypothesis, neuroscientists tacitly assumed that the recorded single neurons had several unrecorded partners and that their mean collective behavior was more reliable than the information available from a single neuron.[12] To test the true collective behavior of neurons and prove or dispute Hebb's assembly idea, researchers had to wait until simultaneous recording of large numbers of neurons became possible.

Population Vector

Apostolos Georgopoulos and his colleagues at the University of Minnesota were interested in the neuronal control of movement direction. They trained a monkey to move its arm to one of eight possible targets and observed a striking relationship between the discharge activity of single neurons and the direction of the monkey's arm movement. Many neurons in the motor cortex had a preferred direction of reach. That is, they fired action potentials maximally when the monkey's hand moved in a particular direction, less so when it moved toward a neighboring target, and not at all when the hand moved opposite to the preferred direction (Figure 4.1). Although the experimenters recorded only one neuron at a time, they collected the activity of many neurons, one after the other, and analyzed them as if they had been recorded simultaneously. This cavalier simplification was possible because the monkey's motor task was very stereotypical.

To assess the contribution of all neurons to any given action, Georgopoulos formulated a *population vector* hypothesis. In such a vector, the contributions of neurons with different preferences are summed to produce a final movement command. Each neuron fires the most spikes when the arm moves toward its preferred target, but neurons with nearby preferred directions can also support the same direction less strongly, so the final vote is calculated by vectorial summation of preferred directions of individual neurons weighted by their firing rates. By examining the firing rates of many direction-tuned neurons in a given time window, the population vector model can precisely describe the resulting movement direction. Moreover, when the monkeys were required to translate locations of visual stimuli into spatially shifted reach targets, the population vector accurately predicted mental rotations.[13] In recent years, using

---

12. A similar sentiment was also expressed by von Neumann (1958), who suggested that multiple neurons redundantly work together and that their joined activity corresponds to a bit of information.

13. Georgopoulos et al. (1986, 1989). An implicit prediction of the population vector idea is that neurons are relatively equal in their contribution to command muscle movements when

Chapter 4. Neuronal Assembly

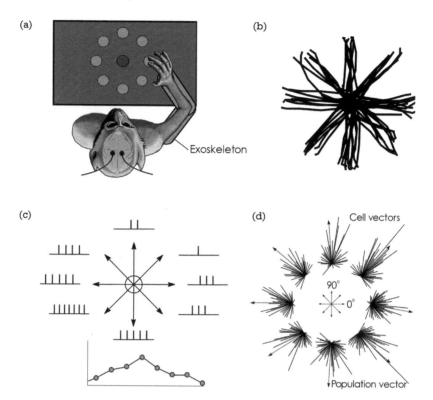

Figure 4.1. Population vector coding of reach movements. A: Monkey reaches to one of eight possible reach directions. B: Trajectory of arm movements. C: Spiking activity of an example neuron. Note very different firing rates in different reach directions. D: Accurate coding of eight movement directions by the population vector computed from the discharge rates of many neurons recorded in the motor cortex. Each one of the eight clusters consists of the weighted preferred directions of the neurons (*solid lines*) and the calculated neuronal population vector (*dotted arrows*). The arrows in the center diagram indicate the direction of the instructed movement.
Figure courtesy of Apostolos Georgopoulos.

simultaneous recording of many neurons, improved versions of the population vector method have been used to decode movement direction in real time. It became an indispensable tool in so-called *brain–machine interface* experiments, allowing both monkeys and permanently paralyzed human patients to control

the movement is in the neuron's tuning direction. However, this assumption turned out to be incorrect as the contribution of individual neurons is strongly skewed (Chapter 12). For a contemporary approach to the population coding of arm movement, see Churchland et al. (2012).

computer cursors or robotic arms via thoughts; that is, by the voluntary production of spikes in the appropriate sets of neurons.[14]

Relationship Between Population Vector and Cell Assembly

There is a clear parallel between cell assemblies and population vectors. However, in contrast to the mathematically defined population vector, the cell assembly concept is only loosely described as representing "things" through excitatory connections. In line with the prevailing framework of cognitive psychology, the assembly concept has primarily remained a sensory input-driven model of brain activity. The strength of Hebb's hypothesis is its generalizability, but this is also its major weakness. The idea that changing constellations of neuronal firing patterns underlie our cognitive capacity is not different from the layman's view and is hardly falsifiable since there is no alternative hypothesis. What makes Hebb's concept distinct and specific are his suggestions that cell assemblies are formed by experience-driven synaptic changes and that unique assemblies with assumed boundaries can represent separate objects and thoughts.

However, the absence of a rigorous definition raises many questions. Who has more cell assemblies, Albert Einstein, William Shakespeare, you, or me? If cell assemblies are created by training, as is typically the case in artificial neuronal networks, does the brain start with no assemblies and accumulate them over the years? How would the brain's dynamic landscape look without any experience? What happens to cell assemblies during sleep? How big is a cell assembly? As the average path length in the brain is only a few synapses,[15] every input-triggered ignition would spread to the entire brain through the excitatory connections. Because Hebb's original assembly concept does not involve inhibition, it is not clear what mechanisms would mark the boundaries among the neuronal groups. Ideally, these questions would be addressed through

---

14. Using the concept of the population vector, the pioneering Miguel Nicolelis of Duke University and John Donoghue of Brown University have dreamed up a spectacular chapter of translational neuroscience by creating brain–machine interface devices that can read intentions and translate them into the movements of artificial actuators (Nicolelis and Lebedev, 2009; Hochberg et al., 2012). Georgopoulos was a student of the legendary neurophysiologist Vernon Mountcastle, who coined the term "cortical columns" (Mountcastle, 1957). In turn, a prominent student of Georgopoulos, Andrew Schwartz, now at Pittsburgh University, perfected population vector analysis for a more efficient two-dimensional control of a robotic arm in a tetraplegic patient (Collinger et al., 2013). The brain–machine interface field is a rapidly moving translational application of the cell assembly coding ideas.

15. Sporns (2010).

experiments in the learning, behaving animal as opposed to subjective reasoning. However, even when the best tools are available, experiments must be guided by a hypothesis that can be rejected or retained with some confidence.

CELL ASSEMBLIES FOR A PURPOSE

In my view, a fundamental problem with Hebb's "representational concept" is that it presupposes that the same inputs always mobilize the same set of neurons because this framework suggests that objects in the world should correspond to neuronal responses in the brain. We can call it the "neuronal correlate of x" approach. According to the representational outside-in framework, the best strategy by which to understand mechanisms of perception and to identify input-representing cell assemblies is to present various stimuli to the brain and examine the spatiotemporal distribution of the evoked neuronal responses.[16] However, as discussed in Chapters 1 and 3, it is a questionable strategy to identify the neuronal correlates of human-invented mental constructs on the presumption that they must have clear boundaries that correspond to their neural representation. Correlating spiking with external stimulus features is informative to the experimenter, but such correlation is not available to neurons in the brain (Chapter 1). If cell assemblies are meaningful events for neuronal computation, they should have predictable consequences for their downstream partners. Finally, because the brain is also active during sleep, cell assemblies are expected to be as important during sleep as in the waking state (Chapter 8). The outside-in, representational framework is mute about the existence and function of cell assemblies during offline modes of brain operation.

In Chapter 3, I already expressed doubt that the representational strategy can objectively identify cell assemblies because the brain's fundamental priority is not to faithfully "represent" the surrounding world but to simulate practically useful aspects of it based on prior experience and select the most advantageous action in the current situation. From this perspective, an objective definition of the cell assembly requires two related key conditions: a reader classifier and a temporal frame. Let me elaborate on this bold statement a bit. Inspired by the population vector concept, I suggest that a cell assembly can only be defined from the perspective of downstream "reader" mechanisms because the biological relevance of a particular constellation of active neurons (i.e., a presumed cell assembly or assembly sequence) can only be judged from its consequences. In my world, the term "reader" refers to a mechanism that can use the inputs

---

16. Engel et al. (2001); Hebb (1949); James (1890); Milner (1996); von der Malsburg (1994); Hubel and Wiesel (1962); Rieke et al. (1997).

it receives to respond one way or another. The reader mechanism[17] can be a muscle, a single neuron, groups of neurons, a machine, or even a human observer who interprets the meaning of the inputs. To be meaningful, the same constellation of the assembly should always lead to a similar action, whether it is at the level of a single neuron, neuronal groups, muscles, or hormonal release. The assembly has to have a consistent consequence. This action outcome is what makes any neuronal coalition into a meaningful assembly.[18]

Reading the impact of a cell assembly requires a temporal integration mechanism. Neurons come together in time; that is, they synchronize to achieve an action that is not possible for single members alone. Whether events are synchronous or not can be determined only from the point of view of an observer.[19] By extension, I suggest that a neuronal assembly can only be defined from the perspective of a neuronal reader mechanism.

Reader Mechanism–Defined Cell Assembly

All the preceding discussion is based on reasoning and speculation. However, the only way forward for science is by quantitative measurements; in this case, examination of the temporal relationships of spikes among putative assembly members. By the 1990s, my group and other laboratories worked out methods to record sufficiently large ensembles of neurons so that we could address a critical question: What determines the precise timing of a neuron's spikes? We hypothesized, as did Hebb, that the recorded neurons take part in different assemblies, but not all assembly members are active on each occasion. We also

---

17. Synonyms for "reader" can be terms like "observer," "classifier," "integrator," or "actuator," depending on whether one's background is physics, biology, computational science, or engineering. In the present definition, they all refer to the same thing.

18. At the most complex level, such "caused" effects may be plans, memories, decisions, or thoughts (Chapter 13). Berkeley's dictum *"Esse est percipi"* ("To be is to be perceived") was a response to British empiricist philosophy. Berkeley famously asked "if a tree falls in a forest and no one is around to hear it, does it make a sound?" (Berkeley, 1710/2010). Galileo Galilei, the most prominent champion of Aristotelian logic, posed a similar question: "if ears, tongues, and noses be taken away, the number, shape, and motion of bodies would remain, but not their tastes, sounds, and odors. The latter, external to the living creature, I believe to be nothing but mere names" (1623/1954). Perhaps nothing is more foreign to contemporary neuroscience than Berkeley's subjective idealism, yet the idea that in the absence of a "reader-actuator" *meaning* does not exist well resonates with an engineering approach to brain function. Although the metaphor of the reader/interpreter/actuator admittedly has an element of spookiness, it captures the key features of control theory of dynamical systems: a goal or desired output.

19. This is a classic problem in relativity theory (see Chapter 10).

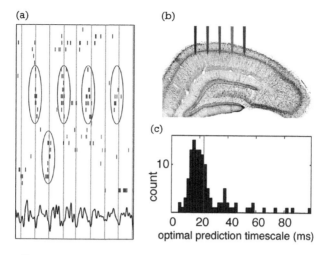

Figure 4.2. Cell assembly: the fundamental unit of neural syntax. A: A raster plot of a subset of hippocampal pyramidal cells that were active during a 1-sec period of spatial exploration on an open field, ordered by their temporal relationship. Each line corresponds to a neuron, and each tick is an action potential in that neuron. Top four ellipsoids indicate a repeatedly active cell assembly, which alternates with another one (*bottom*). Bottom trace shows local field potential dominated by theta and nested gamma oscillatory waves. Vertical lines indicate troughs of theta waves. B: Recording sites from the hippocampal CA1 pyramidal layer in the rat. C: Spike timing of a randomly chosen member neuron of a cell assembly is predictable from peer activity. Distribution of time scales at which peer activity optimally improved spike time prediction of a given cell. The median optimal time scale is 23 ms (*vertical line*).
Modified with permission from Harris et al. (2003).

reasoned that members of the assembly should work together within a measurable time window (Figure 4.2). Like members of an orchestra, neurons in a circuit can effectively time their actions relative to others.[20] So we tried to determine the time window within which neurons can best predict the timing of each other's spikes. By varying the analysis window experimentally, we found that the best prediction of the spike timing of single hippocampal neurons from the activity of their peers was when the time window varied between 10 and 30 ms.[21]

This is an important time window in neurophysiology because many physiological variables share it. First and foremost, the membrane time constant (τ) of

---

20. This project was led by Ken Harris et al. (2003), then a postdoc in my lab. See Truccolo et al. (2010).

21. Jensen and Lisman (1996, 2000); Harris et al. (2003); Harris (2005); Kelemen and Fenton (2010); Lansner (2009).

cortical pyramidal cells is exactly in this range, and it determines their integration ability.[22] Discharges of upstream neurons within this window can successfully trigger an action potential in the downstream reader neuron. Generating spike responses in reader neurons is the main reason for a cell assembly to come together. Therefore, from the point of view of a single reader neuron, all neurons whose spiking activity contributes to its own spike can be regarded as a meaningful assembly. Other upstream neurons that fire outside this critical time window (i.e., nonsynchronously) can only be part of another assembly. Thus, by monitoring spiking activity of reader neurons, one can objectively determine whether the upstream neurons are part of the same assembly and serve the same goal (the discharge of the reader neuron) or belong to different assemblies (Figure 4.2). Members of an assembly can project individually to hundreds or thousands of other neurons. Therefore, each neuron can be part of many assemblies, determined largely by its number of targets. These other potential constellations become an assembly when they activate another reader neuron.

In my definition, principal cells do not need to be anatomically connected to each other to form an assembly. It is also inconsequential whether the upstream assembly member neurons are neighbors or reside in multiple corners of the brain. For example, a coactive group of thalamic neurons that lead to the discharge of their target cortical neurons is a meaningful cell assembly. However, by Hebb's definition, these neurons would not be regarded as an assembly because they are not connected to each other by excitatory synapses. In contrast, viewed from the reader neuron definition of the cell assembly, the synchronous activity of these thalamic neurons within the membrane time constant of their downstream cortical neurons renders them an effective assembly of neurons. Therefore, this definition of the cell assembly is closer to the concept of the population vector than to Hebb's definition. However, if the presynaptic neurons are also connected, their synchronous discharge may strengthen their synaptic communication, as predicted by Hebb's plasticity

---

22. The membrane time constant (τ) is the product of the resistance r_m and capacitance c_m of the membrane ($\tau = r_m c_m$) and represents a duration within which the membrane potential relaxes to approximately 37% (= e^{-1}) of its initial value after a voltage step. Neurons with longer τ can integrate postsynaptic potentials for longer time periods (Johnston and Wu, 1995; Koch et al., 1996). The time constant can vary depending on the state of the network in which the neuron is embedded (Destexhe et al., 2003). During intense synaptic activity, such as hippocampal sharp wave ripples, the neuron's input impedance decreases, accompanied by a shortened time window of input integration, corresponding to the fast ripple waves (6–7 ms; Buzsáki, 2015).

rule, and such coalitions may enhance their co-firing in the future.[23] This is a bonus but not a requirement.

Gamma Wave Frame of the Cell Assembly

Another argument for the physiological importance of the cell assembly's ephemeral lifetime is that its 10- to 30-ms time window is similar to the duration of fast synaptic signaling mechanisms. Both excitatory and inhibitory receptors work in this temporal range.[24] The temporal interaction between the opposing excitatory and inhibitory postsynaptic effects gives rise to an oscillating tug of war and forms the basis for one of the best known brain rhythms, the *gamma oscillation*.[25] The time scale of gamma waves also corresponds to the temporal window of spike timing-dependent plasticity, a mechanism that can modify the synaptic connections between neurons.[26] The temporal similarities among the membrane time constant, fast inhibitory and excitatory signaling, the plasticity window, and the period of gamma oscillations make a strong case for the potential importance of upstream cell assemblies acting within this time window, which we discuss further in the next sections.

Cell Assembly Is a Letter in the Brain's Vocabulary

To be effective, cell assemblies acting within single gamma waves (10–30 ms epochs) must mobilize enough peer neurons so that their collective spiking activity can discharge the target (reader) neuron(s). Whether different constellations of spiking upstream neurons are regarded as parts of the same or different assemblies can only be specified by downstream reader neuron(s). Because of the all-or-none spike response of the target neuron, the cell assembly

---

23. Hebb did not consider inhibitory connections. However, inhibitory neurons interconnecting with excitatory pyramidal cells are very effective in routing spikes in neuronal circuits (Fernández-Ruiz et al., 2017).

24. These are AMPA receptor-mediated excitatory postsynaptic potentials (EPSPs) and $GABA_A$ receptor-mediated inhibitory postsynaptic potentials (IPSPs).

25. Buzsáki et al. (1983); Bragin et al. (1995); Kopell (2000); Whittington et al. (2000); Csicsvari et al. (2003); Bartos et al. (2007); Atallah and Scanziani (2009); Colgin et al. (2009); Buzsáki and Wang (2012); Schomburg et al. (2014); Bastos et al. (2015); Lasztoczi and Klausberger (2016). Layer 4 cortical neurons show a resonance and subthreshold oscillations in this time period (Pedroarena and Llinás, 1997), which make them especially sensitive to inputs at this frequency.

26. Magee and Johnston (1997); Markram et al. (1997); Bi and Poo (1998, 2001).

defined by this reader neuron denotes a discrete, collective unitary event, which we can term "fundamental cell assembly"[27] or, by analogy to written language, a "neuronal letter." Several of these gamma assemblies can be concatenated to comprise a neural word (Figure 4.2; Chapter 6).

Although I have written so far from the perspective of a particular reader neuron, in fact, there are no special, dedicated reader neurons. Every neuron can be a reader, and every reader can be part of an assembly, much like members of an orchestra who both produce actions and respond to others' actions. I simply use the term "reader" as a metaphor to refer to a classifier-actuator mechanism, which sends an action in response to a particular constellation input pattern. Separation of the reader mechanism from the assembly concept is needed only for an objectively quantified definition of neuronal alliances serving well-defined goals. In interconnected local networks, the reader is both an observer-integrator and a contributor in the sense that it generates a measurable and interpretable output. In the simplest case, the output is binary, such as an action potential. In other cases, the reader mechanism may be a population burst or a sequential spiking pattern. Therefore, the reader-centric perspective of assembly organization provides a disciplined framework to uncover the mechanisms that enhance the relationship between upstream firing patterns and the readers of those patterns. This is how the spiking of neurons *becomes* information.

How Big Is a Cell Assembly?

Do neuronal assemblies resemble quartets, chamber orchestras, or philharmonic orchestras? In Hebb's definition of cell assemblies, membership is defined by connectedness through excitatory synapses. However, as discussed earlier, neither the sufficient nor the total number of assembly members can be determined without first knowing the timeframe and the goal. For me, the question of the Hebb's assembly size, for example, underlying a visual percept is an unanswerable, ill-posed question. If you catch a glimpse of a couple kissing at a party, the stimulus may transiently activate only a few thousand neurons. On the other hand, if one of the couple is your partner, it may mobilize half of your brain, even though the number of retinal neurons conveying the two sensations is pretty much the same.

27. Again, my definition of the cell assembly is different from other uses. For example, Valentino Braitenberg suggested that, at each moment in time, there is only one assembly active in the entire cortex. His formulation was motivated by a search for the subjective unity of conscious experience. However, to become aware of a sensory experience requires engagement of the appropriate brain networks for at least 500 msec (Libet, 1985, 2005).

Chapter 4. Neuronal Assembly

Because the reader-centric definition of the cell assembly depends on integrator/actuator mechanisms, it also provides an objective way to approach the question of assembly size. If the goal of an assembly is to discharge a downstream pyramidal cell in vivo, the number of neurons whose spikes can be integrated in approximately 20 ms (i.e., in one gamma cycle) can quantitatively define the size of the effective assembly. Approximately 1% of hippocampal pyramidal cells fire in a gamma cycle, and 15,000–30,000 pyramidal cells provide input to each pyramidal cell. Thus, an estimated 150–300 pyramidal cells firing within a gamma cycle may comprise an assembly,[28] although these numbers can vary extensively if the synaptic strengths are stronger or weaker. In the olfactory bulb, less than 10% of sharply tuned reader-classifier mitral cells are responsible for generating discrete and defined outputs.[29]

An inherent difficulty in determining the size of a neuronal assembly is that, without an explicit goal, it is not possible to quantitatively define which neurons belong to the primary assembly and which represent feedback activation of assembly members or newly recruited assemblies serving other goals. Although many neurons can contribute to a cell assembly, the contribution of individual members is not equal, as in an orchestra. The absence of the first violinist is not the same as the absence of a second violinist. Analogously, activity in just a few strongly firing neurons can be as informative about a particular outcome as several dozens of other neurons simultaneously recorded from the same

---

28. The assembly size, of course, can vary extensively depending on the nature of the integrating reader neuron, synaptic strength, and synchrony of the active upstream population. Under special conditions, when the inputs converge on the same dendritic branch and neurons fire synchronously in less than 6 ms, as few as 20 neurons may be sufficient to initiate a forward-propagating dendritic spike (Losonczy and Magee, 2006). These conditions may be present in the hippocampus during sharp-wave ripples (Csicsvari et al., 2000) and in the geniculocortical system during visual transmission (Wang et al., 2010). If the synapse is strong, a single spike or a burst of spikes in a single presynaptic neuron may be sufficient to discharge a postsynaptic neuron (Csicsvari et al., 1998; Constantinidis and Goldman-Rakic, 2002; Henze et al., 2002; Bartho et al., 2004; Hirabayashi and Miyashita, 2005; Fujisawa et al., 2008; English et al., 2017). In a different approach to estimating the minimum size of a cell assembly to effectively substitute for the effect of a sensory input, channel rhodopsin-2 (ChR2)-expressing neurons in the motor cortex were directly stimulated by light. The mice reported sensing a stimulus when single action potentials were evoked synchronously in approximately 300 neurons. Even fewer neurons (~60) were required when the light induced a train of spikes (Huber et al., 2008). Yet all these considerations will become obsolete when we learn about the strongly skewed distribution of synaptic weights and firing rates in Chapter 12.

29. Niessing and Friedrich (2010). Under special conditions, stimulation of a single pyramidal cell or interneuron can recruit many target neurons in the circuit. Intense trains of intracellularly evoked spikes in a single motor cortex neuron were sufficient to evoke population activity or reset whisking movement in the rat (Miles, 1990; Brecht et al., 2004; Bonifazi et al., 2009; Ellender et al., 2010; Quilichini et al., 2010).

volume. The consequences of such nonegalitarian contribution of neurons to brain function will be discussed in Chapter 12. For now, we can only conclude tentatively that even a small cell assembly in the cortex may involve tens to hundreds of pyramidal cells and their transient partner interneurons, but how these assemblies are concatenated to neuronal words to lead to a percept or an action depends on additional factors.[30] Before we discuss the ability of brain circuits to combine neuronal letters to words and words to sentences (Chapters 6 and 7), we need to address the expected advantages of such syntactical operations. Do not skip Chapter 5, please.

SUMMARY

Spikes of single pyramidal neurons are rarely effective in discharging a postsynaptic neuron.[31] This makes a lot of sense. If a single neuron could discharge its thousands of postsynaptic targets all the time, spikes would have very little value for encoding neuronal information. To effectively send a message, a single neuron must cooperate with its peers. Such cooperation can be achieved by synchronizing their spikes together within the time window limited by

---

30. We recognize a familiar face in 150 msec or so, and it takes even less time to tell whether we see one or more persons. In this time frame, signals from the retina pass through at least six to eight layers of neurons to the visual cortex, allowing approximately 20 msec processing time for each layer. Knowing the maximum firing rate of neurons (say 100/s), only one or occasionally two spikes can be emitted (vanRullen et al., 2005; Guo et al., 2017) in such a short time (one gamma cycle). In this brief temporal window, the concept of rate coding is meaningless. What matters is how many neurons fire together. From the view of the observing downstream neurons, this togetherness is judged as assembly synchrony.

31. Pioneering studies by Sir Bernard Katz and colleagues established that the strength of a synapse depends on at least three factors: the number of synaptic contacts, the quantal size of the postsynaptic polarization caused by neurotransmitter release from a single synaptic vesicle, and the probability of neurotransmitter release from the presynaptic terminal (Katz, 1966). These parameters vary vastly among the large variety of synapses in the brain. Whereas most cortical pyramidal neurons contact their postsynaptic partners by a single synapse and release a single vesicle (quantum) per spike (Gulyás et al., 1993), some specialized synapses, such as climbing fibers from the inferior olive, make more than 500 synapses onto a Purkinje cell in the cerebellum (Llinás and Sugimori, 1980). The probability of neurotransmitter release is stochastic (Zhang et al., 2009). In the cortex, the probability that a single spike will induce a postsynaptic response is typically very low (0.05–0.5). The probability that a single spike will induce a postsynaptic spike is at least an order of magnitude lower. The distribution of these values across synapses is highly skewed, however (Chapter 12). An important advantage of low release probability is the flexibility it provides. In this way, synapses can act as frequency filters and resonators; they can vary their strength dynamically over both short and longer terms and act as a multiplier/divider gain control mechanism (for a review, see Branco and Staras, 2009).

the ability of the downstream reader neurons to integrate the incoming signals. Therefore, the cell assembly, defined from the point of view of the reader neuron, can be considered a unit of neuronal communication.[32]

Acting in assemblies has several advantages. Simple chains of neurons would be vulnerable to synaptic or spike transmission failures from one neuron to the next, resulting in the loss of neuronal messages. Furthermore, minor differences in synaptic weights between the leading and trailing neurons could divert the flow of neuronal traffic in unpredictable ways in the presence of noise. In contrast, cooperative assembly partnership tolerates spike rate variation in individual cells effectively because the total excitatory effect of the assembly is what matters to the reader. Interacting assembly members can compute probabilities, rather than convey deterministic information, and can robustly tolerate noise. We can call the assembly a unit of communication or a putative "neuronal letter."

---

32. Spike communication allows neurons to affect other neurons, both near and far, through synaptic release of a neurotransmitter. This is only one of many types of neuronal communication. Nearby neurons can affect each other by electric junctions, ephaptic effects, dendrodendritic release of neurotransmitter, and neuron-glia-neuron effects as well.

5

Internalization of Experience

Cognition from Action

I move, therefore I am.

—Haruki Murakami[1]

In the beginning was the act.

—Goethe's *Faust*

Not hearing is not as good as hearing, hearing is not as good as seeing, seeing is not as good as knowing, knowing is not as good as acting; true learning continues until it is put into action.

—Confucius [2]

Theories of cognition have evolved through multiple stages. The first was the outside-in, empiricist view, postulating that the brain is an associational, representational device that analyzes the world around us and makes judgments. Then came the Pavlovian reflex theory, which did not make much space for cognition; everything was a hierarchy of associative reflexes. Similarly, the behaviorist paradigm argued that there is no need to think about cognition as actions can always be explained as a response to immediate external cues. In response to these views, a hard-thinking minority argued that behavior cannot be understood simply as an input–output function and activity

---

1. https://www.goodreads.com/quotes/503630-i-move-therefore-i-am.

2. Confucius's formulation of the problem is also representational. Although it acknowledges the paramount important of action, it starts with the senses.

in the hidden layers of the brain is critical (Chapters 1 and 3).[3] In this chapter I discuss that *internally organized activity* in these hidden layers, detached from immediate sensory inputs and motor outputs, is a necessary condition for cognition.

DISENGAGEMENT OF THE BRAIN FROM ITS ENVIRONMENT

Humans, and likely other animals, can imagine into the future and recall the past; we spend a great deal of time on such mental activities. I suggest that these activities arise from the brain's ability to break free from external control. The core idea is that cognition depends on prior action-based experiences of the world, which allow internally generated sequences to test "what if" scenarios and anticipate the possible consequences of alternative actions without actually taking them. This process then helps to select future overt actions. To unpack my hypothesis, consider the three possible brain network scenarios shown in Figure 5.1.

In a small nervous system (left panel), the connections between output and input networks are short and simple. This hypothetical small brain's main goal is to predict the consequences of the animal's actions. Such a brain can make experience-based predictions in a simple environment and at a relatively short time scale. These predictions are generated through learning rules that either apply across generations (evolutionary adaptation) or within the lifetime of the organism. As a result of stored past experience, organisms can deal with a future occurrence of similar situations more effectively. This is the fundamental organizing principle of adaptive systems, of which the brain is a prime example. For example, animals learn to predict the occurrence of food or an unpleasant air puff after some training.[4]

---

3. Tolman's cognitive map theory (1948) was a major departure from the outside-in framework. As opposed to trial-and-error or "instrumental response" solutions, Tolman (1948) speculated that rodents use a mapping strategy (i.e., a "cognitive map") created through exploration to solve maze problems. His thinking fueled John O'Keefe's pioneering physiological experiments (O'Keefe and Dostrovsky, 1971; see Chapter 6).

4. According to the Pavlovian theory, the initially "neutral" conditional stimulus becomes a "substitute" for the unconditioned stimulus after their systematic pairing. Thus, the neutral stimulus will also induce salivation because neurons "representing" the conditional stimulus will become connected to the neurons representing the unconditioned signal. In turn, the latter neurons trigger an inborn stimulus-response reflex (outside-in). One of Pavlov's students, Piotr Stevanovitch Kupalov, was already unhappy with this explanation (Giurgea, 1974) and, along with my mentor, Endre Grastyán, argued that the conditional signal does not become a substitute for the unconditional signal but a new goal, which acquires significance for the animal through active investigation of the conditional signal (Brown and Jenkins, 1968; Grastyán and

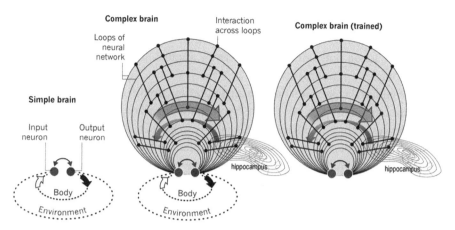

Figure 5.1. Externally driven and self-organized brain operations. A: Small brains contain simple neural networks. The output moves sensors, which scan the surroundings or the body so that the brain can predict the consequences of its action based on prior experience (phylogenetic or ontogenetic) in similar situations. B: In more complex brains, multiple interacting loops of increasing lengths and layers improve prediction of more elaborate events in more complex environments and at longer time scales by creating an efficient memory mechanism (*postdiction*). C: After extensive training, the loops can sustain self-organized, long-lasting neuronal sequences without reliance on external cues and can, therefore, support cognitive operations such as memory, planning, and imagination. Learning and disengagement from the sensors take place in parallel. I show them separately here only to illustrate the principle.
Modified from Buzsaki (2013).

More complex brains are organized in a "multiple loop" pattern. These parallel loops, each with increasing levels of complexity of wiring and temporal dynamics, are inserted between the output and input (middle panel). For example, in mammals, the most direct circuit between sensors and muscle activation is the monosynaptic spinal cord/brainstem connection. More complex loops connecting these same inputs and outputs include the thalamocortical system and bidirectional communication between the neocortex and several subcortical structures. In addition, a side loop of the hippocampal system is dedicated to generate sequential neuronal activity such that content in the thalamocortical loops is organized into orderly trajectories, allowing simultaneous postdiction (memory) and prediction (planning) operations (as elaborated in Chapter 13). Large and small brains share the same main goal: to predict the future

---

Vereczkei, 1974; Kupalov, 1978; for a review, see Buzsáki, 1983). Under this hypothesis, action is essential because it is the motor output that adjusts the sensors to optimally sense the conditional stimulus so that it becomes a meaningful signal.

consequences of their actions. However, the multiple interacting loops and the hippocampal postdiction-prediction system allow more sophisticated brains to predict action outcomes more effectively at much longer time scales and in more complex environments.

Efficient predictive operations in a complex environment require the storage of a vast amount of past experience to compare the current situation to similar occurrences in the past, along with the ability to evaluate and weight the importance of a range of possible actions. Many of these operations require brain computation to continue long after the disappearance of transient sensory inputs and the completion of actions. Such persistent activities are needed to compute the importance of the items perceived and the actions performed. Learning, a result of both phylogenetic and ontogenetic accumulation of experience, makes the longer, more complex neuronal loops "smarter" or more effective by enabling them to interpolate and extrapolate the events encountered by the organism.

Self-organized activity also allows brain networks to disengage from inputs and overt actions (right panel). I propose that, in a trained brain, communication from the action systems to sensory and higher order areas via the expansion of corollary discharge circuits (Chapter 3) allows for such operations.[5] In other words, the corollary discharge mechanism supplemented with a memory loop can substitute for the feedback provided by the body and environment. This virtual action is not translated to muscle movement or heart rate change, yet it is interpreted by the brain as if those actions could have occurred. Such internalized actions—we can call them simulations of real actions—allow the brain to inspect its own computations; evaluate internally generated, imagined scenarios; and estimate the consequences of potential actions, all without overt inputs or outputs (Figure 5.2).[6]

For a brain network, there is no difference between sensory inputs or activity conveyed by the same upstream neuronal group in the absence of external inputs (see Figure 1.2). Without external constraints, disengaged processing in the brain can create an internalized "virtual world" and new knowledge through vicarious or imagined experience, tested against preexisting and stored knowledge. This process—which most scientists and philosophers would call cognition[7]—provides dramatic advantages in predicting the consequence of

---

5. Mátyás et al. (2010); Mao et al. (2011).

6. This process is expected to develop during the course of evolution and should be distinguished from the Taoist idea of *wu-wei*, the "nonaction with action" or "unremitting alertness" and from the Buddhist internalized focusing, all of which are practiced during individual lives.

7. Merleau-Ponty (1945/2005), Gibson (1977), and their intellectual heirs developed ideas, similar to those I outlined here, which have come to be known as "embodied cognition" (Thelen, 1989; Gregory, 1980; Maturana and Varela, 1980; Beer, 1990; Brooks, 1991; Varela et al., 1991;

Chapter 5. Internalization of Experience   105

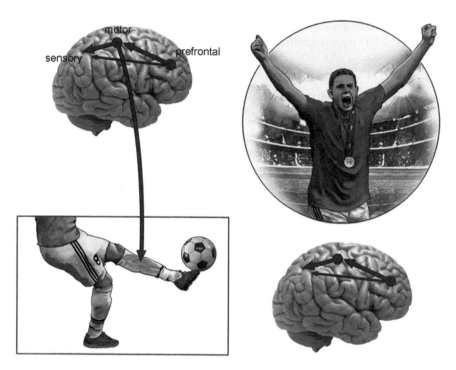

Figure 5.2. The brain can disengage from immediate control of the body. *Left*: Whenever a motor command is sent to the muscles, several other brain regions are informed that an output signal was sent (*arrows*). Prefrontal neurons activate motor neurons and both area sends a corollary discharge (Chapter 3) to sensory areas. *Right*: After the corollary discharge system is calibrated by experience, involving feedback from the body and environment, activation of the command signal can simulate real actions by activating the same target circuits but without sending signals to the muscles. Such disengagement from immediate muscle control ("internalized or simulated action") is hypothesized to be the neuronal basis of imagination and planning. A "thought" might be regarded as a buffer for delayed action.

actual behavior in complex environments and at long time scales. Of course, the validity of these newly created ideas should be tested eventually in real-world interactions. Until then, the internal content remains a belief. In short, cognition is time-deferred action.

An important aspect of the framework just presented is that most brain structures have two uses, which I detail in the coming examples. At times, they are connected to sensory input or motor output, directly or indirectly, and vary

Jeannerod, 1994; Clark and Chalmers, 1998; O'Regan and Noë, 2001; Llinás, 2002; Noë, 2004; Buzsáki, 2006; Prinz et al., 2013; Goodrich, 2010). While I share the general philosophy of the situated, embodied mind (i.e., that brains gain their smartness from the body, environment, and other brains), my approach is from a neurocircuit perspective.

their cell assembly contents at the pace of changing sensory inputs. At other times, they rely largely on their internal dynamics, often maintained by brain rhythms (Chapter 6).[8]

I do not claim that processes like input-dependence, self-organized activity, and output disengagement developed sequentially or independently. On the contrary, these mechanisms are strongly intertwined, and the degree to which they dominate in each brain depends on the complexity of the circuits, the availability of external cues, and likely other factors. Thus, the emergence of cognition is gradual and quantitative rather than step-like and qualitative. Let me illustrate the "emergence of cognition" hypothesis by concrete examples. The text will inevitably get a bit technical here and there, but I try to keep it simple.

INTERNALIZATION OF HEAD DIRECTION SENSE

The Aristotelian inheritance in cognitive neuroscience is perhaps most evident when people talk about our five senses. The eyes, ears, nose, skin, and tongue contain sensors for seeing, hearing, smelling, touching, and tasting.[9] Any sense beyond these classical ones is still referred to as a "sixth sense." But there are more, including our senses of temperature, pain, balance, motion, and head direction. Of these, the sense of head direction is perhaps the simplest as it has only a single dimension: the direction the head points to.

Neuronal substrates of the head direction sense were discovered by James Ranck at the State University of New York Downstate in 1984.[10] He observed that neurons in the post subiculum, a structure that is part of the hippocampal system, fire robustly when the animal's head points in a specific direction, and he called them *head direction cells*. For example, when the rat faces a corner of the testing box, one set of neurons become active. When the rat turns its head to right, another set of head direction cells will turn on their activity. For each of the 360 degrees, a different set of neurons fires with equal probability (Figure

---

8. We have already discussed a striking example of this process, the "internalization" of early spindles, which depends on external stimuli before it becomes a spontaneously occurring sleep spindle.

9. The five senses or "wits" were originally derived from the faculties of the soul, described in the three-book treatise, *De Anima*, by Aristotle. Interestingly, he first lumped taste together with touch before treating it as a separate sense.

10. Jim Ranck debuted his discovery at small meeting that I organized for Endre Grastyán's sixtieth birthday in Pécs, Hungary, in 1984 (Ranck, 1985). All participants, including John O'Keefe, the discoverer of hippocampal place cells, immediately recognized the importance of Jim's unexpected finding. For a comprehensive review, see Taube (2007).

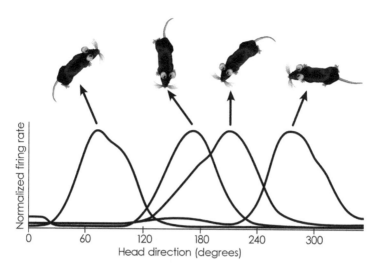

Figure 5.3. For each head direction, a subset of head direction cells become active. Firing patterns of four head direction cells are shown, each firing maximally when the mouse's head faces particular directions.

5.3). The head direction *system* consists of many neurons in a serially connected brain network, including the brainstem, mammillary bodies, anterodorsal thalamic nucleus, post subiculum and entorhinal cortex, lateral septum, and parietal cortex. This widespread representation of head direction in so many brain structures indicates that this sense is of fundamental importance for the brain.

Researchers have tacitly assumed that head direction neurons are largely controlled by peripheral inputs, similar to other sensory systems. Indeed, numerous experiments in multiple laboratories show that head direction neurons are controlled by vestibular, proprioceptive, visual, and ancillary inputs. Such responses, of course, are expected in a sensory system because neurons in all sensory systems respond to the relevant sensory modality. But I often wondered what such "sensory" neurons do when there is no sensation.[11] If they are active without their dedicated outside inputs, do they show any organized, cooperative pattern?

To address this question of how head direction neurons work together, we had to wait until the technology matured to allow recording from a sufficiently large population of neurons. We then compared their dynamic interactions in the tiny

---

11. Computational models have already postulated that head direction neurons with similar preferred directions fire together. According to these models, the temporally engaged group of head direction neurons (i.e., a "hill of activity" or "an activity packet") moves on a virtual ring as the animal turns its head (Redish et al., 1996; Burak and Fiete, 2012; Knierim and Zhang, 2012).

anterodorsal nucleus of the thalamus during various behaviors.[12] By training an artificial neuron network, we were able to reconstruct the animal's head position from the population interactions of the recorded neurons. When the mouse's head was turning from left to right, this behavior was reliably reflected in a sweep of neuronal firing sequences as each head direction neuron is tuned to a preferred direction. There was excellent agreement between the measured (real) head direction and the (virtual) direction reconstructed from the neuronal firing.

The exciting part of the experiment began when the mouse fell asleep. Because during sleep all peripheral inputs are either absent or constant, one might expect that head direction neurons will be silent or just fire randomly. Yet, to our surprise, not only did head direction neurons continue to be active, but they replicated the sequential activity patterns seen in the waking mouse. Neurons coding for nearby orientation continued to fire together, while neurons with opposite head direction tuning remained anti-correlated. During rapid eye movement (REM) sleep, when brain electrophysiology is strikingly similar to its awake state, the neural patterns moved at the same speed as in the waking animal, recapitulating the activity seen when the mouse moved its head in different directions during waking. In essence, from the temporally shifting activities of cell assemblies, we could determine the mouse's virtual gaze of direction during REM sleep. During non-REM sleep, the activity was still perfectly organized, but now the virtual head direction drifted ten times faster than it did in waking and in REM sleep.[13] Because no head movement accompanied these changes in the sleeping animal, the temporal organization had to rely on internal mechanisms rather than external stimuli.

The observation that the neuronal code of the head direction sense is preserved during sleep suggests that the system is maintained by a self-organized mechanism.[14] Of course, this self-organized dynamic is expected to remain active in the waking animal and should contribute to some function. For example, sensory inputs can be combined with the internally generated prediction of head direction to rapidly adapt to reconfigured environments. In case of ambiguous or conflicting signals, self-organized mechanisms may generate the brain's "best guess" by interpolating across input signals or extrapolating the correct position of the head direction vector from limited or ambiguous sensory information.

12. Adrien Peyrache was the key player in these experiments (Peyrache et al., 2015).

13. The several-fold faster virtual head direction–related neuronal activity during non-REM is in sync with the previously described faster dynamics of non-REM sleep in cortical networks (Buzsáki, 1989; Wilson and McNaughton, 1994; Nádasdy et al., 1999).

14. "Self-organized" is the term most often used in physics and engineering; it refers to autonomously generated patterns, "changing from unorganized to organized," or any change with decreasing entropy (Winfree, 1980).

Overall, these experiments illustrate how internally organized mechanisms can improve the brain's interpretation of the external world. Internal computation continues even without external cues and can perform both interpolation and extrapolations (also called *prediction*; Chapter 2). When I close my eyes, my head direction sense persists. But the self-organized head direction system needs to be calibrated by active exploration to exploit its potentials.

FROM PHYSICAL NAVIGATION TO MENTAL NAVIGATION

When you emerge from a subway station in an unfamiliar city with a map in your hands, the first thing you need to do is properly orient the map. North on the map should orient to north in the world. Second, you have to locate yourself—that is, find your subway station. In addition, the vertical and horizontal grid lines on the map (a grid) can then help you to find museums, your hotel, and other places of significance. Third, you need distance calibration to determine whether your target is within walking distance or if you would be better off taking a cab. Now you can use the grid to identify any location or the relationship between two arbitrary locations. The alternative to this map-based navigation would require you to memorize a long list of street names, number of steps, turns to be taken, and so on. This latter strategy works well for moving from one place to a few other places. However, when the goal is to get from anywhere to anywhere else, maps are superior to route descriptions.

To construct a map, someone needs to explore every single locale of the environment. Each location can then be uniquely described by the intersection of horizontal and vertical coordinates on the map. The advantage of this approach is that it allows us to calculate the most efficient route between any two points, thus minimizing distance or time, including potential road blocks or shortcuts. Physical maps are a prime example of the "externalization" of brain function (Chapter 9). Instead of carefully exploring and memorizing the surroundings in each brain, as other animals must do, humans can navigate effectively and without prior exploration as long as they can read a map prepared by someone else.[15] Without speech or other direct communication, maps provide clues about landmarks, distances, and specific locations.

---

15. The everyday use of maps is rapidly shrinking in modern societies. With the wide availability of global positioning system (GPS), travel and goal finding have reverted back to the locale (list) system. No familiarity with the real world or abstract maps is needed; just follow the verbal commands. While the immediate convenience of GPS-assisted navigation is clear, its exclusive use may deprive people of their spatial sense. Imagine the impact of an attack on or

An entire system in the brain, consisting of several interactive structures, is dedicated to the cognitive map that supports spatial navigation. Head direction neurons provide orientation guidance, the map grid is represented by entorhinal "grid cells," and self-localization is assisted by hippocampal "place cells." The source of distance calibration is still debated, but self-motion remains a leading candidate mechanism.[16] The positive impact of practice in navigation on the hippocampal system has been demonstrated by a series of celebrated studies by Eleanor Maguire and colleagues at University College, London. These researchers showed, using structural imaging of the brain, that the posterior (tail) part of the hippocampus in taxi drivers who spent at least three years learning their way around London was larger than in matched controls. Thus, taxi drivers' brains grow on the job.[17] This is likely the case whenever navigation skills are increased by practice.

Grid Maps in the Entorhinal Cortex

Much like the line grid on a physical map, spatial firing patterns of grid cells provide a metric for the neural representation of space. However, instead of the square grid of typical city maps, grid cells tile the two-dimensional surface of the environment in a honeycomb fashion. This pattern becomes apparent to

---

failure of GPS navigation satellites. Some people would not even be able to find their way home. For this reason, there is an effort to construct "dead reckoning" devices which do not rely on satellites but track the exact route of travelers (e.g., combat soldiers) and can assist in guiding them back to home base.

16. A significant milestone in our understanding of cognitive representation of space in the brain was the treatise by O'Keefe and Nadel (1978). Although there are hundreds of reviews on the navigational aspects of the hippocampus system, this volume remains the most eye-opening, thought-provoking, and comprehensive treatment of the subject. It has remained the de facto reference for almost everything we have claimed about the hippocampus over the past decades. On October 6, 2014, *The New York Times* heralded that John O'Keefe, Edvard Moser, and May-Britt Moser received the Nobel Prize for Medicine and Physiology for the discovery of a brain GPS that "helps us know where we are, find our way from place to place and store the information for the next time." The three scientists' discoveries "have solved a problem that has occupied philosophers and scientists for centuries—how does the brain create a map of the space surrounding us and how can we navigate our way through a complex environment?" Because the neuronal mechanisms of spatial navigation have been detailed in several excellent reviews (O'Keefe and Nadel, 1978; O'Keefe, 1991, 1999; O'Keefe and Burgess, 1996; Burgess and O'Keefe, 2011; Buzsáki and Moser, 2013; Moser et al., 2008, 2014, 2017; Hasselmo and Stern, 2015; Mehta, 2015; Redish, 2016; McNaughton et al., 2006; Connor and Knierim, 2017), I will only briefly summarize the key papers of the large literature on navigation and the hippocampal-entorhinal system here with a bias to help expose my own views.

17. Maguire et al. (2000).

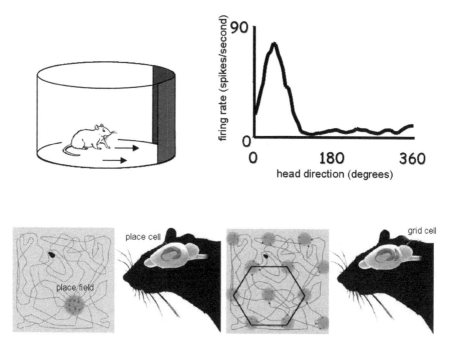

Figure 5.4. Neuronal components of navigation. *Top*: Tuning curve of a head direction neuron from the presubiculum while the rat's head is facing east. *Lower left*: Singular place field of a hippocampal pyramidal neuron during exploration of a rectangular platform. Lines indicate the rat's path; circular area is where the place cell fired most (place field). *Lower right*: Grid cell firing (from the medial entorhinal cortex) while the rat has explored the rectangular platform. Note increased firing rates at the apexes of a hexagon.

the experimenter when the spikes emitted by a single grid cell are projected onto the floor of the experimental apparatus. Think of a TV screen, where a stationary-looking picture is formed by the illumination of many points in fast succession, compared to the slow temporal resolution of the eye. For grid cells, the moving beam is the rat's movements. To see the hexagonal pattern, we have to wait patiently until the animal passes through each pixel of the environment a few times. When enough spikes are emitted during a long exploration, the hexagonal grid firing pattern becomes clear (Figure 5.4).

Neighboring grid cells have a similar pattern but with a slight spatial offset. These cells are arranged as semi-independent modules or clusters along the dorsal-to-ventral axis of the medial entorhinal cortex, where grid scales vary from small to large. This arrangement is like overlaying multiple maps of the same environment with low-, intermediate-, and high-resolution on each other. When the firing patterns of many grid cells are superimposed, they precisely tile the entire floor of the testing apparatus. Their combined activity amounts to a sequence of population vectors or a trajectory (Chapter 4) that describe the animal's position with high precision at any time because each position evokes a unique firing pattern among grids of different sizes and locations.[18] This self-independent or so-called *allocentric* representation describes the relationship of each position to any other position: a map.

Place Cells and Place Fields

In contrast to the entorhinal grid patterns, hippocampal neurons display highly flexible spatially tuned patterns, called *place fields*. Typically, in a given environment, most pyramidal neurons are silent, and the minority forms a single place field: that is, the neuron becomes active only when the animal reaches a particular location in the testing apparatus. This response is why John O'Keefe named them *place cells*. The size of the place fields is small in the dorsal part of the hippocampus and grows in size toward the ventral pole, in register with the grid size increase of the topographically connected medial entorhinal cortex. Because of the discrete place fields of hippocampal neurons, they carry much more specific information than the entorhinal grid cells. A large population of place cells with unique locations can provide a reliable map and faithfully track the animal's position in the environment.[19] Surprisingly, though, in contrast to the entorhinal grid map, the hippocampal map is not static. If the animal is placed in a different environment, a different set of place cells becomes active. Neighboring place neurons in one environment may have a very different spatial relationship with one another. For example, a pair of place cells representing adjacent pieces of the floor in one apparatus may become silent in another, or just one of them may remain active but at a different position. If both of them continue to fire, the distance between their preferred firing locations may be different in the two mazes. Thus, each environment

---

18. The discovery of grid cells by the Moser group was a spectacular moment in neuroscience (Hafting et al., 2005). Its beauty has captured the attention of many researchers outside spatial navigation and recruited many new investigators into the field. Initially, it appeared that mapping and navigational functions, attributed to the hippocampus, resided in the entorhinal cortex, freeing up the hippocampus for other functions. The pendulum has swung back and forth a few times over the past decade, intensifying the debate and forcing us to think in a larger context. Despite numerous models and experiments, there is no agreed-upon explanation of how the grid pattern is supported by the circuits in which they are embedded (i.e., the superficial entorhinal layers, the presubiculum, and parasubiculum). It is becoming clearer that functions do not reside in this or that structure but emerge through their interactions.

19. The ability of hippocampal place cell assemblies to track the animal's position in the environment was first demonstrated in a landmark paper by Wilson and McNaughton (1993).

is represented by a unique combination of active place cells and place fields.[20] Even seemingly minor changes in the testing apparatus or the variables of the task may strongly alter the firing pattern of the hippocampal place cells.

The spatial layout of individual place cells (i.e., the map) is not related to the location relationships of place cells in the hippocampal circuitry. Two neighboring pyramidal neurons are just as likely to represent adjacent or distant patches of the environment. Instead, place cells dynamically and relatively randomly reconfigure under various conditions. Largely because of this arrangement, the densely connected hippocampal recurrent collaterals can generate discrete maps individualized to the many environments an animal visits in its lifetime.[21] To envision such relationships, imagine that many place cells on a piece of paper are interconnected with rubber bands. When you crumple the paper, the different place cells will occupy a new set of displacement relationship to each other. These distances on the surface of the wrinkled paper may be thought of as the synaptic strengths between them. Now you can flatten the paper and crumple it again a bit differently; this gives another set of relationship, and so on. Thus, the same set of neurons can give rise to many different constellations of relationships, each of them representing a map.

Of course, a map is just a static positioning tool, a reference frame, in which the spatial relationships among landmarks assist in defining the animal's own location in the environment. This relationship is described eloquently by O'Keefe[22]: "Each place cell receives two different inputs, one conveying information about a large number of environmental stimuli or events, and the other from a navigational system which calculates where an animal is in an environment independently of the stimuli impinging on it at that moment.... When the animal had located itself in an environment (using environmental stimuli) the hippocampus could calculate subsequent positions in that environment on the basis of how far, and in what direction the animal had moved in the interim.... In addition to information about distance traversed, a navigational system would need to know about changes in direction of movement either relative to some environmental landmark or within the animal's

---

20. The unique firing patterns in different contexts is a segregation mechanism (also often called "orthogonalization"; Marr, 1969), which is believed to be accomplished by interactions between the numerous independently modulated granule cells and the small number of mossy cells whose axons massively project back to the granule cells (Senzai and Buzsáki, 2017). The pattern segregation and integration mechanisms and the assumed role of attractor dynamics in these processes are discussed by McNaughton and Morris (1987), Treves and Rolls (1994), and Knierim and Neunuebel (2016).

21. Samsonovich and McNaughton (1997).

22. Chapter 11; p. 499 in Andersen et al. (2007).

own egocentric space." However, a map is useful for the animal only after it has already learned the significance of the relationships among the landmarks though active exploration.

As the Crow Flies: Navigation and Self-Motion

Bruce McNaughton and Carol Barnes at the University of Arizona took an important step in linking the static representation of maps to animal navigation through carefully designed experiments and reinterpretation of existing data. Their first key insight was that head direction information is an integral part of the navigation system. The second insight was that self-motion is the primary source of spatial information leading to place-specific firing.[23] In other words, visual landmarks and other stimuli in the environment gain their meaning by action (Chapter 3), although they did not explicitly spell out this link.

Several experiments support the primacy of action in navigation. First, the firing frequency of place cells depends on the animal's speed. Second, on a running track, a robust place field observed on the left-to-right journey is often absent on the return run. That is, place fields are not absolute descriptors of the x-y coordinates of the environment; they also depend on context, such as which direction the rat is facing and how head direction orients the entorhinal-hippocampal map. Third, under certain circumstances, the same physical space can be represented by different sets of place cells depending on other contextual variables. For example, when a rat is placed into an otherwise familiar environment in total darkness, new place fields can emerge and persist in subsequent light. Thus, place fields can arise independently of the visual landmarks, although landmarks can control the expression of place cells. Fourth, changing the features of a landmark without changing its position can affect the map. For example, after a white cue card in a symmetric apparatus (cylinder) is replaced with a black one, the place cells stay put relative to the radial axis of the cylinder. However, after several trials in which the white and black cards are alternated, the place-field distributions of several neurons become largely uncorrelated across the two conditions as the animal learns to distinguish the

---

23. McNaughton and his long-term collaborator Carol Barnes were postdoctoral fellows in O'Keefe's lab. They provided firm experimental evidence and drew attention to the importance of local cues and body-derived signals in self-localization in navigation (McNaughton et al., 1996). On computational arguments, Redish and Touretzky (1997) also emphasized the need for a path integrator independent of the cognitive map. See also Gallistel (1990).

two cues.[24] Fifth, when a rat is repeatedly disoriented (by turning it by hand multiple times before placing it into a cylinder with few landmarks), its place fields tend to destabilize. Sixth, when the distance between the start and goal boxes changes, only a fraction of neurons remain under the control of distant room cues. Many of them realign as a function of the distance from either the start box or goal box.[25]

The list goes on. The profound implications of these observations led McNaughton, Carol Barnes, and their colleagues to postulate a hypothetical "path integrator" system, a sort of a body cues feedback-supported guiding mechanism, based on information from local visual and somatosensory cues, proprioceptive feedback from the body (e.g., muscles, tendons), the number of steps taken, vestibular input (translational and rotational head accelerations), and, likely, corollary discharge from self-motion activity. Integrating self-motion allows animals to move through space even without prominent landmarks while keeping track of their starting location. The path integrator system can operate without a prior spatial reference or even in complete darkness by calculating the distance traveled and the turns the animal made.

Dead reckoning, another name for path integration, was one of the earliest forms of marine navigation. From a home base, the captain calculates the movement of the ship based on heading, speed, and time. Multiplying speed by time provides the distance traveled. The sequence of distances and heading direction changes is noted on a chart. This route chart can always be used to calculate the straight path to return home. Dead reckoning was the main navigation strategy for sailors in the Mediterranean Sea for centuries before Portuguese sailors introduced maps and landmark (more precisely "skymark" or celestial) navigation.[26]

---

24. Several experiments convincingly demonstrate that distant visual cues can have a large impact on place cells. For example, when the walls of the behavioral apparatus are rearranged, the size and shape of place fields often follow such modifications by enlargement, elongation, or even splitting the place field into two (Muller and Kubie, 1987; Kubie et al., 1990; O'Keefe and Burgess, 1996; Wills et al., 2005; Leutgeb et al., 2005).

25. Redish et al., 2000; Gothard et al., 2001; Diba and Buzsáki, 2008).

26. Christopher Columbus used dead reckoning to discover the New World and safely return home. The term "path integration" (*Wegintegration*) was introduced by Mittelstaedt and Mittelstaedt (1980) to reflect the idea that a continually updated representation of direction and distance from a starting point is a reflection of signals generated by self-motion (also known as *idiothetic*) as opposed to the intermittency of navigation via landmarks, which may not be sensed continuously. These authors also wrote that the updating of position is much more accurate during active motion as opposed to passive shifts. This is familiar to us: the passenger's sense of the surroundings is always poorer than that of the driver. Before the work of the Mittelstaedts on gerbils, Wehner and Srinivasan (1981) reported that desert ants can return directly to the nest after an outward journey of hundreds of meters, using the shortest path. See also Whishaw et al. (2001).

Navigation by path integration is intuitively appealing.[27] Not only sailors but also animals can find their way without landmark cues. A gerbil mom can reliably retrieve her pups from within a circular arena and return to her nest at the arena border in complete darkness using an internal sense of direction. When the platform is slowly rotated so that the mom does not notice it, she carries the pups to the original spatial position even though the nest is no longer there. Yet path integration presents some challenges. Every change in speed and direction contributes to error accumulation. As each estimate of position is relative to the previous one, these errors quickly accumulate.

The exact brain networks involved in path integration have been debated from its conception. Initially, the hippocampus was the suspect, based on both recording and lesion experiments. With the discovery of grid cells, the entorhinal cortex became the favorite structure. Although the majority of grid cells are present in layer II of the dorsomedial entorhinal cortex, many neurons in deeper layers respond to head direction, and a subset of these neurons also show typical grid responses.[28] The running speed of the animal correlates with the firing rate of some place cells and most inhibitory interneurons. Thus, the conjunction of positional, directional, and translational information in the same entorhinal groups of neurons may enable the updating of grid coordinates during self-motion–based navigation—as the argument goes. However, such correlation alone may not be sufficient because entorhinal neurons receive information about head direction and speed from multiple sources. In support of this idea, damaging the anterodorsal nucleus of the thalamus, a main hub of the head direction system, leads to the disappearance of grid dynamics. Furthermore, position, speed, and head direction information, and even distance, are present in hippocampal and the adjacent subicular neurons as well. Neurons in the parietal and retrosplenial cortex may also be involved in computing the animal's motion relative to environmental and body cues, suggesting that path integration involves multiple systems.[29]

---

27. A blind child who is walked from a base to three objects (landmarks) in a large room can immediately find the routes between the three objects on her own, even though she has never experienced those paths. To commute between any two targets, the child must encode the three training paths by dead reckoning, store these paths, and combine two correct stored paths to deduce a new one, a demonstration that a cognitive map can be constructed solely on the basis of dead reckoning navigation (Landau et al., 1984; Wang and Spelke, 2000).

28. The context-invariant firing properties of grid cells led to the suggestion that they are part of a path integration spatial representation system (Hafting et al., 2005; McNaughton et al., 2006).

29. Alyan and McNaughton (1999); Cooper and Mizumori (1999); Leutgeb et al. (2000); Whishaw et al. (2001); Etienne and Jeffery (2004); Parron and Save (2004); McNaugthon et al. (2006); Sargolini et al. (2006); Samsonovich and McNaughton (1997); Winter et al. (2015);

A navigation system is only as useful as its ability to navigate the body. How does the abstract hippocampal cognitive map instruct the body to choose the right path in this map? The most dense projection of the hippocampus converges on a subcortical structure called *lateral septum*, which in turn projects to the hypothalamus, the motivational mover of the brain. Indeed, spatial information about the animal's whereabouts is also present in neuronal assemblies of lateral septal neurons.[30] Therefore, the lateral septum is an interpreter and translator of the abstract map to motivated action.

Allocentric map representation and self-referenced or *egocentric* path-integration route information work together. The environmental conditions determine which strategy dominates. In cue-rich environments, representations can be updated frequently by changes in the configuration of sensory inputs. In environments with few stationary landmarks or in complete darkness, path integration is the default mode. Do these map and path integration mechanisms need different neuronal networks, or are they different manifestations of the same or overlapping neuronal substrates? We can only guess that navigation effectively exploits both systems. When the goal is to generate a universal solution (getting from anywhere to anywhere else), maps are superior to route descriptions provided that landmarks are available. Planning a route within a map requires creating a mental image of the self and its position on an *allocentric* map. We can gain some insight into how this may happen by looking at how autonomous robots learn to navigate.

Navigating Robots

Like animals, navigating robots also face the problem of locating themselves in the environment. The breakthrough in autonomous robot navigation came from the recognition that perfecting the robot's sensors does not help much with navigation. Sensors are useless unless the robot can actively move them while keeping track of the motion-induced sensory feedback from the environment. Once mapping and robot localization were combined into a single estimation problem, progress accelerated. This insight is nicely expressed

---

Acharya et al. (2016). People can still learn simple and distinct routes after damage to the hippocampus. However, the hippocampus is indispensable when delays are inserted between the legs of a journey or when novel routes between separate path representations need to be calculated (Kim et al., 2013).

30. Tingley and Buzsáki (2017).

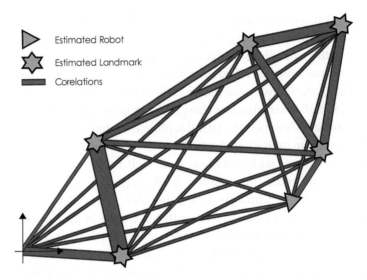

Figure 5.5. Spring network analogy. Springs connect the landmarks and describe their correlations. As the vehicle moves back and forth through the environment, spring stiffness or correlations increase (links become thicker). As landmarks are observed and estimated locations are corrected, these changes are propagated through the spring network. These continuously updated correlations are learned or 'internalized' by the robot. Such correlations can be effectively constructed in recurrent networks, such as the hippocampal CA2/3 systems.
Reprinted with permission from Durrant-Whyte and Bailey (2006).

by the acronym for the solution: simultaneous localization and mapping (SLAM), a combination of motor-driven path integration and sensor-dependent landmark detection (Figure 5.5). The movement trajectory of the mobile robot and the locations of all landmarks are continuously estimated and updated without the need for any a priori knowledge. In the process, the robot can construct a map of an environment by fusing observations from different locations while using the map to deduce its location. The estimation accuracy of the robot's current position is bounded only by the quality of the actively constructed map.[31] The more the robot moves around, the better the estimation of the relationships among the landmarks becomes. Thus, performance improves with exploration. At the conceptual level, autonomous robot navigation is now considered a solved problem, although numerous practical

---

31. The *spring map model* is reminiscent of the "cognitive graph" idea of Muller et al. (1996), in which the springs are replaced by the synaptic strengths between hippocampal CA3 neurons.

issues remain in realizing more general solutions, increasing speed and accuracy, and generalizing to multiple environments.[32]

This spectacular progress in autonomous robot navigation may soon make them part of our everyday lives. A variety of mobile robots have been tested indoors and in outdoor environments where not all landmarks are fixed. Robotic cars must constantly update their position relative to other moving objects. Recent mobile robots can even (relatively) autonomously navigate through busy urban city centers and interact with pedestrians and pets.[33] These technical developments have gone a long way from Grey Walter's *Machina speculatrix* (Chapter 1), but fundamental ideas about ambulation-induced knowledge collection have remained the same.

If machines with a few microprocessors or insects with tiny brains can solve the navigation problem, why do mammals need a complicated navigation system with millions of interactive neurons? To be fair, autonomous robots are not (yet) perfect, and unmanned cars and trucks strongly depend on global positioning system (GPS; i.e., human made) assistance. Most insects do their business in relatively constant environments so they do not need to construct many independent maps. Their seemingly astonishing navigation abilities depend on rigid computational rules. Ants and bees can proceed directly from one familiar food goal to another by selecting the appropriate long-term vectors and combining them with one another.[34] However, they consistently fail in situations that require explicit comparison between separate maps. The mammalian navigation system not only represents more environmental details,

---

32. Solutions to the probabilistic SLAM problem involve finding an appropriate representation for both the observation model (landmark navigation) and the motion model (dead reckoning navigation) that allows efficient computation of the prior and posterior distributions (Durrant-Whyte and Bailey, 2006). The graph-based SLAM is modeled as a sparse graph, where nodes represent the landmarks and each instant pose state, and edges correspond to either a motion or a measurement event. The advantage of graphical SLAM methods is their ability to scale the environment, so they can generalize from small to related larger environments (see also Brooks, 1991). Many novel nonlinear control systems have been developed recently that have found practical applications ranging from digital "fly-by-wire" flight control systems for aircraft, to "drive-by-wire" automobiles, to advanced robotic and space systems (Slotine and Li, 1991).

33. Unmanned navigation with robotic cars has been studied intensively. These efforts have been largely promoted and financed by the Defense Advanced Research Projects Agency (DARPA; Seetharaman et al., 2006; Urmson et al., 2008; Rauskolb et al., 2008), Google (Google, Inc., 2012), and the European Land Robot Trial (ELROB). It is hard to keep up with the progress in this fast-moving field (Schneider and Wildermuth, 2011; Adouane, 2016).

34. The spatial cognitive abilities of invertebrates, rodents, and humans are reviewed by Wehner and Menzel (1990). For an opposing view, see Gould (1986).

but it can flexibly combine and compare any of those details to symbolize and categorize events, objects, and living things.[35]

INTERNALIZING ROUTES AND MAPS FOR MENTAL NAVIGATION

To investigate how such combinatorial abilities may have emerged in mammals, we must recall the *dictum* "structure constrains function," which is especially true in the hippocampal system. Nothing about its architecture and connectivity suggests that it is evolved for spatial navigation but nothing else. The expansively parallel granule cell interface and the strongly recurrent CA2/3 excitatory system, two very different architectural designs, appear ideal for segregating and integrating places and objects. However, segregation and integration are fundamental operations that could benefit numerous brain processes.[36] Thus, the hippocampal system, evolved initially for spatial navigation, may be repurposed for other tasks as well.

One Circuit, Many (Apparent) Functions

The hippocampus-entorhinal system has a topographically organized bidirectional communication with the large neocortex. In a rodent, a major fraction of the neocortex deals with controlling motor outputs and processing sensory inputs. In primates, in contrast, a large part is dedicated to computing more complex functions.[37] The nonsensory representation in the hippocampus has progressively increased with the enlargement of the neocortex during mammalian evolution (Figure 5.6). Therefore, computation of spatial information might be only a part-time job for the hippocampus in animals with

---

35. Etchamendy et al. (2003) showed that mice with hippocampal damage cannot flexibly compare previously learned relationships between the arms of a maze and reward, whereas intact mice can.

36. Similar is different! Whether something is deemed similar or different can be judged from a particular point of view (a "classifier") by an observer. Without an a priori goal, no meaningful classification is possible, or the classification process may turn up numerous minute differences whose meaning cannot be interpreted without a goal. The hippocampus is such a classifier, and the meaning of its computed results (i.e., the segregated items returned to its senders) is interpreted by the neocortex.

37. Because neuroscience's philosophy followed the footsteps of British empiricism (Chapter 1), it is not surprising that these nonallied, higher order cortical areas are still referred to as "association" cortex.

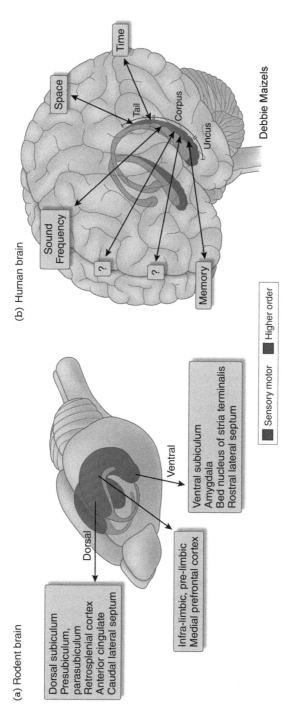

Figure 5.6. Homologous regions of the hippocampus in the human and rat brains. The ventral quadrant of the rodent hippocampus became disproportionally enlarged in primates to keep up with the increasingly larger share of higher order neocortex. Only the relatively small tail part of the primate hippocampus communicates with visuospatial areas. This tail is homologous to the rodent dorsal-intermediate hippocampus. The different connections to and from the segments of the septotemporal axis are shown. Most recordings and manipulations in the rodent brain have been done in the dorsal hippocampus. Adapted from Royer et al. (2010).

larger brains.[38] This idea suggests that whatever information the hippocampus receives from whichever parts of the neocortex, the same general computation will be performed. In other words, the hippocampus is "blind" to the modality and nature of the inputs. It processes the messages from the neocortex in the same way irrespective of their origin and returns its judgment to the source.[39] Whether it appears to process spatial, temporal, or other information is determined largely by the experimental setup.

If the hippocampus served only navigation, it would remain a mystery how I can jot down these lines without any help from environmental cues in Narita Airport, Japan, where I am currently waiting for my connecting flight. These very same structures are also responsible for seemingly separate functions, such as memory, planning, and imagination all of which must be supported by self-generated cell assembly sequences rather than rely on external or body-derived stimuli. As I explain later, these cognitive functions can be thought of as mental navigation that emerged from mechanisms which were initially introduced for spatial navigation in the physical environment.

---

38. The septal (dorsal) and midtemporal thirds of the rodent hippocampus receive visuospatial and other sensory inputs from the dorsal entorhinal cortex, whereas the temporal pole communicates mainly with hypothalamic, amygdalar, and prefrontal areas (Risold and Swanson, 1996; Amaral and Lavenex, 2007). In the primate brain, the ventral pole becomes disproportionally enlarged to keep up with the large fraction of the nonallied neocortex and forms the uncus and body. Only the relatively small tail part of the primate hippocampus communicates with visuospatial areas. This tail is homologous to the rodent dorsal hippocampus, and most recordings and manipulations in the rodent brain have been done here. The relationship between the size of the hippocampus and neocortex is unusual in cetaceans. In dolphins and several other aquatic mammals, the hippocampus is embarrassingly smaller than would be expected from its enlarged neocortex partner. Another paradox is that these mammals have little or no postnatal neurogenesis of granule cells (Patzke et al., 2015). One argument is that landmark (or "sea-mark") navigation is not very useful in the vast waters of the ocean. All you see most of the time is water anyway, so there is no need for a cognitive map. Thus, the hippocampus "shrinks." In marine parks, the dolphin pools are separated by an underwater net, which rises above the water only a foot or so. I have long wondered why they never jump over the net. Are they not curious enough to investigate what is over there? Or do they have a different concept of space than terrestrial mammals (Gregg, 2013)?

39. For example, subgroups of ventral CA1 pyramidal cells project relatively selectively to the prefrontal cortex, amygdala, and nucleus accumbens. These subgroups become activated contingent on the nature of the task (e.g., anxiety or goal-directed behavior; Ciocchi et al., 2015). However, the hippocampus runs the same "algorithm" in each of these behaviors. See Aronov et al. (2017); Terada et al. (2017).

Forms of Mental Navigation

Starting with the famous patient HM, who had both hippocampi surgically removed in an attempt to alleviate his intractable epilepsy, a consensus emerged that the hippocampus and associated structures are responsible for making memories.[40] How are memories related to spatial navigation, another postulated hippocampal function? John O'Keefe, the preeminent promoter of the cognitive map theory, speculated that "addition of a temporal component to the basic spatial map in the human provides the basis for an episodic memory system."[41] On the one hand, this suggestion implicitly implies that the hippocampus in rodents and in humans does different things and that rodents lack a "temporal component."[42] On the other hand, it is not at all clear how adding time to a cognitive map, an allocentric construct that depends on external drive, can provide a mechanism for the most egocentric thing we possess, our collection of episodic memories.

Perhaps if the navigation system can disengage from environmental or body cues, its internalized performance could support memories. Analogous to spatial navigation, two forms of hippocampus system-dependent memories can be distinguished. These are personal experiences (episodic memory or memory for specific instances) and memorized facts (semantic memory or memory for statistical regularities). We are consciously aware of both types and can verbally declare them, so together they are called *declarative memories*. Terms such as "episodic-like" or "flexible" are used to refer to analogous memories in non-human animals.[43]

---

40. The amnesic syndrome following bilateral hippocampal lesion was presented by Scoville and Milner (1957). Independent of this celebrated study, Nielsen (1958) concluded after studying a large number of cases that different brain systems are involved in different memories. One system supports the life experiences of the individual and involves a time component (*temporal amnesia*). The other type of memory corresponds to acquired knowledge, facts that are not personal (*categorical amnesia*). See also Milner et al. (1998); Corkin (2013). Several comprehensive reviews have been written on this topic (Squire, 1992a, 1992b; Cohen and Eichenbaum, 1993; Milner et al., 1998; Eichenbaum, 2000).

41. O'Keefe (1999).

42. This argument is fueled by Tulving's (1983, 2002) repeated assertion that only humans have episodic memories because only humans have the ability to "feel" subjective time. His famous patient KC could not recollect any personally experienced events, circumstances, or happenings. KC could not perform mental time travel but could use a map. His semantic knowledge base was relatively normal, and he knew many facts about himself.

43. "Semantic" is from the Greek word *semantikos*, meaning "significant." "Episodes" are combination of things, from the Greek *epeisodion*, meaning "linking songs" or "odes" in Greek tragedies. Generalization across episodes means to leave behind the concrete in order to abstract the common links. The distinction between episodic and semantic memory was made by

Episodic Memory

Personal memories are important episodes of our lives that give rise to the feeling of the self and are the sole source of individuality.[44] To re-experience such first-person (egocentric) context-dependent episodes, we need to project ourselves back in time and space. Endel Tulving of the University of Toronto, Canada, coined the term "mental time travel" for this hypothetical brain operation. Mental time travel allows us to be aware of the past and the future. Going backward in time is known as *episodic recall*, while traveling into the imagined future can be called *planning*. Some people might protest, of course, that the past and future are vastly different, and this distinction should also apply to memory and planning. Indeed, these categories are typically dealt with in different chapters of neuroscience handbooks and assigned to different brain structures and mechanisms. But the brain may not see things this way (Chapter 10).

My first proposal to address these problems is that the brain mechanisms that evolved initially for navigation in physical space by dead reckoning are basically the same as those used for navigation in "cognitive space" to create and recall episodic memory. My second, related proposal is that the neural algorithms evolved to support map-based navigation are largely the same as those needed to create, store, and remember semantic knowledge. My third proposal is that generation of semantic (allocentric) knowledge requires prior self-referenced episodic experience, akin to map creation by dead-reckoning exploration (Figure 5.7).[45]

Tulving (1972) on theoretical grounds and has been supported by numerous lesion, physiological, and imaging data (Squire, 1992a).

44. Becoming an "in-dividual" means acquiring the single sense that my self cannot be divided into parts or subsystems.

45. Although today it is well accepted that the hippocampal system is essential for memory (Milner et al., 1998) and spatial navigation, views regarding the fundamental role of the hippocampus were diverse not so long ago. Here is a quote from the leader of the cognitive map theory, John O'Keefe (1999) "I reiterate the basic tenet of the cognitive map theory that the processing and storage of spatial information is the primary and perhaps exclusive role of the hippocampus in the rat, and that the data that appear to contradict this have been misinterpreted." Others also postulated that the hippocampus supports different functions in rodents and humans (e.g., Tulving and Schacter, 1990). The idea that the evolutionary roots of episodic and semantic memory systems are the dead reckoning and landmark-based forms of navigation, respectively, are discussed in Whishaw and Brooks (1999); Buzsáki (2005); Buzsáki and Moser (2013). The relationship among place cells and grid cells is not confined to the spatial domain and physical navigation but may represent a common organizational structure of all two-dimensional spaces that are needed for various cognitive computation such as memory, inference, generalization, imagination, and learning (Constantinescu et al., 2016).

Chapter 5. Internalization of Experience

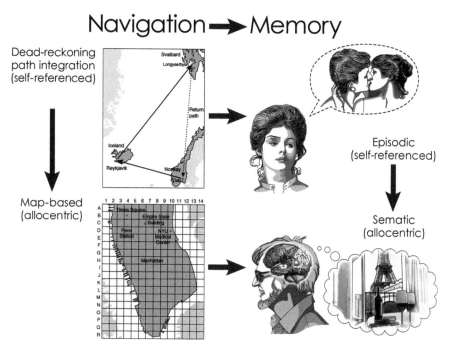

Figure 5.7. Relationship between navigation and memory. *Left*: Path integration (dead reckoning) is based on self-referenced information: keeping track of travel distances (elapsed time multiplied by speed) and the direction of turns. Calculating positions relative to the start location allows the traveler to return to the start along the shortest (homing) path (dotted arrow). *Bottom left*: Map-based navigation is supported by the relationships among landmarks. A map is constructed by exploration (path integration, as indicated by the large arrow from dead-reckoning to map). *Top right*: Episodic memory is "mental travel" in time and space referenced to the self. *Bottom right*: Semantic memory is the explicit representation of living things, objects, places, and events without temporal or contextual embedding. Semantic knowledge can be acquired through multiple episodes with common elements (as indicated by the arrow from episodic to semantic).
From Buzsaki and Moser (2014).

Semantic Memory

In contrast to the observer-dependence of egocentric episodic memories, semantic knowledge is observer-independent.[46] Semantic memory defines objects, facts, and events of the surrounding world independently of temporal

---

46. Philosophers usually refer to observer-independent things as the *objective reality*, referring to things in the physical world. In addition, many other human-created ideas also fall into the semantic category, such as names, events, concepts (such as space and time), and imagined things.

context, similar to how place cells and grid cells explicitly define position coordinates in a map. The deep relationship between physical and mental navigation is also illustrated by the practice of Greek orators. They memorized their speeches by imagining that they walked through different rooms of their house, with each room containing a particular topic. Professional jazz musicians also imagine a landscape of notes and navigate through a preplanned path in a concert.

Unlike the one-trial acquisition of episodic memory of a unique event, *semantic information* is typically learned after repeated encounters with the same thing or event. Initially encoded in episodic memory, information gradually loses its contextual features to become generalized and explicit, reminiscent of the place fields, which are omnidirectional; that is, a place cell fires in its place field irrespective of from which direction the animal approaches it.[47] For example, if you are lucky and discover something, that episode remains precious for the rest of your life. However, if my laboratory and others confirm your discovery by reaching the same conclusion from different angles, it becomes a fact: explicit knowledge understood by everyone in the same general way, irrespective of the conditions of the discovery and confirmation.

Not everyone agrees that episodic experience is necessary to encode semantic information. Indeed, in humans, speech and other externalized functions of the brain allow rapid acquisition of semantic knowledge without the need for personal experiences (Chapter 9). A caregiver can provide the names of objects, which the child then learns. Patients born without a hippocampus can acquire facts and use them efficiently to communicate. However, even though social interaction can provide shortcuts to knowledge, this ability does not undermine the evolutionary origin of episodic and semantic memory from dead-reckoning and map-based navigation, respectively.

---

47. For further discussion of the relationship between episodic and semantic memories, see Tulving (1972); McClelland et al. (1995); Nadel and Moscovitch (1997); Eichenbaum et al. (1999); Manns et al. (2003); Hasselmo (2012). Place cells explicitly define the coordinates of a map. Similarly, neurons in the hippocampal and entorhinal cortex in epileptic patients implanted with depth electrodes respond selectively to pictures of objects or persons, largely independently of their physical characteristics, and even to their names (Heit et al., 1988; Quian Quiroga et al., 2005), which are the defining criteria of explicit knowledge. Experiments on amnesic patients suggest that acquisition of semantic knowledge is possible when brain regions required for episodic memories are damaged. Patient KC was unable to recollect any personal experience, yet he could learn new facts and relations (Hayman et al., 1993; see also Schacter et al., 1984; Shimamura and Squire, 1987). However, his learning was very slow, likely because of the lack of the boosting effect of the episodic system. Three young adults who became amnesic at an early age acquired nearly normal semantic knowledge of world facts and relations despite a severe episodic memory deficit (Vargha-Khadem et al., 1997).

Do Memories Migrate?

The relationships among events and objects, known as *semantic proximity*, shares many features with distance relationships in landmark navigation. Models of semantic relatedness use a metric based on topological similarity, just like the spatial relations among landmarks in the cognitive map.[48] There is a general agreement that declarative memories depend on the entorhinal cortex–hippocampal system, although debate persists about the fate of consolidated semantic information. One controversy is whether the neural circuits involved in creating semantic and episodic memories overlap or completely separate.

Another controversy is whether memories remain in the circuits where they were created or move gradually from the hippocampal system to the neocortex over time, so that retrieval of consolidated memories no longer requires the structures that originally created them. The latter idea likely derives from the initial description of the cognitive symptoms of patient HM. According to early analyses, he could not learn or remember new declarative memories but could recall virtually anything that happened before his brain surgery. However, subsequent work on HM and similar patients revealed that, after bilateral hippocampal damage, they cannot narrate and lack the ability to "travel mentally" either to the past or the future. Their imaginative and planning skills are as strongly impaired as their ability to place themselves into a spatiotemporal context. Instead, they use their intact logic and semantic knowledge to invent scenarios that might sound reasonable.[49] For example, if you asked about the house where the amnesic patient grew up, you might be impressed with the recalled details and perhaps events. However, the patient remains unable to narrate a single personal episode, such as a dinner with the family or a fight with siblings, in the proper sequence. In other words, these patients cannot recapitulate or reimagine a situation by embedding themselves into it.

A more recent view, known as the "two-trace" model, suggests instead that although episodic memories are transferred to the neocortex, a copy of

---

48. Trope and Liberman (2010). The spatial origin of our semantic classification of explicit knowledge is also reflected by the vector distances used for defining relationships among words (Talmy, 1985; Navigli and Lapata, 2010). Such distance representations are also used in "mind maps" and "concept maps." See https://wordnet.princeton.edu/.

49. Such confabulation is a core symptom of Korsakoff's syndrome, originally described in people with chronic alcohol abuse. Degeneration of the mammillary bodies, due not only to alcohol but also to vitamin B (thiamine) deficiency, drug abuse, or head trauma affecting the diencephalon, as well as to bilateral damage to the temporal lobe, can result in an inability to faithfully recall one's past (Bayley and Squire, 2002; Downes et al., 2002; Squire et al., 2004; Gilboa et al., 2006). The memory troubles of the character Leonard Shelby in Christopher Nolan's movie *Memento* (2000) pretty accurately illustrate the symptoms of retrograde amnesia.

the memory is retained in the hippocampus. In support of this view, functional magnetic resonance imaging (fMRI) experiments in human research participants consistently show hippocampal activation during memory recall even for events acquired decades earlier.[50] An alternative explanation is that the memory trace itself no longer resides in the hippocampus, but that hippocampal networks help to coordinate the sequential recall of items stored in the neocortex, thus providing an egocentric perspective for the items and events of the episode (see Chapter 13).

Overall, comparison of spatial and mental navigation indicates how disengagement of navigation networks from their external dependence could transform them into internalized memory operations.[51] This recognition simplifies our understanding of neuronal mechanisms of apparently different functions. While spatial navigation, memory, planning and imagination are distinct terms, their neuronal substrates and neurophysiological mechanisms are identical or at least similar. Viewed from the brain's perspective, these distant chapters of behavioral neuroscience and psychology can be unified.[52]

INTERNALIZING ACTIONS: THE "MIRROR" NEURON SYSTEM

Another spectacular example of the disengagement of the brain from its sensory input and motor reafferentation dependence is the mirror neuron system. Like neurons in the hippocampal-entorhinal system that are active during both encoding and recall of experience, many neurons in the neocortex become active not only when we act intentionally but also when we interpret other people's or even our pet's intentional actions. They serve dual functions.

---

50. Support for the transfer of memory trace can be found in Squire and Alvarez (1995); McClelland et al. (1995); Eichenbaum (2000); Frankland and Bontempi (2005). Many experiments supporting the two-trace model are reviewed in Nadel and Moscovitch (1997).

51. A striking example of how externalized spatial memory may be the precursor to internal memory is the demonstration that the brainless slime mold *Physarum polycephalum* can avoid areas it has previously explored. Drawn into a trap, it can avoid it as if it has constructed a map of the environment (Reid et al., 2012).

52. Functions are often repurposed as conditions change in evolution. The transformation of physical to mental navigation is only one striking and well-researched example. Neuronal mechanisms, such as the grid cell dynamic that supports allocentric (map-like) navigation might also underlie more complex functions, such as logical inference and reasoning (Constantinescu et al., 2016). Grid cell mechanisms could support periodic encoding in multiple cortical areas for abstract categorization and general knowledge organization.

Chapter 5. Internalization of Experience

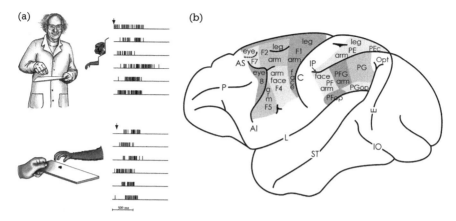

Figure 5.8. Mirror neurons recorded from the prefrontal area of the monkey. A: The neuron is equally active when the monkey reaches for the raisin (*bottom*) or when observes the experimenter reaching for raisin (top). Vertical ticks, action potentials of the neuron. B: Shaded areas indicate brain regions from which mirror neurons were recorded.

The serendipitous discovery of these "mirror neurons" is arguably one of the most celebrated moments of cognitive neuroscience in the past century. Giacomo Rizzolatti and his colleagues at the University of Parma, Italy, observed that some neurons in the premotor cortex (called F5) of macaque monkeys fired not only when the monkey performed a particular action, such as picking up a peanut with its fingers, but also when it watched another monkey or the experimenter performing the same action (Figure 5.8).[53] Observing the robust responses of these neurons leaves no doubt that they are special, and they represent a wide range of actions. In a small subset of mirror neurons, the correlation between the observed and executed actions is strong, both in terms of general action (e.g., grasping) and the executed action (e.g., precision grip). However, most neurons in the premotor area display a reliable correlation between the visual action they respond to and the motor response they organize.[54]

---

53. The discovery of mirror neurons illustrates the importance of a prepared mind in research. There was no a priori hypothesis being tested since there was no way of knowing whether and where these neurons existed. The original observations were published by di Pellegrino et al. (1992), and the term "mirror neuron" was coined by Gallesse et al. (1996). For a comprehensive overview of the mirror neuron system, see Rizzolatti and Craighero (2004). The existence of mirror neurons has also been used as a physiological support for the "theory of mind"; that is, the brain's ability to detect other people's (or animal's) intentions, desires, and fears from their actions.

54. Disengagement of motor action from cortical activity can occur at a remarkable speed. In a brain–machine interface experiment, monkeys were trained to move a lever initially by hand and subsequently by producing neuronal activity in the premotor cortex, which was read out

Mirror neurons catapulted to fame after their indirect demonstration in the human brain. College students watched experimenters make finger movements and in some trials made the same finger movements themselves while their brains were monitored by fMRI. During both observation-execution (action) and observation-only conditions, activity was seen in two areas: the inferior frontal cortex (opercular region) and the rostral-most part of the right superior parietal lobe. Similar activations were observed when participants imagined movements instructed by symbolic cues. Although both brain areas participate in action imitation, their roles are somewhat different. The opercular region represents the observed action as a motor goal or plan, without defining the precise details of the movement. In contrast, the right parietal lobe represents the kinesthetic details of the movement (e.g., how much the finger should be lifted and in what direction). If these areas respond equally well to real and observed or imitated actions, how do people know if they are the actor or the imitator of a movement? One possible answer is that the signal in the parietal cortex reflects a corollary discharge copy of the planned action (Chapter 3), creating some firing pattern differences between real and internalized actions that can inform the brain that "it is my body that is moving."[55]

Extracellular recordings of neurons in human patients shed some light on such subtle differential patterns. A significant fraction of neurons in the supplementary motor area and even in the hippocampus respond while patients either execute or observe hand-grasping actions and facial emotional expressions. Some of these neurons were excited during action execution but inhibited during action observation. These findings are compatible with a mechanism that differentiates the motor and perceptual aspects of actions performed by self or others.[56]

---

by a computer algorithm and sent to the actuator that controlled the lever. After a few hundred trials, the monkey's arm ceased to move and the lever was operated exclusively by neuronal activity (or, if you wish, the monkey's will; Nicolelis and Lebedev, 2009).

55. Iacoboni et al. (1999). These and other experiments on overt action and imagination are haunted by the possibility that the motor pathways and muscles are also activated in some attenuated form. The motor system is strongly connected to the rest of the brain, and, therefore, neuronal patterns underlying cognitions and emotions are often expressed in muscular activity even when it is inconsequential. Just watch how fans of football behave during an exciting match. They mimic the action of the players in a subliminal form. Before the study of Iabobini et al. (1999), Rizzolatti and his colleagues examined hand muscle twitching and associated electromyograms while the participants watched the experimenter grasp objects. Often they found a good match between electromyogram patterns when the research participants watched the experimenter or actually grasped objects themselves (Fadiga et al., 1995).

56. Mukamel et al. (2010).

The original mirror neuron studies examined the similarity of neuronal patterns between performing actions and watching others perform actions. Subsequently, this idea spread to other areas of neuroscience, as researchers investigated how such neurons can encode other people's feelings, intentions, and emotions. When we see a tarantula crawling on someone else's chest, are the same type of neurons in the same brain regions active in both the observed person and the observer? fMRI experiments suggest so, at least in the secondary somatosensory area.[57] The mirror neuron system may link an action and its meaning or intent, whether the act is committed by yourself, by others, or even by simulation in the mind. The actions of these neurons may form the neurophysiological basis of some aspects of our social interactions.

The brain's disengagement from overt movement has important consequences. First, during internalized states, the motor system puts the action planning in a "motor format" (Chapter 11) so that it looks like a real action when related to recorded neuronal activity even when it is not accompanied by overt action. From this perspective, imitation can be regarded as a transformation of observed action into an internal simulation of that action. Second, the corollary discharge signals (Chapter 3) to parietal cortex, resulting in covert "actions," would allow the brain to evaluate the potential consequences of future plans. Without any overt behavior, the mirror and corollary discharge systems can then inform the brain's owner that she is the agent of this covert activity.[58]

SOCIAL INTERACTIONS ARE ACTIONS

One of the most elaborate corollary discharge systems relates to eye movements (Chapter 3), which allow a newborn to begin to discover the world long before reaching and crawling develop. Faces and biological motion are especially effective in triggering eye movements in visual animals such as humans. There are at least two ideas to explain this. One view is that these cues are somehow

---

57. Keysers et al. (2004). Both feeling disgusted and watching an actor with a disgusted look activate the anterior insula (Wicker et al., 2003).

58. Dysfunction of the mirror neuron system has been suggested to underlie the derailed social interactions and symptoms of motor speech problems in autism (Iacoboni and Dapretto, 2006). The mirror neuron concept offers a potential framework for other dysfunctional states as well, such as misattribution of actions, hallucinations, delusion of agent influence, and experience of alien control in psychotic patients. The system of mirror neurons has been hypothesized to form a "core" network, including the primary temporal, posterior parietal, and prefrontal cortex, which becomes activated during action execution and simulated states, including observed and imagined actions (Wolpert et al., 1995; Jeannerod, 2001).

special for the brain, so they automatically trigger a response.[59] Another view is that eye movements reflect internally driven seeking for ethologically relevant patterns, so the brain can make adaptive use of those cues.[60] Although the distinction may seem subtle, the answers may determine how we search for and evaluate the underlying mechanisms. Monitoring eye movements in the laboratory tells us about the brain's engagement with the world. This simple method does not require verbal communication.

Alteration of eye movement patterns can be diagnostic for neuropsychiatric diseases, most prominently autism. Autism is usually defined not by its underlying mechanisms but by impaired social communication and repetitive behavior. One of the first things parents discover is that autistic children rarely solicit or reciprocate eye contact with caregivers, unlike normal infants. This difference becomes apparent by two to six months of age.[61] There is nothing wrong with the eye movements in affected children; they simply look elsewhere.

There is no large gap between autism and many 'odd' but healthy people.[62] Instead, there seems to be a continuum of sociality reflected in the amount of social visual engagement, which is largely determined by genetic factors. When identical twin toddlers viewed movies of other toddlers interacting, they tended to fixate their eyes on similar things. Moreover, they shifted their eyes at the same time in the same directions. In other words, their eye searches were remarkably concordant. Such joint patterns were much more rare in sex- and age-matched nonidentical twins or nonsiblings, showing that it is not only the movie scenes that synchronized the eye movement patterns in identical twins but also some internal programs. Heritability of eye search is most prominent in gaze toward the eyes and mouth. Toddlers with autism rarely focus on these critical parts of the face, so they do not learn about the social consequences of their actions from the smiles and grimaces of other children and caregivers who try to engage them.[63]

---

59. This may be similar to the released action patterns in many animals (Tinbergen, 1951).

60. Scarr and McCartney (1983); Hopfinger et al. (2000); Treue (2003).

61. Jones and Klin (2013). In contrast to the active search idea, Markram et al. (2007) proposed that the amygdala is hyperreactive because people with autism find looking into the eyes of others to be stressful.

62. Some claim that scientists belong to the autistic end of the spectrum. Instead of making eye contact, scientists look at each others' shoes. Mathematicians, on the other hand, look at their own shoes.

63. Constantino et al. (2017).

Investigate the White of the Eyes

These findings, admittedly selected with some bias from an enormous literature with various and often conflicting views, suggest that internally driven seeking of face cues is heritable. The neuronal mechanisms responsible for social visual engagement are not well understood, but, if I may speculate a bit, the amygdala is likely to be a key structure. The amygdala and its partners comprise a neural system involved in the perception and expression of emotions[64] and the retrieval of social cues from facial expressions.[65] Nearly all social interactions involve emotions. Patients with complete bilateral amygdala damage have difficulty judging people's trustworthiness based on their appearance, compared to people with intact brains or control patients without amygdala damage. People with bilateral amygdala damage also cannot recognize fearful faces.[66]

These results, along with numerous experiments in animals, have been interpreted to indicate that the amygdala is the seat of emotions, particularly fear. But this textbook teaching may be misleading. First, the impairment of emotional face judgments does not extend to verbal descriptions of people. Second, in contrast to the laboratory, where emotions are tested by viewing still faces, patients with amygdala damage do not necessarily misjudge untrustworthy persons in real life. Third, they easily recognize fear expressed via vocal communication. Fourth, when they are instructed to focus on the eyes in pictures, their judgment becomes close to that of control subjects.

---

64. Adolphs et al. (1994, 1998); Calder et al. (1966). Emotion is one of those constructs on William James's shortlist. This charged term has two aspects. One relates to actions and bodily expressions, which are perceptible and measurable by external observation; for example, breathing changes, flushing, crying, or other facial changes. The other aspect is an internal experience (or *qualium*), which is only available to the participant who has the emotion (Damasio, 1995). Only the observable aspects can be objectively studied (LeDoux, 2014). Brain imaging studies also point to the key role of amygdala in visual processing of emotional faces. Fearful faces induce positron emission tomography (PET) scan changes in the left amygdala, left pulvinar, left anterior insula, and bilateral anterior cingulate gyri (Morris et al., 1998). I cannot do justice here to the vast amount of thinking in this area of research, but I can recommend great summaries (Damasio, 1994; LeDoux, 2015). Classic sources on emotions are James (1890), Papez (1937), and MacLean (1970).

65. Perceptual models of autism suggest that the underlying problem is socioaffective oversensitivity.

66. Neuronal recordings from the amygdala in epileptic patients with comorbid autism show that the spike responses of neurons do not differ from those of a nonautistic group in general, nor does the responsiveness to whole faces. Yet some neurons in the autistic patients showed abnormal insensitivity to the eyes and abnormal sensitivity to the mouth (Rutishauser et al., 2013).

Both the clinical findings and the animal experiments are compatible with the alternative interpretation that the amygdala's main behavioral function is to direct the gaze to cues that are ethologically relevant for social communication. For primates, these cues are mainly aspects of the face (including the ears in monkeys). Physiological experiments support this idea.

There is a general consensus that the inputs that are physiologically relevant to neurons should induce robust firing. However, neurons in the amygdala are infamous for firing at extremely low rates under most circumstances, including fear conditioning experiments. Perhaps the experimental conditions that lead to fear learning are not really what amygdala neurons care most about. But similar to other neurons in the brain, amygdala neurons can also become very active under the right circumstances. In an illuminating experiment, neuronal activity from the monkey amygdala was recorded while monkeys viewed videos of natural behaviors in unfamiliar monkeys. When the recorded monkey made eye contact with the video monkey, the firing rates of many neurons increased several-fold. When the gaze of the movie monkey pointed instead to the peripheral parts of the recorded monkey's retina, only small responses were observed. A critical observation was that the firing latency of these gaze contact neurons varied between 80 and 140 ms, which is tens of milliseconds shorter than their response latencies to visual stimuli. This experiment therefore suggests that amygdala neurons do not respond to a particular visual pattern but may be involved in selecting which parts of the face should be consulted when examining strangers.[67] I interpret these findings to mean that emotions, which have traditionally been approached via sensory inputs, can be reevaluated when viewed from the action side of the brain. The disengagement hypothesis may be used as guidance here as well.

One of the oldest and most influential theories of emotion was put forward by William James and the Danish physiologist Carl Lange, known today as the James-Lange theory of emotion.[68] The theory argues that emotions occur as a result of physiological reactions to body signals that we normally do not detect, such as changes in heart rate and respiration. We run away not because we fear the bear but we fear because we run away from it. In William James's

---

67. Mosher et al. (2014). A dominant monkey initiates eye contact by staring at the eyes of others and waiting for them to return direct gaze. In contrast, submissive individuals avoid direct eye contact altogether or engage in it only briefly (Redican, 1975).

68. James (1884); Lange (1885/1912). Walter Cannon (1927) criticized the James-Lange theory, but he examined only sensory signals conveyed by the vagal nerve and spinal cord. More recent theories argue that emotions are cognitive interpretations, and the brain produces meaning from the autonomic responses in combination with prior experience and social cues (Barrett-Feldman, 2017; LeDoux, 2014, 2016; LeDoux and Daw, 2018).

words, "My thesis, on the contrary, is that the bodily changes follow directly the perception of the exciting fact, and that our feeling of the same changes as they occur is the emotion." Not surprisingly, James's explanation is "outside-in"; it is the body-derived signals whose conscious interpretation by the brain produces fear. The theory has been criticized multiple times over the past century, yet its appeal persists.

Emotion expert Joe Ledoux, working at New York University, recently discussed the distinction between brain mechanisms that detect and respond to threatening signals and those that give rise to conscious fear. Whereas the former can be studied in animals, such as pairing a place with an electric shock and examining neuronal responses during the process, fear is a subjective feeling in the conscious domain that cannot be pinned down by simply correlating neuronal activity with overt responses to those signals. I agree. However, we can attempt to draw a parallel between mechanisms of cognition and mechanisms of emotions. As we discussed earlier in this chapter, cognition is a disengagement of the brain from the motor action system. Similarly, emotion can be conceptualized as disengagement from the autonomous action system. In both cases, the fundamental physiological mechanism is the corollary discharge system; in case of emotion, it is the reafference from the brain's control mechanisms of visceral systems to the amygdala and neocortex. Although the phylogenetic roots of emotions may well be based on overt feedback from the autonomous system, more complex brains can disengage from it so that internalized feelings can occur in the absence of blushing, sweating, or pupil or respiration changes. In summary, internalization of the overt sensory feedback in the James-Lange theory can be an alternative way of conceptualizing and researching emotions.

LANGUAGE IS BASED ON INTERNALIZED ACTION

In social communication, individuals try to understand the gestural and vocal actions and intentions of others. If the mirror neuron system represents the actions and emotions of conspecifics, we can speculate that a similar framework may underlie the neurophysiological mechanisms of language, linking the gestural communication system of the limbs, ears, and face muscles to verbal communication. This may be the evolutionary reason that language is inherently linked with gestures. Look at anyone in a cell phone conversation or watching TV during a plane flight. There is no need for waving hands, smiling, or other gestures when the other party cannot reciprocate them. The brain just cannot help it and mobilizes the body anyway.

Providing anatomical support for the gestural communication hypothesis, area F5 in non-human primates and Broca's speech area in people are homologous brain regions. Broca's area may have evolved atop the mirror system for actions and provided a seed for generating and reading motor sequences of speech. Similarly, the parietal mirror system could be a precursor of the Wernicke's "perceptual" speech area. Through such evolutionary development, the combinatorially open repertoire of manual and body gestures could have provided scaffolding for the emergence of protospeech vocalizations, leading eventually to language with syntax and compositional semantics. The observation–execution matching system could have linked action to social communication, culminating in linguistic syntax, so that an effective sender-to-receiver relationship could develop between actor and observer.[69]

In line with this hypothesized evolutionary path, action, as symbolized by the verb in language, is the architectural centerpiece of a sentence. You cannot master a language without learning the meaning of verbs. Verbs express intentions and encode what is happening in motion. As engines of syntactic action, they can have a large entourage in a sentence. A verb can specify the action or event, describe the trajectory of an action with respect to some reference spatial point, indicate whether an action is completed or ongoing, and select the semantic roles of the participants in an action. Verbs in many but not all languages have a tense, referring to past, present, or future. They express not only the intent but also the mood of the speaker. For example, ambulatory movement can be described by a variety of verbs that refer to different speeds, moods, intentions, directions, and patterns of walking—in essence allowing the listener to visualize the action.[70]

The process that gives meaning to verbs and relates them to motor actions and real-world experiences is called "semantic grounding" (Chapter 3): an initially meaningless sound pattern acquires meaning to the listener. After this learning process, whenever the pattern recurs in the brain, it can bring back the simulated or constructed thing even when the thing (or episode) is not present.

---

69. Similar matching mechanisms could be involved in the recognition of phonetic gestures in humans (Gallesse et al., 1996). This idea was further developed by Rizzolatti and Arbib (1998) and Arbib (2005) and has its antecedents in the "motor theory of speech perception" (Lieberman et al., 1967), which postulates the necessary access to the motor templates of words in their perception.

70. For example, walking, running, approaching, leaving, hopping, heading, trotting, trudging, passing, hitching, and fishtailing refer to different movement patterns, reflecting mood, speed, and direction. In action-oriented languages, such as Hungarian, more than 200 verbs describe the walk-run dimension.

The specific neural mechanisms underlying semantic grounding of nouns and words in general are not known but they are based on our interaction with the environment in unique ways. For example, "up" and "down" have positive and negative meanings, respectively, in most languages, likely because of a number of motor actions that lead to reward require an erect posture, combating the gravitation forces we live in. "She woke up"; "he fell asleep"; "I'm feeling down"; "his income is high, mine fell"; "things in the lab are going downhill"; or "she couldn't rise above her emotions." Numerous examples in language are based on action metaphors. "Digesting these pages" "he killed my brilliant idea"; "Einstein's theory gave birth to a new era."[71]

Early learning in human infants often takes place in the context of active exploration of objects. "Infants do not simply look passively at the jumble of toys on the floor but rather use their body—head, hands and eyes—to select and potentially visually isolate objects of interest, thereby reducing ambiguity at the sensory level."[72] Of course, the process of "reducing ambiguity" is the idea of action-assisted grounding or, as Wittgenstein noted it more eloquently: "Our talk gets its meaning from the rest of our activities." Pointing by hand to objects or just gazing at them in humans is an efficient gesture to invite joint attention between communicating individuals.[73]

Psychophysical, lesion, recording, and transcranial magnetic stimulation (TMS) studies suggest that motor cortex and somatosensory cortex are involved in linking action words to their direct meaning, whereas the inferior frontal cortex, temporal lobe, and inferior parietal cortex encode more abstract features of action words.[74]

As in spoken language, auditory feedback when playing a musical instrument is necessary to learn how action leads to the desired effect. Copies of the planned action from premotor cortices are sent before the sound feedback arrives. Once this corollary system has been trained, sounds can be imagined

---

71. Lakoff and Johnson (1980).

72. Yu and Smith (2012). Pulvermüller (2003) also points out the connection between meaning and action. Reading the word "kick" activates the same part of motor cortex as an actual kick.

73. Pointing and gazing are examples of externalization of brain function in humans (Chapter 9). No such behaviors have been observed in other species, except in domesticated dogs. Miklósi et al. (2003) suggest that reading intention of the other is a result of human–dog co-evolution and is absent in wolves and even non-human primates. However, active communicative systems exist in all animals. Rats communicate the safety of food by breath. Once a "demonstrator" rat conveys information about a novel edible food, naïve "observer" rats exhibit substantial enhancement of their preference for that food even if they did not observe the demonstrator's act of eating (Galef and Wigmore, 1983).

74. Damasio et al. (1996); Tomasino et al. (2008); Pulvermüller (2013).

even in their absence. This may explain why many deaf people can speak, sing, or even compose music if they lose their hearing after acquiring language, although they have difficulty regulating the volume and pitch of sound. Beethoven became virtually deaf by the age of forty-four, yet he continued to compose extraordinary music for the rest of his life.[75] In contrast, no composer born with hearing loss is known. The corollary feedback loop of the action system must be calibrated before it can be deployed in a useful way.

To recapitulate, the mirror neuron system allows us to read the intentions of others by interpreting body language. The expansion of the action system to spoken language permitted us to establish an extensive and effective communication system by interpreting speech, which can be viewed as a metaphoric form of action. The invention of language accelerated externalization of brain function, creating a collective species memory (Chapter 9).

To enable cognitive operations, the brain should be equipped with mechanisms to separate and combine neuronal messages, similar to the syntactical rules of language. How neuronal networks support these operations is the topic of the next chapter.

SUMMARY

We discussed several examples of how disengagement of brain networks from their external inputs can be useful for cognitive operations. The key physiological mechanism of this scenario is a corollary discharge-like system that allows the brain to interpret the activity of action circuits even in the absence of overt movement and sensory feedback from muscles. Within such an internalized world, brain networks can anticipate the consequences of imagined actions without the need to act them out. Instead, the outcomes can be tested against previously acquired knowledge, which creates new knowledge entirely through self-organized brain activity.

For the sake of simplicity, I portrayed externally and internally driven activities as separate processes that emerged sequentially during evolution. In reality, these steps are only abstractions. The logic of this framework is not that the networks of a small brain depend entirely on external inputs and larger brains switch to internal operations. Instead, the two processes are inherently intertwined, and the same circuit can perform both input-dependent and

---

75. Several other composers continued to create music after complete hearing loss. Bedrich Smetana created perhaps his most famous work, *The Moldau*, after his hearing had left him at age fifty.

input-disengaged operations.[76] For example, the head direction system can momentarily disconnect from environmental and body-derived cues to extrapolate the imagined head direction via internal mechanisms. Spatial navigation, especially dead reckoning, assumes some kind of memory. Larger brains may have more capacity for internal processing. Yet, even in a complex brain, memory recall can be improved by external information. When a narrative gets stuck, a simple cue comes in handy, which is why people prepare a list of keywords for talks. In other words, even small brains have elements of internal operations ("cognition"), but, as the complexity of neural networks increases, the share and efficacy of internalized computation also increases. Therefore, complex neuronal networks with numerous interactive loops in larger brains can predict the consequences of the brain's actions over a much longer time scale and in more complex environments than smaller brains.

---

76. In fact, as discussed in Chapter 3, self-organized brain activity may be the fundamental brain operation that can generate useful motor outputs even without sensors (e.g., baby kicks). The modification of such self-organized patterns to external perturbations can provide "meaning" to the actions by matching preexisting neuronal patterns to action–perception (Chapter 13; Buzsáki, 2006). Experiments in developing ferrets show that the similarity between spontaneous and visually evoked activity in visual areas increases progressively with age, suggesting that *internal models* result from the adaptation of self-organized brain patterns to the statistics of the surrounding world (Fiser et al., 2004). In a similar vein, construction of an internal representation of the self as a special entity can be accomplished by reprocessing internal data simultaneously with information about the environment.

6

Brain Rhythms Provide a Framework for Neural Syntax

Rhythm imposes unanimity upon the divergent, melody imposes continuity upon the disjointed, and harmony imposes compatibility upon the incongruous.

—Yehudi Menuhin[1]

Language is a process of free creation; its laws and principles are fixed, but the manner in which the principles of generation are used is free and infinitely varied.

—Noam Chomsky[2]

I would define, in brief, the poetry of words as the rhythmical creation of beauty.

—Edgar Allan Poe[3]

I was mesmerized when, as a middle school student in Hungary, I first heard Morse code in a ham radio course I attended. Sending signals through the ether with a transmitter, the instructor told us that somebody from New Zealand had responded to his call. The operator on the other side of the world informed us about the sunny weather there, the receiving conditions, the details of his rig, and the antenna type he used. My surprise only increased

---

1. https://en.wikiquote.org/wiki/Yehudi_Menuhin.
2. http://www.famousphilosophers.org/noam-chomsky/.
3. Poe, EA (The Poetic Principle; CreateSpace Publishing, 2016).

when I figured out that our instructor did not speak either English or Maori, and presumably the operator in New Zealand did not understand Hungarian. Instead, I was informed, they "spoke" to each other with dots (short pulses) and dashes (long pulses), through a universal language in which all words consist of three letters, called *Q-code*. To be part of such a conversation, all I had to do was to learn Morse code, memorize the Q language, learn a bit about electronics, pass exams, get a license, build a transmitter and receiver, and set up a wire antenna between the chimneys of our house and the neighbor's. Then I could communicate with any ham radio around the globe. That is exactly what I did, and the problem of coding has bugged me ever since.

In learning Morse code, the biggest hurdle is to recognize letters in the sea of dots and dashes separated by short spaces. Separation of messages is the most important thing in any coding system, be it speech, written language, computer language, or spike transmission in the brain. In Morse code, letters are separated by silent periods of the duration of a dot, whereas words have boundaries of at least two dots' length: [.. _ _ _ .._ _._ ...._ ....] contains information (translation: "it makes sense"), whereas [..._.._._........_....] the exact same sequence of dots and dashes, but without spaces between the characters, is just nonsense noise. Similarly, if this entire page were a single long word, it would be very difficult to decipher the content, although not impossible because unique combinations of thirty letters are much easier to decode than combinations of only two characters. The situation with neuronal messages is even more complicated because action potentials are all pretty much the same: just dots.[4]

Human language is also produced by combining elements. For instance, linking letters in particular sequences produces words, which have new qualities not present in the individual letters. Such combinations are regulated by *syntax*, a set of rules that govern the transformation and temporal progression of discrete elements (such as letters or musical notes) into ordered and hierarchical relations (words, phrases, and sentences, or chords and chord progressions) that allow the brain to interpret their meaning. Grouping or

---

4. In principle, single spikes and bursts of spikes could be conceived of as dots and dashes. Indeed, it has been suggested that the postsynaptic neuron can effectively discriminate between single spikes and burst events (Lisman, 1997). Ham radio operators are not exceptional among neuroscientists. I have met a few of them, including the late W. Ross Adey (UCLA), John Hopfield (Princeton University), Terry Sejnowski (Salk Institute), and Fritz Sommer (UC Berkeley). I also had radio contacts with a few famous people, including Yuri Gagarin (UA1LO), Senator Barry Goldwater (K7UGA), King Hussein of Jordan (JY1), and his Queen Noor (JY1H). The aviator Howard Hughes was also a ham radio operator (W5CY). Q-code is also used in maritime communication. In addition to Q-code, hams use a more specialized code, which includes some arbitrary signals (e.g., 88 = love and kisses), while others are simplified English words (e.g., vy = very, gm = good morning, r = are).

chunking the fundamentals by syntax[5] allows the generation of virtually infinite numbers of combinations from a finite number of lexical elements using a minimal number of rules in music, human, sign, body, artificial, and computer languages, as well as in mathematical logic.[6] With just thirty or so letters (or sounds) and approximately 40,000 words, we can effectively communicate the entire knowledge of humankind. This is an extraordinary combinatorial feat. Syntax is exploited in all systems where information is coded, transmitted, and decoded.[7] In this chapter, I suggest that neuronal rhythms provide the necessary syntactical rules for the brain so that unbounded combinatorial information can be generated from spike patterns.

A HIERARCHICAL SYSTEM OF BRAIN RHYTHMS: A FRAMEWORK FOR NEURAL SYNTAX?

Researchers who study neuronal rhythms or oscillations have created an interdisciplinary platform that cuts across psychophysics, cognitive psychology, neuroscience, biophysics, computational modeling, and philosophy.[8] Brain rhythms are important because they form a hierarchical system that offers a

---

5. Segmentation or chunking has been often considered as a basic perceptual and memory operation. According to Newtson et al. (1977), segmentation is an interplay between perceptual, bottom-up processes, and inferential, top-down processes. The observer has a preformed anticipatory schema that helps to select features that are characteristic for anticipated events or activities and their change. Recognition of event boundaries partitions a continuous stream of events into discrete units and forms the basis for a redefinition of the search and extraction of new information. The "preformed anticipatory schema" may correspond to the system of brain rhythms. See also Shipley and Zacks (2008).

6. Several previous thinkers have considered the need for brain rules to support syntax in language; for example Port and Van Gelder (1995); Wickelgren (1999); Pulvermüller (2003, 2010). Karl Lashley (1951) was perhaps the first who specifically thought that sequential and hierarchical neuronal organization must be behind "fixed action patterns" or "action syntax," behaviors that are elicited by ethologically relevant cues but proceed according to species-specific rules even without prior experience (Tinbergen, 1951).

7. In a strict sense, "syntax" (from Greek *syntaxis*) means "arrangement," and in language it roughly refers to the ordering of words (i.e., what goes where in a sentence). It does not provide any information about meaning, which is the domain of semantics. Sometimes syntax is used synonymously with *grammar*, although grammar is a larger entity that includes syntax. Grammar sets the rules for how words are conjugated and declined according to aspect, tense, number, and gender.

8. I devoted a book (*Rhythms of the Brain*, 2006), a monograph (Buzsáki, 2015), and several reviews (e.g., Buzsáki, 2002; Buzsáki and Draguhn, 2004; Buzsáki and Wang, 2012) to this topic. See also Whittington et al. (2000), Wang (2010); Engel et al. (2001); Fries et al. (2001); Fries (2005).

syntactical structure for the spike traffic within and across neuronal circuits at multiple time scales, and their alterations invariably lead to mental and neurological disease.[9]

The transmembrane potential of neurons can be measured by electrodes, which can report neuronal activity with submillisecond time resolution. Electric current flows from many neurons superimpose at a given location in the extracellular medium to generate a potential with respect to a reference site. We measure the difference between "active" and reference locations as voltage. When an electrode with a small tip is used to monitor the extracellular voltage contributed by hundreds to thousands of neurons, we refer to this signal as the *local field potential* (LFP). [10] Recordings known as the *electrocorticogram* (ECoG) are made with larger electrodes placed on the brain surface or its surrounding dura mater to sample many more neurons. Finally, when even larger footprint electrodes are placed on the scalp, we call the recorded signal an *electroencephalogram* (EEG). Each measure refers to the same underlying mechanisms and neuron voltages, but because of the different sizes of the monitoring electrodes used in LFP, ECoG, and EEG recordings, they integrate over increasingly larger numbers of neurons, from a few dozens to millions. The magnetic field induced by the same transmembrane activity of neurons is known as the *magnetoencephalogram* (MEG).[11]

The patterns measured by LFP, ECoG, EEG, or MEG are very useful for the experimenter because they provide critical information about the timing of cooperating neurons. Like noise measured in a football stadium, these techniques cannot provide information about individual conversations but

---

9. In the brain (and other systems), hierarchy is defined as an asymmetrical relationship between forward and backward connections.

10. The term LFP is an unfortunate misnomer, since there is no such a thing as "field potential." What the electrode measures is voltage in the extracellular space (Ve; a scalar measured in volts). The electric field is defined as the negative spatial gradient of Ve. It is a vector whose amplitude is measured in volts/meter. All transmembrane ionic currents, from fast action potentials to the slowest voltage fluctuations in glia, at any given point in brain tissue, superimpose to yield Ve at that location. Any excitable membrane (spine, dendrite, soma, axon, and axon terminal of neurons) and any type of transmembrane current contributes, including slow events in glia to pericytes of capillaries. Ve amplitude scales with the inverse of the distance between the source and the recording site. The LFP and EEG waveforms, typically measured as amplitude and frequency, depend on the proportional contribution of the multiple sources and various properties of brain tissue. A major advantage of extracellular field recording is that, in contrast to other methods used to investigate network activity, the biophysics related to these measurements are well understood (Buzsáki et al., 2012; Einevoll et al., 2013).

11. Hämäläinen et al. (1993).

can precisely determine the timing of potentially important events, such a goal in the stadium or the emergence of a synchronized pattern in the brain. Monitoring LFPs or magnetic fields is the best available method for identifying brain state changes and faithfully tracking dynamic population patterns such as brain rhythms.

A System of Rhythms

There are numerous brain rhythms, from approximately 0.02 to 600 cycles per second (Hz), covering more than four orders of temporal magnitude (Figure 6.1).[12] Many of these discrete brain rhythms have been known for decades, but it was only recently recognized that these oscillation bands form a geometric progression on a linear frequency scale or a linear progression on a natural logarithmic scale, leading to a natural separation of at least ten frequency bands.[13] The neighboring bands have a roughly constant ratio of e = 2.718—the base for the natural logarithm.[14] Because of this non-integer relationship among the various brain rhythms, the different frequencies can never perfectly entrain each other. Instead, the interference they produce gives rise to metastability, a perpetual fluctuation between unstable and transiently stable states, like waves in the ocean. The constantly interfering network rhythms can never settle to a stable attractor, using the parlance of nonlinear dynamics. This explains the ever-changing landscape of the EEG.

12. Rhythms are a ubiquitous phenomenon in nervous systems across all phyla and are generated by devoted mechanisms. In simple systems, neurons are often endowed with pacemaker currents, which favor rhythmic activity and resonance in specific frequency bands (Grillner, 2006; Marder and Rehm, 2005).

13. Penttonen and Buzsáki (2003); Buzsáki and Draguhn (2004). The multiple time scale organization of brain rhythms is similar to Indian classical music, where the multilevel nested rhythmic structure, explained by the concept of *tāla*, characterizes the composition. *Tāla* roughly corresponds to a rhythmic framework and defines a broad structure for repetition of musical phrases, motifs, and improvisations. Each *tāla* has a distinct cycle in which multiple other faster rhythms are nested (Srinivasamurthy et al., 2012).

14. The constant e is sometimes called Napier's constant (see Chapter 12) and is perhaps the most famous irrational number with the exception of π. Yet the e symbol honors the German mathematician Leonhard Euler who defined it. If you ever wondered why naming things and getting credit for discoveries are often an idiosyncrasy of history, read these entertaining accounts about credit assignment debates in mathematics: Maor (1994); Conway and Guy (1996); Beckmann (1971); Livio (2002); Posamentier and Lehmann (2007).

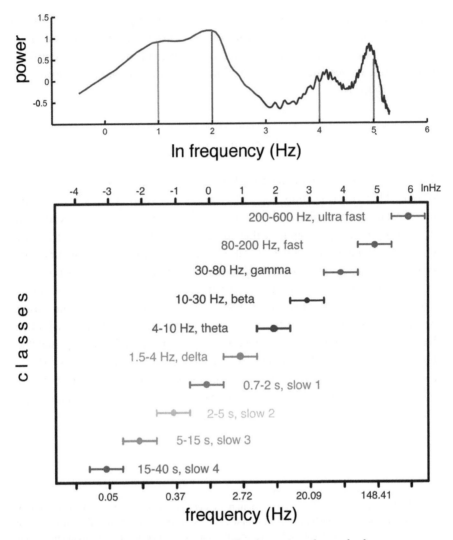

Figure 6.1. Multiple oscillators form a hierarchical system in the cerebral cortex. *Top*: Power spectrum of hippocampal electroencephalogram (EEG) in the mouse recorded during sleep and waking periods. Note that the four peaks (1, 2, 4, 5), corresponding to the traditional delta, theta, gamma, and fast ("ripple") bands, are multiples of natural log integer values (ln frequency). *Bottom*: Oscillatory classes in the cerebral cortex show a linear progression of the frequency classes on the log (ln) scale. In each class, the frequency ranges ("bandwidth") overlap with those of the neighboring classes so that frequency coverage is more than four orders of magnitude. Modified with permission from Penttonen and Buzsáki (2003).

Traditionally, the frequency bands are denoted by Greek letters.[15] Several rhythms can temporally coexist in the same or different structures and interact with each other. Their constellation determines various brain states, such as stages of sleep and the arousal levels of waking. The different oscillations generated in cortical networks show a hierarchical relationship, often expressed by *cross-frequency phase modulation* between the various rhythms. This term implies that the phase of a slow oscillation modulates the amplitude of a faster

---

15. The Greek letters do not indicate a logical order of frequencies but roughly the order of their discoveries. Hans Berger (1873–1941), the discoverer of the human EEG, first observed 8–12 Hz rhythmic patterns above the occipital cortical area when his study participants closed their eyes, and he called them alpha waves (Berger, 1929). In their absence, the smaller-amplitude, faster vibrations, present when the eyes were open, were named beta waves (13–30 Hz). The largest amplitude waves are present during deep sleep and are known as delta waves (0.5–4 Hz). We already discussed theta oscillations (4–10 Hz), most prominent in the hippocampus. The story of gamma oscillations is really fascinating. Perhaps Jasper and Andrews (1938) used the term "gamma waves" first for frequencies between 35 and 45 Hz. The idea that this "40-Hz" oscillation is a "cognitive" rhythm likely originates from Henri Gastaut (Das and Gastaut, 1955), who described 40-Hz rhythmic trains in the scalp EEG of trained yogis during the *samadhi* state. Banquet (1973) also observed 40-Hz bouts during the third deep stage of transcendental meditation. In normal participants, Giannitrapani (1966) found increases in 35–45 Hz immediately before they answered difficult multiplication questions. Subsequently, Daniel Sheer popularized "40 Hz" in his many papers on biofeedback (e.g., Sheer et al., 1966; Bird et al., 1978), which he thought was basically a high frequency "beta" mechanism. The phrase "gamma rhythm" became popular in the 1980s. According to the late Walter Freeman "I coined it in 1980, when the popular term was '40 Hz'.... At that time Steve Bressler was beginning graduate work in my lab, so I assigned him to a literature search to document the inverse relation of OB [olfactory bulb] frequency and size. Berger had coined alpha. Someone else coined beta (I don't remember who), and Berger coopted it. In analogy to particle physics, the next step up would be gamma. I found no prior use of 'gamma' in EEG research, so Steve and I called it that. We sent the manuscript to Mollie Brazier, long-time editor of the *EEG Journal*. She wrote that she would submit our term to the Nomenclature of the International EEG Society. A month later she reported back that the committee had refused to endorse the usage, so she wouldn't publish the paper unless we took it out. Steve needed a publication to get support, so we complied. When the article appeared (Bressler and Freeman, 1980), there it was: 'Gamma rhythms in the EEG,' in the running title. I'd forgotten to take it out, and nobody noticed, least of all Mollie. So that's how it first appeared in print. I continued to use it in lectures, insisting that gamma is a range, not a frequency, and it caught on. Now, like numerous successful developments in science, few people know where it came from" (cited from my e-mail correspondence with WF, April 3, 2011). Subsequently, Steve Bressler sent me a copy of his original manuscript (he kept it for 30 years!). Voila, its title reads "EEG Gamma Waves: Frequency Analysis of Olfactory System EEG in Cat, Rabbit, and Rat." In the Introduction, they acknowledge that "The label 'gamma waves' was used by Jasper and Andrews (1938)," but Walter did not remember this detail. Putting aside the unavoidable errors of source memory, we should thank Walter Freeman for reintroducing gamma oscillations. *Gamma rhythm* catapulted to its current fame after Wolf Singer and his colleagues suggested that it could be the Rosetta Stone for the problem of perceptual binding (Gray et al., 1989).

rhythm, meaning that its amplitude varies predictably within each cycle. In turn, the phase of the faster rhythm modulates the amplitude of an even faster one and so on. This hierarchical mechanism is not unique to the brain. Spring, summer, fall, and winter are four phases of a year that "modulate" both the amplitude and duration of day length. In turn, the phases of the day are correlated (i.e., phase-locked) with the alignment of the sun and moon, which modulate the tidal magnitude of the oceans.

Because of cross-frequency coupling, the duration of the faster event is limited by the "allowable" phase extent of the slower event. The fastest oscillation is phase-locked to the spikes of the local neurons and, because of cross-frequency coupling, to all slower rhythms of the hierarchy. For example, the ultrafast oscillatory "ripple" waves (5–7 ms or ~150–200/s) in the hippocampus are phase-locked to the spikes of both pyramidal cells and several types of inhibitory interneurons, and the magnitude of the short-duration ripple events (approximately 40–80 ms) is modulated by the phase of thalamocortical sleep spindles. The spindle events, in turn, are phase-modulated by cortical delta oscillations, which are nested in the brain-wide infra-slow (slow 3) oscillations (Figure 6.2).[16] The nested nature of brain rhythms may represent the needed structure for syntactic rules, allowing for both the separation of messages (e.g., gamma oscillation cycles containing cell assemblies; see Chapter 4) and their linking into neuronal words and sentences.

PUNCTUATION BY INHIBITION

The excitatory pyramidal neurons (also called principal cells) are considered to be the main carriers of information in the cortex. Their potential runaway excitation is curtailed by inhibitory interneurons: the 15–20% of cortical neurons that contain the inhibitory neurotransmitter gamma aminobutyric acid (GABA). The main function of these neurons is to coordinate the flow of excitation in neuronal networks. There are several different classes of inhibitory

---

16. Sirota et al. (2003). There are many examples of cross-frequency coupling across structures and species (Buzsáki et al., 1983; Soltesz and Deschênes, 1993; Steriade et al., 1993a, 1993b, 1993c; Sanchez-Vives et al., 2000; Bragin et al., 1995; Buzsáki et al., 2003; Csicsvari et al., 2003; Chrobak and Buzsáki, 1998; Leopold et al., 2003; Lakatos et al., 2005; Canolty et al., 2006; Isomura et al., 2006; Sirota et al., 2008). In fact, nobody has ever seen an exception. For excellent reviews, see Jensen and Colgin (2007); Axmacher et al. (2010); Canolty and Knight (2010). Yet caution should be used in the quantification of cross-frequency phase coupling as waveform distortion and nonstationarities can often yield spurious coupling (Aru et al., 2015). One way to avoid artifactual coupling is to first establish that two independent oscillations exist in the examined network.

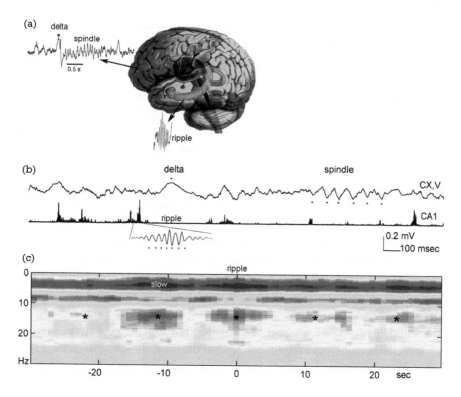

Figure 6.2. Hippocampal sharp wave ripples, neocortical slow waves and sleep spindles are often temporally coupled. A: Schematics of events and structures. B: Example traces of neocortical (CX, V) local field potential and power of fast oscillations (100-200 Hz) in the hippocampal (CA1) local field potential in the rat. An example of the fast local oscillation (ripple) is also shown. C: Hippocampal ripple peak-triggered (time zero) neocortical spectrogram. Note increased correlation of power in the slow oscillation (slow) and sleep spindle (10–18 Hz) bands with hippocampal ripples. *Ultra-slow (~0.1 Hz) comodulation of neocortical and hippocampal activity. Reproduced from Sirota et al. (2003).

interneurons with specialized functions. One group selectively innervates the axon's initial segment of pyramidal cells, where the action potential is produced, and can influence the timing of spikes. A second group innervates the region surrounding the cell body and can electrically segregate the dendrites from the axon. A third group is dedicated to dendritic inhibition; their main function is to attenuate or "filter out" excitatory inputs onto different dendritic segments. Dendrite-targeting interneurons can also effectively short-circuit entire dendritic branches or trees and thus dynamically alter the biophysical properties of the inhibited principal neurons. Yet another group exclusively inhibits other interneurons. Interneurons collect excitatory inputs from the surrounding pyramidal cells and respond equally effectively to distant excitatory neurons or to subcortical neurons carrying various neuromodulators. In turn, the axons of most interneurons arborize locally, providing inhibition to the surrounding pyramidal cell population. Finally, a smaller important group of inhibitory cells project their axons to distant structures, hence their name: *long-range interneurons*.

Although there is no agreed job description for each of these interneuron types, their overall task is akin to traffic controllers in a big city. The ability to stop or slow excitation and route the excitatory traffic in the desired direction is an important requirement in complex networks. To be effective, the various traffic controllers should be temporally well-coordinated for each given job.[17] Inhibition of excitatory neurons can be conceived as the punctuation marks of a neural syntax that can parse and segregate neuronal messages.

The segregating or gating effect of neuronal oscillations can be illustrated by considering a single neuron whose membrane potential is fluctuating around the action potential threshold. The outcome of afferent excitation of a neuron depends on the state of the neuron. If the membrane potential is close to the threshold, a very small amount of excitation is enough to discharge the cell. However, when afferent excitation arrives at the time of hyperpolarization (i.e., when the membrane potential is more negative than at rest), the input may be ignored. Because axons of the interneurons target many principal cells, inhibition can effectively synchronize the action of the principal cells. If the discharge of interneurons is temporally coordinated—for example, by oscillatory mechanisms—many pyramidal cells in the network can produce synchronous output and exert a stronger effect on their downstream targets compared with their noncoordinated or asynchronous firing. In sum, the concerted action of interneurons can route the excitatory information at the right time and in the right direction.

Inhibition Creates Oscillations

Inhibition is the foundation of brain rhythms, and every known neuronal oscillator has an inhibitory component. Balance between opposing forces,

---

17. There are several reviews on cortical interneurons (Freund and Buzsáki, 1996; Buzsáki et al., 2004; McBain and Fisahn, 2001; Soltesz, 2005; Rudy et al., 2011; Klausberger and Somogyi, 2008; DeFelipe et al., 2013; Kepecs and Fishell, 2014). Long-range interneurons are also found in the cortex and may be related to other inhibitory neuron types with long axons, such as the medium spiny neurons of the striatum and GABAergic neurons of the ventral tegmental area, substantia nigra, raphe nuclei, and several other brainstem areas.

such as excitation and inhibition, can be achieved most efficiently through oscillations. The output phase allows the transmission of excitatory messages, which are then gated by the build-up of inhibition. Oscillations are energetically the cheapest way to synchronize the constituents of any system, be it mechanical or biological. This may be a good reason for brain networks to generate oscillations spontaneously when no useful external timing signals are available. Inhibitory interneurons can act on many target neurons synchronously, effectively creating windows of opportunity for afferent inputs to affect inhibition-coordinated local circuits. In summary, oscillatory timing can transform both interconnected and unconnected principal cell groups into transient coalitions, thus providing flexibility and economical use of spikes.

BRAIN-WIDE COORDINATION OF ACTIVITY BY OSCILLATIONS

In contrast to computers, where communication is equally fast between neighbors and physically distant components, axon conduction delays in the brain limit the recruitment of neurons into fast oscillatory cycles. As a result, infra-slow waves are of tsunami size and involve many neurons in large brain areas, whereas superfast oscillations are just ripples riding on the vortex of the slower waves and contribute mostly to local integration. The consequence of such cross-frequency relationships is that perturbations at slow frequencies inevitably affect all nested oscillations.

An important utility of cross-frequency phase coupling of rhythms is that the brain can integrate many distributed local processes into globally ordered states. Local computations and the flow of multiple signals to downstream reader mechanisms can be brought under the control of more global brain activity, a mechanism usually referred to in cognitive sciences as *executive, attentional, contextual,* or *top-down control*.[18] These terms reflect the tacit recognition that sensory inputs alone are not enough to account for the variability of network activity, so some other sources of drive must be assumed. These other sources can be conceived of as a supervising signal coming from preexisting wiring and the brain's knowledge base, which can provide the necessary grounding for the meaning of the sensory input (Chapter 3).

Global coordination of local computations in multiple brain areas through cross-frequency coupling can ensure that information from numerous areas

---

18. Engel et al. (2001); Varela et al. (2001).

is delivered within the integration time window of downstream reader mechanisms. A useful analogy here is the cocktail party problem, when many people speak at the same time. Comprehending a conversation is much easier when the listener can also see the actions of the speaker. Movements of the lips and facial muscles involved in speech production precede the sound by several tens of milliseconds, and their patterns are characteristically correlated with the uttered sounds. Thus, visual perception of the sound production information can prepare and enhance sound perception. Visual information reaches the auditory areas of the brain just in time to be combined with the arriving auditory input. Indeed, watching a silent movie with human speakers activates auditory areas of the listener's brain.[19] Understanding someone with a foreign accent is difficult initially, but the listener can catch up quickly by combining auditory and visual information. A temporal discrepancy between acoustic and visual streams may introduce comprehension difficulties, but people quickly learn to compensate for it.

Input–Output Coordination

Neuronal oscillations have a dual function in neuronal networks: they influence both input and output neurons. Within oscillatory waves, there are times when responsiveness to a stimulus is enhanced or suppressed. We can call them "ideal" and "bad" phases. Oscillation is an energy-efficient solution for periodically elevating the membrane potential close to threshold, thus providing discrete windows of opportunity for the neuron to respond. The physiological explanation for this gating effect is that the bad phase of the oscillatory waves is dominated by inhibition, as discussed earlier, whereas excitation prevails at the ideal phase. The same principle applies at the network level: when inputs arrive at the ideal phase of the oscillation—that is, at times when neurons fire synchronously and thus send messages, they are much less effective compared to the same input arriving during the bad phase of the oscillator, when most neurons are silent.

As discussed earlier, activation of inhibitory interneurons can hyperpolarize many principal neurons simultaneously. Recruitment of inhibitory interneurons can happen via either afferent inputs (feedforward) or via pyramidal neurons of the activated local circuit (feedback). As a result, the same mechanism that gates the impact of input excitation also affects the timing of output spikes in

---

19. Schroeder et al. (2008); Schroeder and Lakatos (2009).

many neurons in the local circuit.[20] Such synchronized cell assembly activity can have a much larger impact on downstream partners than on individual, uncoordinated neurons with irregular interspike intervals. This dual function of neuronal oscillations is what makes them a useful mechanism for chunking information into packages of various lengths.[21]

BRAIN RHYTHMS ACROSS SPECIES

The brain is among the most sophisticated scalable architectures in nature. Scalability is a property that allows a system to grow while performing the same computations, often with increased efficiency. In scalable architectures, certain key aspects of the system should be conserved to maintain the desired function while other aspects compensate for such conservation.

Temporal organization of neuronal activity, as represented by rhythms, is a fundamental constraint that needs to be preserved when scaling brain size. Indeed, perhaps the most remarkable aspect of brain rhythms is their evolutionarily conserved nature. Every known pattern of LFPs, oscillatory or intermittent, in one mammalian species is also found in virtually all other mammals investigated to date. Not only the frequency bands but also the temporal aspects of oscillatory activity (such as duration and temporal evolution) and, importantly, their cross-frequency coupling relationships and behavioral correlations, are also conserved (Figure 6.3). For example, the frequency, duration, waveform,

---

20. The idea that randomly occurring inputs will produce outputs determined by the phase of the oscillating target networks can be traced to Bishop (1933). Bishop stimulated the optic nerve and observed how the cortical response varied as the function of peaks and valleys of the EEG. Many subsequent models of oscillatory communication are based on these observations (e.g., Buzsáki and Chrobak, 1995; Fries, 2005).

21. Brain rhythms belong to the family of weakly chaotic oscillators and share features of both harmonic and relaxation oscillators. The macroscopic appearance of several brain rhythms resembles the sinusoid pattern of harmonic oscillators, although the waves are hardly ever symmetric. An advantage of harmonic oscillators is that their long-term behavior can be reliably predicted from short-term observations of their phase angle. If you know the phase of the moon today, you can calculate its phase a hundred years from now with high precision. This precision of the harmonic oscillators is also a disadvantage, as they are hard to perturb and, therefore, they poorly synchronize their phases. Functionally, neuronal oscillators behave like relaxation oscillators, with a "duty phase" when spiking information is transferred, followed by a refractory period. This refractory phase of the cycle can be called the perturbation or "receiving" phase because in this phase, the oscillator is "vulnerable," and its phase can be reset. Due to the separation of the sending and receiving phases, relaxation oscillators can synchronize robustly and rapidly in a single cycle, making them ideal for packaging spiking information in both time and space.

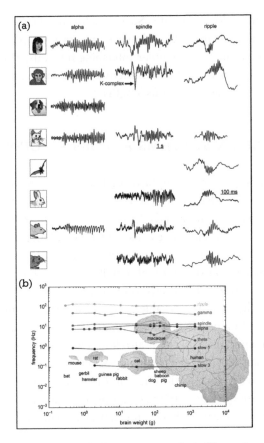

Figure 6.3. Preservation of brain rhythms in mammals. A: Illustrative traces of neocortical alpha oscillations, sleep spindles, and hippocampal ripples in various species. Note the similarity of frequency, temporal evolution, and waveforms of the respective patterns across species. B: Relationship between brain weight and frequency of the various rhythm classes on a log-log scale. Note the small variation of frequency changes despite increases in brain weight of several orders of magnitude.
Reproduced from Buzsaki et al. (2013).

and cortical localization of sleep spindles are very similar in the mouse and the human brain.

On the one hand, this may not be so surprising. After all, neurotransmitters, their receptors, and the membrane time constants of principal cells and interneurons are also conserved, and these properties underlie various oscillations. Thus, irrespective of brain size, the management of multiple time scales in neuronal networks is supported by the same fundamental mechanisms. On the other hand, the speed of communication between areas varies considerably between small and large brains, making the conservation of rhythms unexpected. For example, for coherent perception of multimodal inputs, the results of local computation in the thalamus and several primary sensory cortical areas should arrive within the integration time window of the target associational cortices. The same applies to the motor side of the brain. As discussed in Chapter 3, the fundamental properties of myosin and actin are largely conserved across mammals. Therefore, the motor command computations in the motor cortex, cerebellum, and basal ganglia should be performed in comparable time windows, and the command signals to the spinal cord should be delivered within the same time range in different species. However, the distances of these structures vary by orders of magnitude across species. Thus, all of the timing constraints required for adequate function have to be reconciled with the complexity imposed by the growing size of the brain. This is not a trivial task, given a 17,000-fold increase of brain volume from the small tree shrew to large-brain cetaceans. The constancy of the many brain oscillations and their cross-frequency coupling effects across species suggest a fundamental role for temporal coordination of neuronal activity.[22]

Conservation of Brain Rhythms Across Species

There appear to be at least two mechanisms that allow scaling of neuronal networks while conserving timing mechanisms. The first mechanism compensates for the increase in neuronal numbers and the enormous numbers of possible connections by shortening the *synaptic path length* between neurons, defined as the average number of monosynaptic connections in the shortest path between two neurons. This problem is akin to connecting cities with highways and airlines to obtain an efficient compromise between the length of the connections and the number of intermediate cities required to get from city A to city B. Directly connecting everything with everything else is not an option in most real-world networks as the number of connections increases much more rapidly than the number of nodes to be connected. An efficient compromise can be achieved by inserting a smallish number of long-range short-cut connections into local connections. The resulting "small-world-like" architectural solution allows for scaling while keeping the average synaptic path length similar as brains increase in size.

While such scaling prevents excessive volume growth, longer axons increase the travel time of action potentials. This would pose a serious problem for neuronal communication. Thus, a second mechanism is needed to compensate

---

22. Buzsáki et al. (2013). The Supplementary Material section of this review compares hundreds of papers on various rhythms recorded in a multitude of mammalian species.

for these time delays. This requirement is solved by using larger caliber, better insulated, and more rapidly conducting axons in more complex brains. For example, to achieve interhemispheric gamma band synchrony in the mouse brain, a conduction velocity of 5 m/sec is sufficient. On the other hand, maintaining coherent oscillations in the same frequency range in the human brain, with 70–160 mm between hemispheres, requires much more rapidly conducting axons. Thus, the benefits of timing mechanisms in larger brains can be preserved by adding larger-caliber and more strongly myelinated axons that allow signals to travel longer distances within a similar time window. In the human brain, the great majority of interhemispheric (callosal) axons have small diameters (<0.8 μm) but the thickest 0.1% can exceed 10 μm in diameter. These thick fibers scale best with brain size, whereas the fraction of thinner fibers actually decreases. Although adding some large-diameter axons does increase brain volume and metabolic costs, the resulting volume increase is still considerably smaller than would result from a proportional increase in axon caliber across all neurons. This small added fraction of large-diameter axons may be responsible for the minor variability of interhemispheric conduction delays across species.[23] Although the exact neuron types with large-caliber axons are not known, experimental data indicate that at least some them originate from long-range inhibitory neurons. In turn, theoretical considerations and modeling suggest that long-range interneurons are critical for brain-wide synchronization of gamma and potentially other oscillations.[24]

In summary, preservation of brain rhythms in increasingly larger brains can be taken as supporting evidence for the fundamental importance of timing in brain performance. As the various nested rhythms occur in parallel in multiple brain systems, it is clear that oscillations per se do not serve special biological functions. The meaning of gamma oscillations in the sensory olfactory bulb is different from those in prefrontal circuits serving cognitive functions.

---

23. Aboitiz et al. (2003); Wang et al. (2008). See further discussion on axon diameter in Chapter 12.

24. Densely connected local interneuron networks are supplemented by a small fraction of long-range interneuronal connections, which effectively reduce the synaptic path lengths between distant circuits (Buzsáki et al., 2004) and allow the entrainment of oscillations across distant networks. Long-range interneurons tend to have large-diameter, myelinated axons and can, therefore, provide fast conduits (Jinno et al., 2007). In the brain, most neuronal connections are local but interspersed with long-range shortcuts (Bullmore and Sporns, 2009). This architectural design is reminiscent of mathematically defined "small-world" networks (Watts and Strogatz, 1988). Small-world type networks allow activity to propagate from one neuron (or more likely assemblies of neurons; "node") to distant neurons. Steve Strogatz's book (*Sync*, 2003) is an excellent reference for small-world networks, a term he coined. For connection rules of the brain, see Chapter 12.

Instead, the benefits of a particular oscillation depend on the function of the brain system that supports it.

ENTRAINMENT OF OSCILLATORS: TRACKING TEMPORAL DYNAMICS OF SPEECH

When two oscillators with identical frequency engage each other, the outcome depends on the phase of the two rhythms. In-phase interactions induce resonance and, as a result, amplification.[25] In contrast, opposing phase interactions may annihilate or dampen the rhythm. Oscillators with noninteger relationships induce perpetual interference. This is typical of brain rhythms, and the interference mechanism explains why brain dynamics is constantly changing, similar to the interference of ocean waves. Occasionally, the oscillatory reader mechanism may transiently adjust its phase to the incoming inputs. Such phase adjustment is among the most important flexible features of brain oscillators. This is similar to how musicians in an orchestra keep the beat. If the first violinist is a bit faster, the rest of the musicians adjust the timing of their movements. Furthermore, the phase separation of the maximum and minimum spiking activity of the information-transmitting principal cells make oscillators a natural parsing mechanism. As mentioned earlier, this is an effective separating mechanism of neuronal messages, a fundamental requirement for syntactical operations.[26]

Speech Rhythms Are Brain Rhythms

A striking example of the selection and amplifying properties of brain oscillators is their responses to speech. Rhythms of human speech are remarkably similar

---

25. Single neurons can also exploit resonance and filtering. The leak conductance and capacitance of the neuronal membrane are mainly responsible for the low-pass filtering of neurons (essentially an RC filter). In contrast, several voltage-gated currents, whose activation range is close to the resting membrane potential, act as high-pass filters, making the neuron more sensitive to fast trains of spikes. These resonant-oscillatory features allow neurons to select inputs based on their frequency. Neurons with high-pass and low-pass filtering can be combined to build neuronal networks that can function as band-pass resonators (Llinás et al., 1988; Alonso and Llinás, 1989; Hutcheon and Yarom, 2000). Cortical interneuron classes have a wide range of preferred frequencies (Thomson and West, 2003), and their diverse frequency-tuning properties are important for setting network dynamics.

26. The volume edited by Bickerton and Szathmáry (2009) contains many excellent chapters on syntactical rules related to both language and neuronal activity.

across all spoken languages, and our brains are tuned to efficiently track and extract such information. We can easily recognize a stutterer or a person with a speech impairment in any language because of the altered speech rhythm. The so-called *prosodic features* of speech, such as intonation, stress, and pause, are characteristic of individuals, yet they also share common features among all people, varying between 0.3/s and 2/s (delta band).[27] Syllables also repeat somewhat rhythmically between 4 and 8 times per second (theta band), while phonemes and fast transitions are characterized by a frequency band between 30/s and 80/s (gamma band). Two features of brain oscillations can facilitate the extraction of hierarchically organized continuous speech. First is the correspondence between speech and native brain rhythm frequencies. Their cross-frequency phase coupling can amplify sound features and assist with the segmentation of speech components. Second, the ability of neuronal oscillators to reset their phase can effectively track the temporal features of quasi-rhythmic spoken language.

It is tempting to draw a parallel between neuronal syntax, supported by brain rhythms, and language syntax. Brain rhythms may constrain the way our motor system controls utterance of sound. In return, the "phonetical syntax" of speech, referring to the sequences of sounds, can influence brain oscillations. I make this comparison later with the explicit acknowledgment that there is a difference between speech syntax and syntactic coding of speech. Yet the brain-constructed and human community-constructed syntactic rules often seem correlated.

Experiments using EEG, ECoG, or MEG methods have demonstrated that the edges of the sound envelope of speech can reset the phase of slow brain oscillations in the right temporal and frontal areas and that speech sound envelope fluctuations are correlated with amplitude variations in the delta band. The time-varying dynamics of syllable sequences can be faithfully tracked by the amplitude of theta oscillations and the integrated amplitude of gamma activity. In high-density ECoG recordings from the surface of the superior temporal gyrus (nonprimary auditory cortex) of patients, the spectrotemporal auditory features of continuous speech could be reliably reconstructed from neuronal signals, and many words and pseudowords could be statistically

---

27. Prosodic features describe the auditory qualities of speech beyond individual phonemes. Prosodic phrases are concatenated phonemes, syllables, and morphemes, reflecting suprasegmental features of speech. They are characterized by the pitch and intonation of voice, rhythm, stress, and loudness. A *phoneme* is an abstract sound feature that differentiates between two words (e.g., "house" vs. "mouse"). A *syllable* is a unit of pronunciation with one vowel. A *morpheme* is a word or part of a word and is the smallest grammatical unit that has meaning. For example, information [in-for-ma-tion] has four syllables and one morpheme.

separated. Similarly, when song segments were played to the research participant, the characteristic time-varying sound patterns could be recognized from the smoothed power changes of high gamma oscillations. The requirement for tracking the temporal dynamic of speech by brain rhythms is illustrated by experiments in which test sentences are time-warped. When speech is compressed, but comprehension is maintained, both phase-locking and amplitude-matching between the temporal envelopes of speech and the recorded brain signal are maintained. However, when speech compression is large, brain oscillations no longer effectively track speech, and comprehension is degraded.[28]

An outside-in argument could be that the dynamic relationship between speech and brain oscillations is due to the imposition of the speech patterns on brain patterns. In other words, speech stimuli "train" the brain to effectively segment and parse speech content.[29] This argument has little validity because neuronal oscillations are the same in all mammals, and speech rhythms are the same in every human culture. Instead, I suggest that the coding strategy of the auditory system is matched to internally organized brain rhythms. In support of this view, when rats hear complex sounds, such as music or broadband pink noise, thalamocortical patterns in the superficial layers of the auditory cortex are also segmented into 2–4 frames per second. As in humans, these events reliably reset the phases of the local field potentials at specific times of the acoustic stimulus. Overall, the preservation of brain rhythms across species suggests that human speech was built on preexisting brain dynamics.

28. An important background paper is Shannon et al. (1995), which demonstrated that speech pattern recognition, in general, can use both spectral and temporal cues. Several works have suggested, implicitly or explicitly, that syntactical rules of language are related to neural syntax (Buzsáki, 2010; chapters in the edited volume by Bickerton and Szathmáry, 2009, also discuss such suspected links). For physiological support for the role of brain oscillations in segmentation of speech, see Ahissar et al. (2001); Lakatos et al. (2005, 2008); Howard and Poeppel (2010); Ding and Simon (2012). Valuable ECoG data are typically obtained from patients with brain tumors or epilepsy using subdural electrode arrays, when patients give their informed consent for such tests (Pasley et al., 2012; Sturm et al., 2014).

29. The probability distributions of the amplitude envelope and the time-frequency correlations of all natural sounds are quite similar. Animal vocalizations and human speech are further characterized by the low temporal modulation of most spectral power. The distribution of the amplitude envelopes exhibits characteristic shapes for natural sounds and a relatively uniform distribution for the log of the amplitudes for vocalizations (characterized by the 1/f form, as is the case for the EEG signal). Largely for these reasons, Singh and Theunissen (2003) suggested that the auditory system has evolved to process behaviorally relevant sounds. Under this hypothesis, the statistics of the spectrotemporal amplitude envelopes of the sound should be critical for auditory brain areas to extract sound identity.

Syntactic Segmentation, Grouping, and Parsing by the Brain's Oscillatory System

So far, we have discussed that brain rhythms are efficient for segmenting and parsing continuous natural sounds and that matching between the modulation capacities of the auditory cortex and speech dynamics is a prerequisite for comprehension (although certainly not sufficient). However, the brain's response to the sound envelope is not simply a matter of oscillatory coupling but also involves feature extraction as well. In support of this idea, the magnitude of theta-gamma phase coupling is stronger when the listener understands the speech compared to when the same speech segment is played backward or when the speech components are randomly shuffled.[30] In other words, the variation of cross-frequency coupling magnitude may carry the meaning.

Brain Rhythms at a Cocktail Party

At a cocktail party, where several people may speak at the same time, the speech system helps us by grouping the complex auditory scene into separate "objects." Intelligible speech is extracted by phase-locking the brain's rhythms to the speech of a chosen person. This is a pertinent example of selective gain control (Chapter 11) because the sound emitted by one individual (i.e., the target auditory object) is amplified by phase-resetting and resonance. At the same time, the speech streams from other speakers are filtered and suppressed

---

30. Functional magnetic resonance imaging (fMRI) studies indirectly support the EEG/MEG data. The blood oxygen level-dependent (BOLD) signal in the auditory cortex increases irrespective of whether intelligent or unintelligent (scrambled) speech is presented. In temporal speech areas (such as Broca's area; Brodmann areas 45, 47), coherent information at the sentence level is needed to induced blood flow changes, whereas parietal and frontal areas (Brodmann areas 39, 40, 7, and 22) respond only when intact paragraphs of intelligent speech are presented (Lerner et al., 2011). Two major language-relevant systems can be distinguished in the brain: a dorsal and a ventral system. The dorsal system involves Broca's area (in particular Brodmann area 44), which, together with the posterior temporal cortex, supports hierarchical syntactic computation and comprehension of complex sentences. The ventral system (area 45/47) and the temporal cortex support the processing of lexical-semantic and conceptual information (see e.g., Hagoort, 2005; Berwick et al., 2013). Neurons in the temporal lobe can respond to semantic categories or can be activated by the photograph, sketch, or even name of a familiar person, implying a high degree of semantic abstraction (Quian Quiroga et al., 2005). The speed of word naming depends on "semantic richness." Neuronal assemblies can be combined for many word representations, with semantically related words being encoded by hypothetically overlapping neural ensembles (Li et al., 2006; Sajin and Connine, 2014; Friederici and Singer, 2015).

because their speech arrived during the "ignoring" phase of the rhythms of the listener's brain.

In a real cocktail party, speakers are at physically distinct locations, so you might argue that their spatial position can be triangulated by binaural listening.[31] However, the contribution of spatial localization mechanisms can be excluded in laboratory settings when listeners are asked to selectively attend to one of two competing speakers (typically male and female) heard through a single earphone. Such experiments show that the selection of a speaker relies on the match between the shape of the frequency and the temporal modulation of the chosen speaker's sound and the listener's low-frequency (delta, theta band) MEG/EEG activity. This selective extraction mechanism is achieved primarily by phase-tuning the neuronal oscillators to the prosodic and syllabic tempo of the preferred speaker because varying the sound intensity of the competing speaker, within a reasonable range, does not affect comprehension.

These observations, based on healthy human research participants using noninvasive recording methods (MEG and EEG), are supplemented by high-resolution multielectrode surface recordings (ECoG) from the posterior superior temporal gyrus (a higher order auditory area) in patients with epilepsy. Analysis of the speech spectrograms of a mixture of speakers reconstructed from the envelope of high gamma rhythm (75–150/s) shows that the spectral and temporal features of the ECoG signal triggered by the sound of the attended speaker are more salient than those of induced by competing speakers. These findings demonstrate that not only acoustic features but also intentional speaker selection can be improved by phase-locked oscillations.[32]

---

31. Localization of the sound source also uses timing information. Sounds from a speaker on your right arrive at the right ear first and, after a slight delay, at the left ear. This time difference may be shorter than the duration of a single action potential in the auditory nerve (less than 1 ms). Of course, slight differences in sound intensity are also present between the two ears. Mark Konishi and colleagues at the California Institute of Technology showed that *medial superior olive* neurons in barn owls are selective for timing differences. These neurons are "coincidence detectors": some neurons respond selectively at a 20 μsec time lag, others at 50, 75, 100 μsec, and so on. Their spike output therefore signals the direction of the sound source. Sound intensity differences between the two ears may also assist in computing the distance to the source (Knudsen and Konishi, 1978).

32. Acoustic and semantic features of words are most often assumed to be independent, even though many linguists, starting perhaps with Socrates, have speculated that "good" words have sounds that suit their meaning. Recent large-scale studies rekindled this old debate by demonstrating such a link in several languages. Not only sounds imitating those made by animals (such as "moo") are similar across languages. When people are asked whether "bouba" or "kiki" is the right word for an amoeboid or star-shaped (spiky) figure, speakers of different languages consistently make the right guess (they choose kiki). The word used to denote "sand" often contains the sound /s/ examined across 4,000-plus languages. The vowel /i/ often refers

A computational model (an artificial classifier) trained solely on examples of one of the speakers could recognize both attended words and speaker identity despite the interfering effects of the competing speaker. When behavioral errors occurred, they were reflected in the degradation of oscillatory tuning. Overall, cortical activity does not merely reflect responses to acoustic stimuli but, in addition, it also can identify more complex aspects of speech, including the listener's goal.[33]

In this chapter, we examined how neurons and cell assemblies are governed by their collective behavior, as expressed by brain oscillations. Now it is time to learn how brain oscillations assist with organizing cell assemblies into longer sequences and the functions they support. Curious? Turn the page to Chapter 7.

SUMMARY

From the analogy between speech syntax and brain rhythms, I made the assumption that the hierarchy of brain oscillations may parse and group neuronal activity to decompose and package neuronal information in communication between brain areas. Because all neuronal oscillations are based on inhibition, they can parse and concatenate neuronal messages, a prerequisite for any coding mechanism. The hierarchical nature of cross-frequency-coupled rhythms can serve as a scaffold for combining neuronal letters into words and words into sentences. Brain oscillations, present in the same form in all mammals, represent a fundamental aspect of neuronal computation, including the generation of movement patterns, speech, and likely, music production. Neuronal oscillators readily entrain each other, making the exchange of messages between brain areas effective. I speculate that the roots of language and musical syntax emanate from this native neural syntax since the same brain rhythms that assist

---

to small size, /r/ to roundness, and /m/ to mother or breast even in nonrelated languages (Blasi et al., 2016; Fitch, 2016).

33. Schroeder and Lakatos (2009); Ding and Simon (2012); Mesgarani and Chang (2012); Zion Golumbic et al. (2013). Other investigators emphasize the role of low-frequency oscillators in speech segmentation. Scalp EEG recordings suggest that 4–8 Hz activity in auditory cortex is important for tuning in to the continuous speech content of an attended talker. Interhemispheric asymmetry of alpha power (8–12 Hz) at parietal sites can indicate the direction of auditory attention to a speaker (Kerlin et al., 2010). While the brain efficiently solves the auditory object segregation problem, it still remains a major challenge for automatic speech recognition algorithms (Cooke et al., 2010).

in the generation of speech and music are also responsible for their syntactical segmentation and integration.[34]

Brain waves detected in either LFPs or EEG signals can reveal certain aspects of signal transformation because the experimenter has access to both an external event (speech) and an internal event (brain oscillations). However, because the brain does not use LFPs or EEG for communication,[35] it remains to be demonstrated whether and how brain networks use such packaging mechanisms for their own purposes. The semantic content of information packages can be read out from neuronal spikes alone, but to become information, spike patterns need syntactical rules known to both sender and receiver mechanisms. Brain oscillations represent a candidate cipher because they are present in all the brain regions of all communicating parties.

34. Singh and Theunissen (2003); Buzsáki (2010); Pulvermüller (2010); Giraud and Poeppel (2012).

35. There are exceptions. Electric fields generated by synchronous activity of nearby neurons can affect the membrane potential of the same neurons and increase synchrony further. Such local "ephaptic" effects are especially effective in structures with high neuronal density and regular architectures, such as the hippocampus (Anastassiou et al., 2010).

7

Internally Organized Cell Assembly Trajectories

A city . . . is actually a sequence of spaces enclosed and defined by buildings.

—I. M. Pei[1]

Poets write the words you have heard before but in a new sequence.

—Brian Harris[2]

I thought we'd just sequence the genome once and that would be sufficient for most things in people's lifetimes. Now we're seeing how changeable and adaptable it is, which is why we're surviving and evolving as a species.

—Craig Venter[3]

The annual gathering of the Society for Neuroscience is the world's largest source of fresh ideas on brain science and health. Over the past decade, its Dialogues Between Neuroscience and Society series has featured such luminaries as the Dalai Lama, actress Glenn Close, dancer Mark Morris, and economist Robert Shiller. At the 2006 meeting in Atlanta, Frank Gehry was invited to discuss the relationship between architecture and neuroscience. After the talk, an audience member (actually it was me) asked him, "Mr. Gehry, how

---

1. https://www.brainyquote.com/authors/i_m_pei.
2. As cited in Aggarwal (2013).
3. https://hbr.org/2014/09/j-craig-venter.

do you create?" His answer was both intuitive and funny: "There is a gear [in my brain] that turns and lights a light bulb and turns a something and energizes this hand, and it picks up a pen and intuitively gets a piece of white paper and starts jiggling and wriggling and makes a sketch. And the sketch somehow relates to all the stuff I took in."[4]

Gehry's answer is a perfect metaphoric formulation of the evolving neuronal assembly trajectory concept, the idea that the activity of a group of neurons is somehow ignited in the brain, which passes its content to another ensemble (from "gear to light bulb"), and the second ensemble to a third, and so forth until a muscular action or thought is produced. Creating ideas is that simple. To support cognitive operations effectively, the brain should self-generate large quantities of cell assembly sequences.

Like Gehry, I have known for a long time that the only reason I can write this chapter is because continually changing neuronal assemblies in my brain evolve in a perpetual chain. In fact, this idea is the only current contender to explain internally generated actions and thoughts. Yet I had to devote several decades of my life to tracking down such internally generated cell assembly sequences (or internal sequences for short). In Chapter 5, I outlined a model suggesting that internal sequences are the foundation of cognition. In this chapter, I discuss how such sequences arise in the brain.

CELL ASSEMBLIES (NEURONAL LETTERS) FORM TRAJECTORIES

Morse communication is a bit like human conversation. Speakers take their turns, usually with only one person talking at a time. Decoding is simple if you possess the cipher. But try to decipher the information embedded in a "parallel Morse code" (Figure 7.1). Although many decoding mechanisms, including the visual system, can easily recognize that a picture contains distinct patterns, simply viewing it without a cipher, no matter how long, does not reveal its message. However, if those patterns were converted to sound, most people would instantly recognize it as music, and a sophisticated few as a fragment from Beethoven's *Fifth Symphony*. This recognition occurs because the auditory system has the needed cipher, whereas the visual system does not, except in trained musicians. Parallel streams of musical notes appear complex, but once the reader mechanism is tuned to them, they can convey much richer information. When you examine the right side of the figure, you can see why. Each time slot contains a unique constellation of varying fractions of notes (i.e.,

---

4. Visit http://info.aia.org/aiarchitect/thisweek06/1110/1110n_gehry.htm.

Figure 7.1. The pattern in polyphonic music moves forward on simultaneous multiple scales. *Left*: Ludwig van Beethoven: *Fifth Symphony*, first movement. Graphical score animation. *Right*: Segmentation by time slots. In each slot, the constellation of the pattern is different, although often somewhat predictable.
Courtesy of Smalin's Channel at YouTube.

a population vector; Chapter 4), and the change from one slot to the next is also characteristically different. This is how the sequentially unique population vector of notes allows viewers to decipher the melody (if it is familiar to them) by completing patterns from a small fragment. Cell assembly sequences in the brain are built similarly and may be read in an analogous fashion.

But there are complications. Imagine trying to follow the plot of a movie by examining the variation of only a few pixels on the screen. The changing sequences would provide some information, at least under special conditions, yet such undersampled tunnel vision would make it difficult to enjoy the movie. Morse code is an example of serial communication. In contrast, neuronal communication is more like an orchestra playing polyphonic music than like language[5]. Single neurons form cell assemblies by synchronizing their firing within short (gamma cycle) time windows (Chapter 4). Simultaneous parallel streams offer a much wider bandwidth for neuronal transmission. When performing even elementary actions, such as walking in a park, we coordinate neuronal activity from multiple visual and auditory sources and our past experiences. Messages from the neuronal assemblies in these modalities *become* information only when they are delivered in an organized manner to their target readers. Neuronal networks nearly always use parallel streams, but the transmission of information in interconnected brain circuits is more complex than shown in Figure 7.1.

The evolution of any system can be described mathematically as a motion vector in a multidimensional space. The sequentially visited points in this space are called a *trajectory*. Applying this idea to the brain, a neuronal trajectory corresponds to sequences of population vectors evolving in space and time. Its shape reflects a combination of the input that initiates the sequence and the constraints of the brain networks in which the vector moves. This abstract concept may be easier to understand through specific examples of neuronal assembly sequences and their functions.

Neural Syntax of Birdsongs

The neuronal trajectory underlying birdsong can be conceived of as a neural word.[6] Unlike in evoked neural words, no punctate stimulus is involved in the

---

5. A delightful and accessible introduction to polyphonic music and its relationship to brain activity is by Pesic (2018).

6. Another prominent example of neuronal trajectories has been described in insects. In a series of beautiful experiments, Gilles Laurent at the California Institute of Technology studied the spatiotemporal pattern of evoked neuronal sequences ("neuronal words") in the antennal lobe of the locust in response to odors. When an odor is presented, it induces a transient

generation of birdsong. The presence of a female bird is sufficient to release a song in a male zebra finch (*Taeniopygia guttata*). The song consists of distinct bursts of sounds (syllables), separated by silent intervals (gaps). The syllable sequences are stereotypical and last for less than a second. A critical brain area in song production is the "high vocal center" nucleus. Michale Fee's laboratory at the Massachusetts Institute of Technology demonstrated that the temporal structure of the song is generated by sparse sequential activity of neurons in this area. Approximately 200 neurons form a coherent assembly and emit a single brief burst of spikes at only one time in the song. This may be viewed as a neural letter. The sequential activation of several such letters can be conceived as a word, which is repeated numerous times with slight variations in a singing episode.[7] The impact of the bursts is sensed by neurons in a downstream structure known as the *arcopallium*, which, in turn, drives the motor neurons innervating the *syrinx*, the vocal organ. The vocabulary of each male is limited to a single sentence, which is distinct enough to separate his identity from other conspecifics.

The organization of birdsong provides a simple illustration of how innate neuronal syntax can help to code variations of prosodic content.[8] Learning a song may be a serendipitous process, rather than a well-thought-out algorithm. Young male finches initially "babble" some sounds and progressively refine their motor function from these self-generated syllables and by imitating their father's song. The final product is a song that is both similar to and different from the father's song. Similarly, babbling in human babies may reflect a self-organized intrinsic dynamic. When the uttered sounds resemble a particular word, the happy parents regard them as a real word. They reinforce such spontaneous utterances with a corresponding object, action, or phenomenon, until it acquires a meaning for the baby. Exploitation of the default self-organized

---

gamma-frequency oscillation in the antennal lobe, with different small subsets of antennal lobe neurons firing in each gamma cycle. The odor onset is followed by an evolving sequence of population vectors (a trajectory) lasting for a few hundred milliseconds. Successive presentations of the same stimuli evoke similar neuronal trajectories representing a given odor, whereas different odors are associated with uniquely different neuronal trajectories or words. See Laurent (1999); MacLeod et al. (1998); Broome et al. (2006); Mazor and Laurent (2005).

7. Nottebohm et al. (1976); Fee et al. (2004); Hahnloser et al. (2002); Long and Fee (2008).

8. Birdsongs, in general, are combinatorial but lack semantic composition. In contrast to the single song of finches, nightingales (*Luscinia megarhynchos*) have a repertoire of up to 200 songs, combined from a thousand elements. Studying the neurophysiological basis of such variability in nightingales could provide clues to how assembly sequences (sentences) can be constructed and retrieved in appropriate contexts.

patterns of the brain prior to language is a more effective mechanism of pattern formation than a de novo, blank-slate solution. (Chapter 13).

When zebra finches are raised by Bengalese finch (*Lonchura striata*) foster parents, the juvenile birds imitate the Bengalese finch song syllables. However, the rhythm of the learned song retains the silent gaps characteristic of zebra finch song patterns, which is distinct from the shorter gaps of the Bengalese song. Thus, the syntactic rules are genetically inherited species-specific patterns, just like brain rhythms in mammals, whereas the variable content of the syllables and words can be acquired by experience. The pattern and content are processed by dissociable neuronal circuits in the bird's brain. In the auditory cortex, slow-firing neurons are mainly sensitive to the acoustic features, such as timbre and pitch. In contrast, faster firing, possibly inhibitory, neurons encode the silent gaps and rhythm of the song, and they are insensitive to acoustic features. This division of labor between the inherited temporal patterns that serve as the syntax and the flexible content may be similar to the way human speech is organized.[9]

When we hit a wrong note while singing, we immediately notice the error that deviates from our desired plan. How do we know that we made an error, given that there is nothing intrinsically special about any note? One idea is that an internally generated target pattern is sent to the auditory cortex (i.e., the now familiar corollary discharge; see Chapter 3) to be compared with auditory feedback from the song-generated sound. Similarly, when a juvenile male zebra finch sings, it uses auditory feedback to test if the sounds matched an internal target song. Learning of the song rhythm is assisted by a group of dopaminergic neurons that serve as an error-correcting mechanism. In the laboratory, the experimenter can fool the bird by distorting or displacing some syllables of his song, convincing the bird that he produced the wrong syllable. When his brain detects such a mismatch, the activity of the dopaminergic neurons decreases. Conversely, when the internally generated goal is reached, dopamine signaling increases. This activity therefore is crucial for keeping track of the correctness of the song.[10]

---

9. Araki et al. (2016). As in birdsong, the variance of syllabic durations in human speech is large and skewed, whereas the intersyllabic temporal gaps are shorter and have low variance. There are limits to the similarities between birdsongs and human language, though (Fisher and Scharff, 2009; Bolhuis et al., 2010; Berwick et al., 2011).

10. Gadagkar et al. (2016). These dopaminergic neurons reside in the ventral tegmental area (VTA), and a subgroup of them project their axons to Area X, a structure known to be important in song learning. VTA neurons communicate with a variety of other areas and may serve a similar "error correction" role in those targets as well. Similarly, dopaminergic neurons in the substantia nigra in mice respond robustly to temporal mismatch. When mice are trained to judge the elapsed time between two identical auditory stimuli with random intervals, the

The Grooming Syntax

In most cases, birdsongs are simply repeated syllables or words, so they do not perfectly satisfy the criteria of a neuronal sentence, which requires the combination of sequentially activated different neural words. Several types of stereotypic behavioral patterns, usually called *fixed action patterns*, may serve as examples for neuronal sentences. Fixed action patterns can be elicited by an ethologically relevant cue or can emerge without explicit cues.[11]

A well-studied behavioral sequence in rodents is self-grooming. Unlike repetitive words of birdsongs, grooming is an elaborate, long-lasting behavioral pattern. A typical self-grooming syntactic chain serially links twenty or more behavioral letters and syllables into a serial structure of four distinct phases, each of which may be regarded as a word. Grooming begins with a series of elliptical bilateral paw strokes near the nose (phase 1), followed by unilateral strokes (phase 2). The next phase is a series of bilateral strokes on the head by both paws simultaneously (phase 3), and the chain is concluded with a postural turn and body licking (phase 4). Self-grooming is remarkably similar across mammalian species: compare the preceding sequence to how you clean yourself in the shower.

In contrast to the elaborate behavioral descriptions, physiological correlates of grooming behavior are scarce. Most of our knowledge about the neural circuitry involved in self-grooming comes from lesion studies. Decerebrate animals, whose forebrains are disconnected from the hindbrain and midbrain, can still execute the individual grooming phases, but rarely compile them into the right sequence or complete all four phases. Corticostriatal circuits appear to have an important role in the execution of action sequences. Rats with lesions of the anterior dorsolateral region of the striatum have a permanent deficit in their ability to complete sequential syntactic self-grooming chains. When the damage extends to the midbrain, rats rarely initiate grooming and have

---

activity of dopaminergic neurons tracks the probability of the intervals. If these neurons are optogenetically activated (or inactivated), the animal under- or overestimates the duration by behavioral measures, indicating that dopaminergic neurons are important in the subjective evaluation of elapsed time (Soares et al., 2016; Chapter 10).

11. A fixed action pattern is generally defined as an "instinctive" behavioral sequence, which, when elicited, runs to completion. The behavioral sequences are stereotypical and occur in response to a "releasing signal," even the first time the organism encounters the relevant stimulus (Tinbergen, 1951). Sexual behavior, maternal behavior, and aggression are typical examples. Karl Lashley (1951) was perhaps the first person to suggest that complex sequential behaviors are organized by hierarchically organized neural programs.

difficulties in executing individual phases.[12] As grooming is hard to elicit and occurs at the animal's discretion at unpredictable times, studying the neuronal correlates of such neatly organized behavioral patterns will require stable long-term recordings of neurons in multiple structures. Such data will be essential for a mechanistic description of the neuronal events underlying grooming.

SELF-ORGANIZED CELL ASSEMBLY SEQUENCES

There are two ways to create neuronal sequences. First, temporally changing neuronal activity can reflect outside-in responses elicited by variable external stimuli. When we walk down a path, for example, the resulting changes in environmental stimuli activate different sets of neurons, forming a trajectory. Alternatively, stimuli from the body can "drive" sequentially firing neurons. As I turn my head in the dark, the vestibular system responds, and so my head direction neurons are activated sequentially. Both environmental and body-derived cues count as external from the brain's point of view. Second, the brain's activity can change independently of sensory inputs. Such self-organized activity is the source of perhaps all cognitive operations, including memory, reasoning, planning, decision-making, and thinking (Figure 7.2).

Multiple Time Scale Representations

As discussed in Chapter 5, head direction neurons continue to display organized activity during sleep, when the system becomes disengaged from external inputs. Such internalized activity of hippocampal place cells and entorhinal grid cells is the basis for mental navigation. Several experiments in different laboratories demonstrate that hippocampal neurons do more than simply respond to external cues. One important observation is that a small fraction of hippocampal and entorhinal neurons fire at reliably different rates in the central arm of a T maze depending on where the rat is coming from or heading toward. In the central arm, the animal's behavior is the same, whether it will later

---

12. For excellent reviews on the neuronal mechanisms of grooming and its use as a model of psychiatric conditions such as obsessive-compulsive disorder and autism, see Berridge and Whishaw (1992), Spruijt et al. (1992), and Kalueff et al. (2016). The role of the striatum was demonstrated by Cromwell and Berridge (1996). Mice that lack a gene for a protein called Cntnap4 develop obsessive grooming, which is mainly directed to their peers (Karayannis et al., 2014). This gene is involved in controlling release of gamma aminobutyric acid (GABA) from axon terminals of parvalbumin interneurons and dopaminergic neurons.

Chapter 7. Internally Organized Cell Assembly Trajectories 173

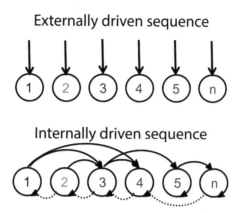

Figure 7.2. Sequential activity of neuronal ensembles (1 to *n*) can be brought about by changing constellation of environmental landmarks and/or proprioceptive information from the body (*top*). Alternatively, sequential activation can be supported by internally driven self-organized patterning (*bottom*).

turn right or left, so such differential firing at the same spatial locations do not qualify for the term "place field." Instead, a mnemonic or planning correlation is a better description.[13]

A second clue comes from the temporal organization of hippocampal place cells. Place fields are relatively large, so the place fields of several neurons can overlap with each other over multiple theta cycles (Figure 7.3). How are the spikes of place cells that represent upcoming locations on the track related to each other? This question is interesting because synaptic integration occurs over tens of milliseconds, while travel between locations takes seconds. During each theta cycle, approximately seven gamma cycles occur and nested within each of them is a cell assembly, representing a spatial position. The spike-timing sequence of neuronal assemblies predicts the sequence of passed and upcoming locations in the rat's path, with larger time lags representing larger distances (Figure 7.3).[14] In other words, if we take a "snapshot" over a single

---

13. These neurons were dubbed "splitter" cells (Wood et al., 2000) or prospective/retrospective cells (Frank et al., 2000). See also Ferbinteanu and Shapiro (2003).

14. This experiment (Dragoi and Buzsáki, 2006) was a logical extension of previous findings by O'Keefe and Recce (1993) and Skaggs et al. (1996). The theta oscillation-based temporal coordination of hippocampal neurons is also related to attractor-based dynamical models (Tsodyks et al., 1996; Wallenstein and Hasselmo, 1997; Samsonovich and McNaughton, 1997). However, the findings of these latter experiments were interpreted to support the idea that spatial cues control hippocampal place cell activity. The theta sequence compression decreases with increasing velocity (Maurer et al., 2012). See also Chapter 11. The sequence model of Jensen and Lisman (1996a, 1996b, 2000, 2005) is even closer to my view. This model suggests

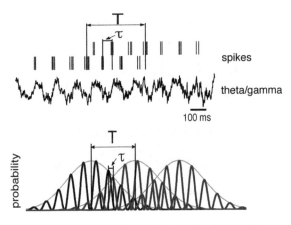

Figure 7.3. Dual time representation of distances in the hippocampus. *Top*: Spiking activity of two place cells and local field potential theta rhythm during maze walking. Temporal duration T is the time needed for the rat to run the distance between the peaks of the two place fields (behavioral time scale). τ, time offset between the two neurons within the theta cycle ("theta time scale"). *Bottom*: Idealized overlapping place fields of three place cells with identical theta oscillation frequency, illustrating the relationship between T and τ. The correlated relationship between distances of place field and theta time scale offset (τ) of many neuron pairs is shown in Figure 2.1.
From Geisler et al. (2010).

theta cycle, the spike sequences correspond to the trajectory of place fields the animal has just passed and is going to visit (Figure 7.4). This is important for several reasons. First, the temporal relationship of cell assemblies over multiple theta cycles is advantageous for strengthening the synaptic connections between the evolving assemblies. Second, these synaptic strengths reflect the distance relationships between place cells. Third, and most important in the present context, these observations challenge the premise that place field spikes are governed exclusively by landmarks or other external cues across theta cycles in an outside-in manner. Instead, such relationships are generated by internal mechanisms that support the packaging of neuronal assembly sequences.

that long-term synaptic plasticity underlies the learning of such sequences, in which theta-nested gamma oscillations play an important role. Diba and Buzsáki (2008) demonstrated that changing the length of the maze track altered many of the firing properties of neurons, including their preferred firing location, peak firing rate, field size, and field overlap. However, the theta-scale timing of place cells remained unaffected, indicating that these parameters set constraints on the mechanisms by which hippocampal networks can represent environments: in larger environments, the spatial resolution is poorer, with larger place fields and larger distances between place fields. The hippocampus thus can "zoom out" or "zoom in" depending on the size of the environment.

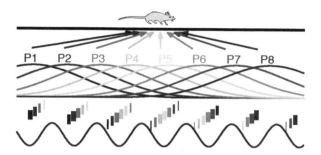

Figure 7.4. Distance coding by within-theta cycle neuronal sequences. Each position (P1 to P8) is defined by the most active cell assembly firing at the trough of the theta cycle. The width of the bars indicates firing rates of the hypothesized assemblies, while the theta timescale temporal differences between assemblies reflect distances of their spatial representations. Because each assembly contributes to multiple position representations in successive theta cycles, multiple assemblies are coactivated in each theta cycle. As a result, the current position/item, represented by the maximally active assembly at the cycle trough, is embedded in the temporal (i.e., theta cycle) context of past and future representations of neighboring assemblies.

Velocity Correction of Distance-Time Compression

A lingering issue is how the theta cycle coordination of cell assemblies is maintained when the rat's running velocity varies. As the rat moves, new place cells join the existing assembly in the theta cycle, while spikes of the place cells whose field the animal just left disappear. Neurons representing upcoming place fields fire at the late phase of the theta cycle, whereas neurons representing recently passed place fields ride on the early phases. Thus, as the animal moves forward, a particular neuron's sequence position moves from the ascending (late) phase to the descending (early) phase of the theta cycle (Figure 7.4).[15] This "one-in, one-out" shifting membership keeps the number of cell assemblies within theta cycles relatively constant in representing the travel path from already past to future locations. Thus, each theta cycle contains a segment of travel (distance) coded as a sweep of time (duration).

How can this distance–duration relationship remain faithful when the animal runs at different velocities? If the rat traverses the place field of a neuron in 1 second and then in half a second in two consecutive trials, the place cell will

---

15. This spike phase shifting is known as *phase precession* (O'Keefe and Recce, 1993). The spike–theta phase relationship correlates reliably with the rat's spatial position on a running track. Because of this relationship, O'Keefe used his discovery to further support his allocentric map-based navigation model. However, the allocentric map does not need time, as discussed in Chapter 5.

be active for 8 and 4 theta cycles, respectively (assuming 8/s theta frequency). The number of spikes within the place field remains the same as the rat's velocity changes. For this reason, the number of spikes per theta wave approximately doubles. Due to the firing rate gain, which reflects stronger excitation of the neuron, as velocity increases, the magnitude of the cycle-to-cycle phase shift increases proportionally. As a result, the velocity gain compensates for the shorter time spent in the place field, leaving the relationship between phase and spatial position relatively invariant (Figure 7.5).[16] If this multiple-step logic is confusing, it is sufficient to remember that the fundamental organization of the hippocampal dynamic is temporal, with the theta cycle as its "scale bar." Because the brain has constant access to velocity from the body and vestibular system, time and distance traveled can be interchangeably calculated (see Chapter 10).

INTERNAL NEURONAL SEQUENCES SERVE COGNITION

As discussed in Chapter 5, the mechanisms for representing a path through an environment are phenomenally similar to those representing sequential items in episodic memory. Both the position-dependent sequential firing of neurons along a linear path and the sequence of arbitrary items in episodic memory tasks are essentially unidimensional. The linking of items in episodic memory, analogous to linking of place cells by theta-gamma coupling, could explain two important principles of memory recall: *asymmetry*, which is the finding that forward associations are stronger than backward associations, and *temporal contiguity*, the finding that recollection of an item is facilitated by the presentation or spontaneous recall of another item that occurred around the same time.[17] In this chapter, I add the information that neuronal networks in the

---

16. Geisler et al. (2007) showed that place cells are velocity-dependent oscillators as their oscillation frequency is determined by the animal's traveling velocity. Every place cell oscillates faster than the ongoing LFP theta, resulting in an interference or phase precession of their spikes (O'Keefe and Recce, 1993). Because the place field span (i.e., the "lifetime" of its activity) is inversely related to the oscillation frequency of the neuron, the slope of the phase precession defines the size of the place field. In other words, neurons that oscillate faster have smaller place fields and display steeper phase-precession slopes. Because neurons in the more caudal (temporal) part of the hippocampus are less sensitive to speed (Hinman et al., 2011; Patel et al., 2012), they oscillate more slowly and therefore have larger place fields and less steep phase-precession slopes (Maurer et al., 2005; Royer et al., 2010).

17. Kahana (1996); Howard and Kahana (2002).

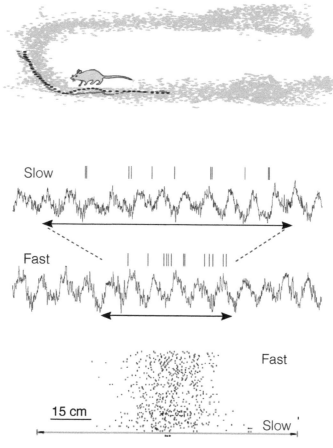

Figure 7.5. Speed-gain compensation for distance-time compression in the hippocampus. *Top*: Trajectories of a rat through a place field on two trials with different speeds. *Middle*: Spikes of one place cell (*vertical ticks*) and the corresponding theta rhythm of the same two trials as on top. Horizontal double arrows indicate the time it took for the rat to run through the place field. *Bottom*: The number of spikes within the neuron's place field is similar on slow and fast run trials. Trials are sorted by velocity from slowest to fastest.
From Geisler et al. (2007).

hippocampus and entorhinal cortex can disengage from the external world and generate their own perpetually changing assembly sequences.

If hippocampal neurons responded exclusively to landmarks or signals from the body in an outside-in manner, the map-based navigation theory (Chapter 5) would predict that a small set of neurons (i.e., place cells) should fire continuously as long as the rat's head remains in the same location. All other pyramidal neurons should remain silent. In contrast, if assembly sequences are generated by internal mechanisms, neuronal activity might instead change continually. My laboratory designed such an experiment to confront these ideas.[18]

Imagine that the animal is "frozen" in place during its travel and yet the theta oscillation is maintained somehow. To achieve this, we trained rats to alternate between the left and right arms of a modified T maze. Rats do this efficiently because finding a water reward at the end of the left arm on one trial informs the animal that a water reward will be available at the end of the right arm during the next trial and vice versa. Such experiments have been conducted on rodents for decades. The novel aspect of our task was the addition of a running wheel at the start of the center arm of the maze. The rat was required to run in the wheel for 10–20 s at about the same speed while facing the same direction on each run. These additional requirements served two purposes. First, the delay between choices is known to make the task hippocampus-dependent. Second, trained behavior guarantees that both the environmental and body-derived cues remain constant during running. Thus, the animal cannot use such signals to help maintain the choice information. Instead, it has to rely on its memory of the previous choice. This task allowed us to distinguish the predictions of the path integration hypothesis from those of internalized memory mechanisms (Chapter 5).

The experiment supported the second alternative (Figure 7.6). Indeed, not a single neuron among the hundreds recorded fired continuously at the same rate throughout the wheel running. Instead, neurons in the wheel were active only for about 1 sec, which is the same "life time" as that of place cells in the maze. Some neurons fired at the beginning of the run, others in the middle, and yet another set toward the end of the run. When enough neurons were recorded simultaneously, every part of the wheel-running epoch was associated with at least one actively firing neuron. Pyramidal neurons fired at the same specific times during wheel running on each trial. In short, the entire journey in the wheel of a single trial is associated with a unique neuronal trajectory of perpetually changing cell assemblies. As the animal's body and head are not displaced while in the wheel, these active neurons do not meet the criteria for place cells. Instead, the evolving neuronal trajectory must reflect some cognitive content.

The internalized cognitive content is information about both the past successful choice or the future goal. This is easy to demonstrate by sorting individual trials according to the rat's future choice of arm (left or right). The neuronal trajectories are uniquely different on left and right choice trials, indicating that the initial condition, set by the correctly identified reward location, determines the pattern of the neuronal trajectory (Figure 7.6). The reader

---

18. Eva Pastalkova, a postdoctoral fellow in my lab, was a key player in these experiments (Pastalkova et al., 2008).

Chapter 7. Internally Organized Cell Assembly Trajectories 179

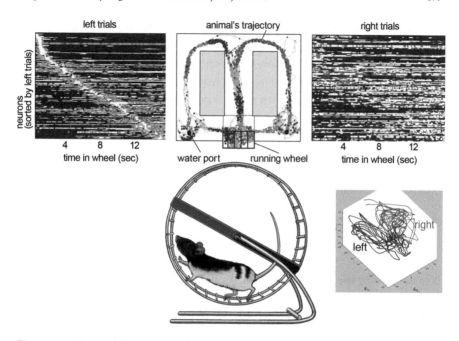

Figure 7.6. Sequential activation of neuronal assemblies in an episodic memory task. *Center*: The rat was required to run in a running wheel during the delay between trials while remembering its last choice between the left and right arms of the maze. It obtained a water reward by choosing the arm opposite to the previous choice. Dots superimposed on the maze (top) represent spike occurrences of simultaneously recorded hippocampal neurons. *Left*: Normalized firing rate trajectory of neurons during wheel running, ordered by the latency of their peak firing rates during left trials (each line is a single neuron). *Right*: Normalized firing rates of the same neurons, ordered the same way as on the *left*, during right trials. We can easily tell the difference between future left or right choices just by looking at the neuronal assembly vector any time during wheel running. *Right bottom*: Neuronal trajectories of the same neuronal population of the prefrontal cortex, reduced to a three-dimensional "state space." Each trial begins and ends at the same initial condition, corresponding to the reward and departs differentially depending whether the rat chooses the left or right arm of the maze.
Sequential activation of neuronal assemblies is from Pastalkova et al. (2008).
Neuronal trajectories courtesy of Esther Holleman.

may rightly ask whether trajectories correspond to the retrospective memory of the previous choice or the prospective future plan. Error trials can help to disambiguate these possibilities. The neuronal trajectories on most error trials are similar to those of correct trials (e.g., when the animal turns left when the right arm is correct, the firing pattern corresponds to left trials) from the very beginning of the wheel running, suggesting that the animal "believes" that its plan was the correct one.

To summarize a bit, these experiments demonstrate that the evolving neuronal trajectory reliably predicts the animal's choice on both correct and erroneous trials tens of seconds before it acts out the plan.[19] These findings support the idea that plans are deferred actions. Another interesting aspect of this experiment is that the firing pattern of the hippocampal neurons, including their sequential organization, firing rates, life times, and the theta time compression, are remarkably similar in the wheel and maze. We can therefore conclude that navigation in the real world and in mental space is supported by identical hippocampal mechanisms,[20] supporting my major thesis in Chapter 5 that the brain's disengagement from the environment can serve cognition.

Neuronal Trajectories Predict Behavior

It may surprise the reader that the neuronal trajectory can reliably predict the animal's choice 15–20 s before its behavioral decision. Yet it can be explained. The population activity on correct trials not only predict the future choice of the animal but can also postdict (i.e., recall) the preceding choice.[21] From the moment of reinforcement, a continuous neuronal trajectory is selected, representing the journey from the water well to the running wheel, the running period, and the traversal of the maze. This trajectory depends on only a

---

19. Prediction of maze arm choice from neuronal activity was most accurate during the first few seconds when the density of neuron firing was highest. This is another similarity with memory, which becomes less accurate over time. The composition of neuronal trajectories during wheel running is unique to the initial condition—that is, trajectories can code for "what." The same sequences also track run distance and elapsed duration, so they relate to *what*, *where*, and *when*, the cardinal features of episodic memory (Chapter 10).

20. The map-based navigation hypothesis also predicts that the phase of spikes will remain fixed if environmental inputs do not change. In contrast to this prediction, every sequentially active neuron displays phase precession during wheel running. Similar to place cells, the theta frequency oscillation of episode cells is higher than that of the field theta rhythm, and the slope of phase precession is inversely related to the length of the episode field (Pastalkova et al., 2008). Numerous control experiments were performed to demonstrate that neuronal trajectories were not driven by optic or haptic flow, steps in the wheel, or the like, but by memory load.

21. Error trials can disambiguate the situation because, during most error trials, the neuronal firing sequence correctly identified the behavioral choice but not the arm from which the rat had come. For example, when the rat erroneously turned left after a previous left turn, the trajectory reliably predicted that the rat would turn left at the T maze intersection and incorrectly encoded its past route (left arm as right arm). However, such disambiguation is information only to the experimenter. From the rat's point of view, the erroneous trial was not an error. One might rightly argue that it acted according to its belief that its choice was correct all the way until the lack of reward revealed that it had made an error.

few conditions: the initial condition, the state of the brain at the moment of the initial condition, and the history of the brain's encounter with the initial condition. After the initial condition is created, the trajectory is similar every time, unless the hippocampus detects something unusual between the expected and the experienced. This neuronal trajectory has interesting qualities. It has no clear break point that would indicate that the past is over and the future is beginning. Beliefs about the future are tied to beliefs about the past. There is no clear demarcation of the "now" or present.[22]

Subsequent experiments showed that theta oscillation is essential for internally organized hippocampal neuronal trajectories as well as for memory-guided spatial behaviors. For example, after local pharmacological inhibition of the medial septal input to the hippocampus, whose integrity is essential for theta oscillations,[23] firing fields in the wheel become disorganized, and neurons fire almost randomly and at a low rate during wheel running. In contrast, the place fields in the maze are largely preserved, presumably because place-cell firing remains under the control of room landmarks. Yet things are not perfect with place cells either because their temporal sequences within theta cycles are completely disorganized. Thus, theta oscillation is a prerequisite for temporal coordination of neurons at the tens of milliseconds time scale.[24]

Overall, these experiments demonstrate that theta oscillations are essential for short-time scale organization of neuronal sequences. Because place fields in the maze are preserved along with their behavioral time-scale sequences, the findings also imply that neuronal sequences can be generated by two fundamentally different mechanisms. First, the internal dynamics of the hippocampus generate sequences under memory load. Second, environmental or other stimuli, by their sequential nature, also generate neuronal sequences at the speed of behavior, which are then compressed into theta time scale

---

22. We can assign an arbitrary segment as "now" by quantifying the number of spikes within the theta cycle (Csicsvari et al., 1998, 1999; Dragoi and Buzsáki, 2006). The assemblies firing on the descending phase of the theta cycle correspond to the past and the assemblies firing on the ascending phase correspond to the future, but the most active assembly occupies the trough of the theta wave (Figure 7.4), which can be designated as representation of the here and now.

23. It has been known since the pioneering work of Petsche et al. (1962) that after destruction of the medial septum, theta oscillations completely disappear from the hippocampus and surrounding structures.

24. Wang et al. (2015). Using calcium imaging to detect the activity of hippocampal CA1 pyramidal neurons, Villette et al. (2015) demonstrated that various lengths of neuronal sequences also emerge in complete darkness without reward and when the treadmill surface offers no external cues. The run distances (or duration) distribute into integer multiples of the span of these sequences. The authors speculate that internally generated hippocampal sequences trigger spontaneous run bouts and track their durations.

sequences. When the internally induced sequence is perturbed in a memory task, for example by some salient cue, the neuronal trajectory "jumps." Then either the original trajectory returns, or a new trajectory emerges.[25] This process is akin to a conversation when one speaker is narrating a story. Another speaker may interrupt it with a question, after which the story can either continue or get diverted into a new direction depending on the nature of the interruption. Thus, the evolution of neuronal trajectory is most often guided by the interaction between internal mechanisms of the self-organized networks and external influences, with the relative importance of the two varying, sometimes substantially, from moment to moment.

Similar Neuronal Trajectories During Learning and Memory Recall

In 2007, I was Visiting Professor at Hebrew University, Jerusalem, where I met Itzhak Fried, a neurosurgeon from University of California at Los Angeles. Itzhak is the de facto leader in chronic physiological recordings from waking epileptic patients. These patients are often implanted with hippocampal-entorhinal electrodes to determine the origin of their seizures. Some of these electrodes can record from single pyramidal neurons, and many of the patients volunteer to participate in cognitive experiments while waiting for seizures to occur for diagnostic purposes. I told Itzhak about our running wheel experiments in rodents, while he described to me how their recorded neurons responded to movie clips.

His team could do things in humans that were not possible with our rodent subjects. Human recordings covered only a few neurons, but the experimenters could show many movie clips in a short period of time. If a neuron responds selectively to only one of the clips, we can assume that it is part of a neuronal assembly sequence that describes the movie clip. In our rat experiments, we had conclusive evidence that the neuronal trajectories encoded a particular mental travel event. However, we could not examine free recall, a critical test in episodic memory, because our rodent subjects could not verbally report their experience. But human subjects could. Free recall is a memory that can be called up consciously from long-term storage without any cue. In Itzhak's experiment, different neurons responded to different famous people's actions or events in the movie clips. After all the clips were shown, the experimenters asked the research participant a critical question: What did you just see? While

---

25. Such jumps in the cell assembly trajectory can occur in a single theta cycle (Zugaro et al., 2004; Harris et al., 2003; Jezek et al., 2011; Dupret et al., 2013).

the participants reported their experience, the same neuron that responded to a particular movie clip (such as Tom Cruise's epic couch jump on Oprah's show) became active a hundred or so milliseconds *before* the participant spoke the actor's name. Their experiment is important for two reasons. First, it demonstrates that similar neuronal trajectories are formed during encoding and recall of an episode, suggesting that the initial condition can be set by voluntary recall. Second, during the encoding process, information flows from the external world to the neocortex and then to the entorhinal-hippocampal system. During recall, and possibly imagination, the direction reverses: activity begins in the hippocampus and propagates to the neocortex.[26]

These neurophysiological experiments echo similar observations in human imaging studies. An advanced technique called *multivoxel pattern analysis* is a conceptual extension of the cell assembly or population vector idea. A voxel, the smallest unit that functional magnetic resonance (fMRI) can resolve, contains about 10,000 neurons. Because many voxels are imaged in each scan, their number of possible combinations is astronomically high. Some of these combinations can be matched to patterns of inputs and potentially movements. In a pioneering experiment, researchers identified population voxel patterns while students studied pictures of celebrities, objects, and places. Later, when participants recalled the learned material, the category-specific voxel constellations reappeared before their overt responses. For example, before the participant recalled the picture of a famous actor, the temporal area of the cortex in charge of face processing was activated together with many other voxels in other parts of the brain, displaying the same pattern as seen during the study phase.[27] These imaging experiments support the view that neuronal assemblies that initially process episodic or semantic information are similar to or at least overlap with those that become active when we recall or imagine the same information.

There is also more direct support for the critical readout mode of the hippocampus for episodic information. In a series of analytical experiments, Susumu Tonegawa at the Massachusetts Institute of Technology tagged neurons that were active in a threat-conditioning task. Hippocampal neurons that responded during the threat triggered the expression of a light-sensitive protein. Later they shined laser light on these same neurons in a different box where no harm was ever inflicted upon the mice. Yet the mice froze in the same way that they did

---

26. Gelbard-Sagiv et al. (2008) According to experiments on monkeys (Miyashita, 2004), two types of retrieval signal can activate cortical representations: one from the frontal cortex for active (or effortful) retrieval (top-down signal) and the other from the medial temporal lobe for automatic retrieval.

27. Polyn et al. (2005).

in the box where they had been shocked.[28] These experiments demonstrate yet again that the neurons activated during learning are related to those needed to recall the experience and initiate appropriate actions.

INTERNALLY GENERATED CELL ASSEMBLY SEQUENCES IN OTHER STRUCTURES

Self-organized neuronal sequences are not just a specialty of the hippocampus. When neurons from the medial prefrontal cortex were recorded during a memory task, similar evolving neuronal trajectories were detected (Figure 7.6). In this experiment, rats were exposed to either chocolate or cheese odor in the waiting area of the maze (instead of the running wheel) and had to learn that the two odors instructed them to turn to the left or right arm of the maze, respectively. As in the hippocampal recordings, the neuronal trajectory was uniquely different in the waiting area before the run and in the central arm of the maze depending on the animal's future choice.[29] Furthermore, the neuronal assemblies lasted for 0.5–2 sec, suggesting that similar mechanisms determine how long a particular neuron or group can be active.[30]

---

28. Liu et al. (2012). In 2010, Michael Häusser's group had already presented a similar experiment using identical molecular biological methods in the hippocampal dentate gyrus at the Society for Neuroscience Meeting, but their experiment was never published. An important critique of these perturbation experiments is that in a threat-conditioning paradigm, very little overlap is sufficient to induce the only relevant behavior: freeze or no freeze. Thus, the labeled and reactivated neuron may not reflect a veridical replication of the experience. Other experiments show that, with time, only a small fraction of neurons activated during the original experience remains the same (Ziv et al., 2013).

29. This match-to-sample test with a delay is known to require both the medial prefrontal cortex and the hippocampus (Fujisawa et al., 2008). Ito et al. (2015) confirmed and extended these findings. In addition to the prefrontal neurons, they also found behavioral trajectory-specific encoding in the nucleus reuniens, a thalamic nucleus connecting prefrontal outflow to the hippocampal CA1 region. They suggested that trajectory-dependent activity arises in the prefrontal cortex and is conveyed to the hippocampus via the nucleus reuniens. However, the future choice of the animal can be also predicted from the firing patterns of granule cells and CA3 pyramidal cells (Senzai and Buzsáki, 2017), which do not receive input from the thalamus.

30. In their classic study, Fuster and Alexander (1971; see also Kubota and Niki, 1971; Funahashi et al., 1989) showed that single neurons in the primate dorsolateral prefrontal cortex fired persistently during the delay period of a working memory task. For several decades, persistent firing of single cells was used as the prime model of bridging delays and keeping information in mind. In our recordings, all such neurons turned out to be fast-spiking inhibitory interneurons. It seems that "persistent" activity is maintained by evolving assemblies rather than by a small group of specialized neurons that sustain their discharge for seconds to tens of seconds.

Yet another cortical structure critically involved in cognition is the posterior parietal cortex. Optical imaging of calcium dynamics, a widespread recording method using laser technology, was used to explore the circuit dynamics in superficial layers of this region. In this test, the mouse's head was fixed in place, but it could run on a large Styrofoam ball to guide itself through a virtual reality environment in which the scenery of a projected movie is controlled by the mouse's movement on the ball. Individual parietal neurons responded with a transient calcium activation, reflecting some degree of spiking, staggered one after another in time to form a sequence of neuronal activation spanning the entire length of a task trial. Distinct sequences of neurons were launched on different trials depending on the animal's goal location. Neuronal assembly sequences have been observed in other structures as well, including the motor cortex, demonstrating that self-organized neuronal trajectories are the norm in brain circuits.[31] Strictly speaking, all the neuronal trajectories discussed earlier and their syntactic combinations can be regarded as "simple" because only the first-order information of sequential activity is used. Whether neuronal syntax also exploits higher order, embedded structures remains to be demonstrated.[32]

HOW DO NEURONAL NETWORKS GENERATE SELF-ORGANIZED SEQUENCES?

Despite the critical importance of internally generated neuronal trajectories to nearly everything we call cognition, the exact mechanisms that generate self-organized sequential activity of neurons are not perfectly understood. The principles are very simple since there are only two essential requirements. The first is a network that contains competitive processes. The second is some kind of short-term accommodation feature. Experimental observation shows that neuronal assemblies typically last for 1–2 sec, which constrains the number of

---

31. Harvey et al. (2012) described internally generated neuronal sequences in the parietal cortex. Such sequences are also prevalent in the motor cortex during planning and execution of voluntary movement (Shenoy et al., 2013) and have been documented in the entorhinal cortex (O'Neill et al., 2017) and the ventral striatum (Akhlaghpour et al., 2016).

32. Gao and Ganguly (2015). In spoken language, first-order or serial syntax refers to the sequence of sounds; that is, how discrete acoustic units are produced and combined to provide diversity ("My father's brother's wife's daughter's dog"; a hierarchy of possessives). This diversity can be enhanced further by embedded, higher order rules ("The dog the cat bit drank the milk"). Such embedding is universal in the world's 6,000–8,000 languages, suggesting that it is based on brain mechanisms with higher order relationships among the neuronal sequence members.

possible mechanisms.³³ This gives some clues regarding the possible physiological mechanisms needed to support neuronal sequences.

Envision a two-dimensional network of neurons (a "neuronal sheet") with relatively symmetric connectivity of both excitation and inhibition. If the excitatory connections are shorter than the inhibitory ones, an excitatory seed of activity will generate a big local "bump" of activity surrounded by inhibition. When the wave of excitation and inhibition is plotted on a graph, it looks like a Mexican hat, with a peak surrounded by a valley. Such connection matrices are often used to model spatially spreading activity. In the simplest of these models, the bump of activity stays stationary or moves only a limited distance. In the parlance of the network modeling literature, the bump is considered to be an attractor that keeps the spread of activity at bay. However, if we add an adaptation component, things change dramatically.³⁴ For example, if the spike threshold of the excitatory neurons increases as the function of spiking intensity, neurons in the regions of high activity (i.e., in the attractor) become increasingly less able to fire over time as their own spiking activity decreases their excitability. The result of this self-increased threshold is that the bump of the activity moves to a competing neighboring region where the spike threshold of the neurons is lower. The speed of the spread of excitatory activity is determined by the time course of the spike threshold adaptation, which is approximately 1 s in cortical pyramidal neurons.³⁵

Now, if such a matrix occupies not a two-dimensional sheet but the surface of a doughnut (technically a toroid), the activity can return to its origin or meander around the toroid forever (Figure 7.7). The exact spatial pattern of the activity (i.e., the trajectory of the neuronal sequences) depends only on where the activity was initiated. If the activity starts at a different location, its trajectory

---

33. When self-sustained activity is not a requirement, sequential activity can be induced in a model chain of neuronal groups connected with fixed synaptic weights. In such "synfire" chains (Abeles, 1991), sequential activity is a consequence of excitatory action going in one direction. Despite the dissimilarity between the model and cortical networks with iterative loops and feedback (Buonomano and Maass, 2009), the synfire chain model generated useful and disciplined thinking about how neuronal activity can spread from layer to layer or region to region (e.g., Ikegaya et al., 2004).

34. The assembly sequence can be viewed as a heteroclinic attractor of dynamical systems, where the noise level is high enough to move the trajectory from one basin to the next by surpassing the energy barriers between the basins (Redish et al., 2000; Rabinovich et al., 2008; Afraimovich et al., 2013). The shift from basin to basin can occur regularly or stochastically, depending on the conditions. If the external cues are strong, they can shift the trajectory into a new direction. Itskov et al. (2011); Howard (2018).

35. Henze and Buzsáki (2001) in vivo and Mickus et al. (1999) in vitro.

Chapter 7. Internally Organized Cell Assembly Trajectories

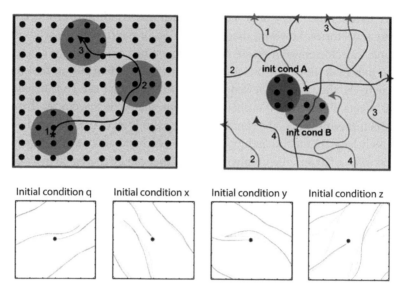

Figure 7.7. Model of self-organized neuronal trajectories. *Top left*: Neurons are arranged on a two-dimensional "sheet" with periodic boundary conditions (toroid arrangement), so that the top and bottom edges and the left and right edges are identical. The trajectories "wrap around" the sheet (arrows "leave" the plane but return on the other side, as would occur on a toroid or a doughnut). Without adaptation, the network activity quickly converges to a "bump attractor" (*shaded circles*), an area which prevents the activity from moving moves away. The center of the bump is marked by the asterisk (position 1). In the presence of adaptation, however, the activity bump moves continuously around the toroid sheet, never settling. The black curve traces the position of the center of the bump traveling through the three attractors. *Top right*: Different initial conditions in the adaptation variables lead to uniquely different trajectories for the bump of activity. Adapted neurons are less likely to fire, so initial condition *A* causes the bump of activity to move initially to the right, whereas initial condition *B* causes the bump of activity to move upward. *Bottom*: Trajectories for four different initial conditions (i.e., different constellation of initiating cell assemblies).
From Itskov et al. (2011).

will be different. In contrast, if the same site is activated repeatedly, the same neuronal trajectory will be induced.

In the running-wheel experiment there were only two initial conditions: the rat returned to the wheel from either the left or the right corridor. In the model, however, activity can begin at any arbitrary location, thus the number of possible trajectories is high and increases dramatically with the size of the toroid matrix. If the system is not strongly affected by noise, the same initial condition can produce reliably similar trajectories for many tens of seconds or minutes. Each of these trajectories can be assigned to a learned event (e.g., an episode in our lives). When very large numbers of initial conditions produce many unique trajectories, however, then the fraction of neurons that participate in multiple trajectories will increase. Such overlapping neurons can be considered as the inevitable noise introduced by the increasing demand on the matrix. Alternatively, they can be viewed as useful linkages that can concatenate different events. Either way, if events can be represented in trajectories involving many neurons over time, the combinatorial possibilities of the network become extremely large. In the rodent hippocampus, which has about a billion synapses and 50 miles of axons, the number of events that can be presented by neuronal trajectories is astronomically high. Along the same lines, the number of chords that can be played on the piano with ten fingers is limited. However, the possible number of variations of melodies that can be played by sequential key strokes over tens of seconds is much, much higher.[36]

Another important source of adaptation is synaptic plasticity. When action potentials arrive at the presynaptic bouton in rapid succession, the probability of neurotransmitter release and the resulting response (excitatory or inhibitory) of the postsynaptic neuron can either increase or decrease. The synapse between pyramidal cells and basket-type inhibitory interneurons is a depressing synapse, meaning that successive action potentials in the pyramidal cell evoke progressively weaker responses in the basket cell. When pyramidal neurons form an assembly, they "enslave" their basket cells, whose function is to suppress competing assemblies by lateral inhibition. However, the hegemony of the dominant assembly can be sustained only transiently because the responsiveness of the basket cells decreases over time. When the assembly-protecting role of these basket cells diminishes, a completing assembly can become the winner for a while, and so on. The duration of the assembly corresponds to the short-term plasticity of the synapse, which is approximately 1 s in vivo.[37]

These simple biophysical adaptation mechanisms explain why the dominance of particular neuronal groups is limited in time and why perpetual activity in the brain is the norm even without any external input. Self-organized

---

36. A few years ago, I visited Endel Tulving in his home in the outskirts of Toronto. I explained that our discovery of internal sequences in the hippocampus (Pastalkova et al., 2008) and prefrontal cortex (Fujisawa et al., 2008) could serve as substrates for episodic memory. We were enjoying tea and biscuits in his backyard while he was feeding his resident raccoon. He briefly shifted his gaze from the raccoon's nose to mine and said, "Did you expect to find something else?" Although he has never done any experimental work inside the brain, I was amazed at how accurately he could predict the physiological operations of the suspected neuronal dynamics. Perhaps he could make such inferences effectively because he had a consistent theory. I also asked him about the definition of an "episode": "Is it a line, a paragraph or chapter in a book, or is it the entire book?" After a short pause, he replied, "All of the above."

37. Royer et al. (2012); Fernandez-Ruiz et al. (2017); English et al. (2017).

activity is the default state of neuronal networks. Identical initial conditions generate similar sequences, whereas different initial conditions give rise to distinct sequences. An important consequence of self-organized activity is that the network's response to inputs can outlast the perturbation. For example, when I observe something unusual in an experiment, it reverberates in my brain for a long while. Many circuits, especially the giant recurrent network of the hippocampus, can generate large numbers of neuronal trajectories without any signals from outside the brain. Even cultured brain tissue in a dish can induce a variety of spontaneous events. That is what neuronal circuits do.[38] They do not just sit and wait to be stimulated.

Based on these examples, now we may consider the possibility that learning is not an outside-in superimposition process where new neuronal sequences are built up with each novel experience. Instead, learning may be an inside-out matching process: when a spontaneously occurring neuronal trajectory, drawn from the available huge repertoire of trajectories, coincides with a useful action, that trajectory acquires meaning to the brain. The richer the experience, the higher the fraction of the meaningful trajectories, but there is always a large reservoir of available patterns. In this alternative model, the number of trajectory sequences would be pretty much the same in a well-experienced brain and a hypothetical naïve brain that has never experienced an input. The advantage of this inside-out solution is that the huge realm of diverse and interdependent trajectories may bring stability to the networks, and there is no fear that novel learning will ever destabilize them.[39] We will return to this topic in Chapter 13.

READING NEURONAL TRAJECTORIES

Neural messages are only as useful as their readability. The biological relevance of the assembly sequences and their alleged "representations" can be verified only through some reader mechanism that differentiates the multiple overlapping sequence patterns. If there exists a reader mechanism that can differentially respond to assembly sequences *a, b, c, d* versus *a, d, c, b*, then such minute differences are significant for the brain. We can be confident that

---

38. Li et al. (1994). Neuronal sequences have been postulated to play a role in nearly all assumed cognitive operations (e.g., memory, planning, decision-making, inference, prediction, imagination, day-dreaming, contemplation). Brain models of cognition (e.g., predictive coding, prospective coding, statistical, probabilistic inference, Bayesian inference, and generative models) also assume their existence (Pezzulo et al., 2017).

39. *Catastrophic interference* is an ever-recurring theme in artificial neuronal networks, in which new learning may erase valuable existing information (Ratcliff, 1990).

mechanisms for sequence readout must exist in neuronal circuits because we can effortlessly differentiate many sound sequences and tell the difference between scratching our palm from left to right or from right to left. Yet, the physiological mechanisms that decipher messages from cell assembly sequences are not well understood. Different reader mechanisms may simultaneously monitor the activity of the same assembly patterns to extract distinct types of meanings. For example, one reader may track spiking intensity of the neuronal population within a given time window, whereas another reader may extract temporal relationships from the assembly members.[40]

In the simplest situation, all the possible temporal patterns of neurons converge onto a large array of separate reader neurons due to the hard-wired features of a circuit.[41] But there are other potential mechanisms for interpreting sequences. Single neurons or even single dendrites may be sensitive to the ordering of excitatory inputs. In this case, sequential activation of neighboring synapses along a dendrite in different directions produces different responses at the soma, provided that some nonlinear properties of the dendrite or local circuit effects assist in discriminating the order of activation.[42] Another potential mechanism exploits the timing effects of lateral inhibition, the kind we discussed earlier in connection with competition between cell assemblies. In this case, upstream inputs nonhomogenously innervate their downstream readers. The neurons that respond to the first input of the upstream sequence will recruit lateral inhibition, which reduces the ability of the remaining

---

40. The weakly electric fish (*Eigenmania*) creates sinusoidal electrical field potentials around its body to detect objects and other fish ("jamming avoidance response"). The frequency difference between electric fields produced by two *Eigenmania* produce systematic phase and amplitude changes that are used to electrolocate nearby conspecifics (Heiligenberg, 1991; Chapter 3). In the hippocampus, the phase of spikes may inform the downstream reader mechanisms about the distance traveled or elapsed time, whereas spiking intensity reflects the animal's instantaneous speed (McNaughton et al., 1983; Hirase et al., 1999; Huxter et al., 2003).

41. MacLeod et al. (1998) showed in the olfactory system of the locust that information encoded in time across ensembles of antennal lobe neurons converges onto single neurons in the mushroom body. These neurons thus can be regarded as readers of population activity sequences of their upstream information-sending neurons.

42. Computational modeling shows that temporal sequences are much easier to read out than rate-dependent changes (Rall, 1964). In experiments, sequential activation of spines from the dendritic branch to the soma or from the soma to the dendrite tip produced directionally sensitive spike responses in cortical pyramidal neurons. The differential responses are due to the interplay between nonlinear activation of synaptic N-methyl-D-aspartate (NMDA) receptors and the impedance gradient along dendritic branches, two fundamental biophysical features common to most neurons (Branco et al., 2010). Although such dendritic mechanisms can discriminate input sequences in the 10–200 ms range, they are unlikely to provide a solution for reading out cell assembly sequences that occur over seconds.

population to respond. The second and third inputs may be able to recruit a few new neurons but with lower probability and strength than the first one. These additional neurons then recruit more inhibition, further decreasing the probability that the remaining neurons will respond to the later arriving inputs in the sequence. Thus, whichever input arrives first has a stronger effect. In this timing primacy model, the important rule is rank order.[43]

After discussing these theoretical possibilities, let's see a concrete example. Place cell assembly sequences in the hippocampus are read out by the target neurons in the lateral septum using a firing rate-independent phase-coding mechanism. These neurons compare the activity of hippocampal CA1 and CA3 place cell assemblies and express their activity ratios by spike preference to the theta cycle. Thus, in the lateral septum, the theta phase of spiking but not the rate in lateral septal neurons varies systematically from the beginning to the end of a journey. A population of septal neurons can precisely identify the past, current, and predicted future positions of the planned trajectory and convey this compressed message to the lateral hypothalamus and locomotion-controlling brainstem areas. This may be the mechanism by which the abstract cognitive map is translated to behavioral action.[44]

Richness of Neuronal Information Depends on the Number of Readers

How many sequential patterns can be read out? In the preceding models, various population patterns evoke spiking responses in one or just a few reader cells. A given reader neuron may learn to respond to a particular constellation of neuronal discharges in the input layer but remain silent when other patterns are present. To provide biological utility to a second pattern, another reader must become selectively tuned to this other pattern. Learning to discriminate numerous patterns requires many selective readers (Figure 7.8).[45] For example, discriminating between two trajectories (assembly sequences) of hippocampal or prefrontal neurons corresponding to two different choices is a relatively simple task. On the other hand, segregating enough trajectories to represent all the episodic memories collected in a lifetime requires complex mechanisms

---

43. Thorpe et al. (2001); vanRullen et al. (2005).

44. Tingley and Buzsáki (2018).

45. Masquelier et al. (2009) presented a reader computational model in line with the framework discussed in this paragraph. In the mushroom body of the locust, the 50,000 reader Kenyon cells, in principle, can respond to 50,000 odorant combinations (Jortner et al., 2007; Perez-Orive et al., 2002).

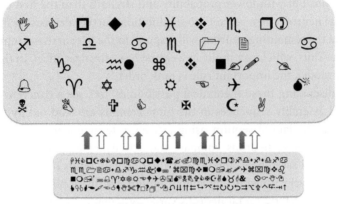

Figure 7.8. Matching preexisting patterns to experience. A sender structure (e.g., hippocampus) can generate extraordinarily large numbers of preexisting patterns even in the absence of any experience (shown here as abstract symbols). Most of these patterns have no meaning to the brain. Yet some of the patterns may acquire meaning by getting linked to some experience-dependent significant event and interpreted as meaningful by the receiving structure (e.g., neocortex). The richness of the information is not determined by the multitude of possible trajectories in the sender but, instead, by the ability of the receiving structure to link the patterns to behaviorally meaningful events through action. The increasing size of the neocortex in higher mammals can read out an ever increasing number hippocampal trajectories.

with many dedicated readers. The neocortex can be conceived of as a giant pattern-segregating structure, with reader mechanisms that learn to classify and segregate overlapping hippocampal output patterns and either lay them down as memories or translate them to plans and overt behavioral responses. However, not all readers are equal. Some may respond to many patterns; we can call them *generalizers*. Others may be selective to just one pattern; we can call them *specialists*. These generalizer and specialist readers occupy a wide distribution with strongly skewed features. We will expand on the significance of such nonegalitarian features of neurons in Chapter 12.

SENDER–RECEIVER PARTNERSHIP IN THE BRAIN

To make sense of the content of neural sequences, both the upstream sender and downstream reader neurons should possess compatible syntactic rules to properly segment and decipher the messages. As discussed in Chapter 6, hierarchically organized brain rhythms are tuned for this purpose. Oscillations not only chunk the messages, but they temporally coordinate the activity of sender and receiver populations to ensure that messages have the correct length and are separated from each other by appropriate time intervals. Transfer of messages from sender to reader is usually considered a unidirectional operation: the source sends the information to an ever-ready recipient network. The brain appears to do things more efficiently. Instead of patiently waiting for messages, the reader actively creates time windows within which the sender is activated and the reader can most effectively receive information.[46] It is a bit like when the boss calls a meeting for 11 AM to gather information about the productivity of the team. The subordinates (senders) show up in the office at the time when the boss (the reader, but also the initiator of the dialogue) offers undivided attention to the information that they provide.

This "call up" is somewhat analogous to the action–perception cycle (Chapter 3). As discussed there, active sensing begins with a motor command informing the sensory system via the corollary discharge mechanisms and adjustment of the sensor. Thus, the temporal windows within which neuronal information can be exchanged are set by the action system. Command signals, such as saccadic eye movements, sniffing, whisking, touching, licking, or twitching of the inner ear muscles can "reset" or synchronize spiking activity in large parts of the corresponding sensory system and enhance the reader (sensory) system's ability to process the inputs.[47]

Neuronal networks in the inner parts of the brain, where movement and sensory inputs are remote, also adapt this reader-initiator principle. The information exchange between the hippocampus and neocortex is a prime example (Figure 7.9). In the waking animal, the hippocampus (the reader) initiates the transfer of neuronal messages via theta phase control of neocortical network dynamics. Hippocampal theta oscillations can temporally bias gamma oscillations at multiple neocortical locations. As a result, neocortical messages contained in the envelope of gamma waves arrive in the hippocampus at the most sensitive (perturbation) phase of the theta cycle so that hippocampal networks can absorb them most effectively. During sleep, the bias works in the opposite direction. Now the dialogue is initiated by the neocortex (the reader) as its slow oscillations bias the timing of hippocampal sharp wave ripples. These are super-synchronized, compressed messages of recently acquired information

---

46. This was first noted by Anton Sirota, a student and postdoctoral fellow in my laboratory (Sirota et al., 2003, 2008); Isomura et al. (2006).

47. See Henson (1965); Halpern (1983); Ahissar and Arieli (2001); and further examples in Chapter 11.

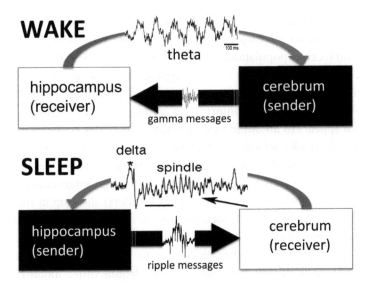

Figure 7.9. "Call up" hypothesis of information exchange. Information transfer in the brain is initiated by slow oscillations produced by the receiver. For example, during waking, hippocampal theta oscillation phase biases neocortical activity (*curved arrow*) and the neocortex (cerebrum) delivers messages, carried by gamma oscillations (*straight arrow*), to the sensitive (receiving) phase of theta rhythm. During non–rapid eye movement (REM) sleep, the neocortex biases the occurrences of hippocampal sharp wave ripples by its slow and spindle oscillations (*curved arrow*). In turn, the hippocampus sends messages to the neocortex in the form of fast ripple messages.

sent from the hippocampus to the neocortex (Chapter 8). This principle appears to be a general mechanism in the brain: the reader initiates the dialogue via a slow oscillation, while the messages are sent via faster barrages, such as gamma or ripple oscillations.[48]

---

48. The reversal of sender and receiver roles across wake and sleep cycles is also supported by combined human resting-state fMRI and electrocorticography. Delta band activity and infra-slow activity propagate in opposite directions between the hippocampus and cerebral cortex in a brain state-dependent manner (Mitra et al., 2016) analogous to the theta–gamma coupling (Figure 7.9). Another excellent demonstration is provided in the visual cortex of macaque monkeys. Here, visual messages from the superficial layer neurons of V1 to the granular layer of V2 are carried by ~4 Hz and gamma-band (~60–80 Hz) oscillations, while feedback from the deep-layer neurons of V2 to the supragranular layer neurons of V1 is mediated by a beta-band (~14–18 Hz) rhythm (Bastos et al., 2015). The notion of a hierarchy rests on the asymmetry between forward and backward connections in the brain. The anatomical hierarchies are thus ordered to the temporal scales of representations, where the slower time scale represents the reader/initiator. Furthermore, information from a higher level brain structure conveys contextual information to the lower level via a slow oscillation (Kiebel et al., 2008).

Both Encoding and Reading Neuronal Information Require Segmentation

There is one more complication in this process. Good reader mechanisms do not instantaneously synthesize information. Instead, they wait with the interpretation until the full stop signal. As in speech, where we must often wait until the end of the sentence to make sense of it, evaluation of neuronal messages takes time. A perfectly synchronized system could offer information from all parts of a network right away, but networks are not synchronized by a global master clock. Instead, the activity travels in neuronal space, much like waves in a pond. For example, theta oscillations in the hippocampus begin to synchronize neuronal activity first at the septal pole and gradually recruit neurons closer to the temporal end of the structure. This sweep of activity takes about 70 ms, corresponding to roughly half a theta cycle.[49] The result of such traveling wave organization is that a static observer neuron downstream from the hippocampus would interpret the temporally synchronous spikes of place cells from the septal pole, mid-segment, and the temporal end as representations of the same position in the outside world. Conversely, place cells that fire at different times within the theta cycle can be interpreted as representations of different positions of the animal's path by the same simple static reader mechanism. Neither interpretation may be right. To illustrate this confusing situation by a metaphor, planes from Paris to New York City depart and arrive at different times (representation of time segments). Conversely, planes arriving close together in time likely come from different cities and continents (representation of locations). Similarly, the reader mechanism looking for representations of the same place in different segments of the hippocampus or in different "time zones" of the theta cycle must take into account the delays introduced by the traveling theta wave mechanisms. Neocortical reader mechanisms of the hippocampus must take into consideration the full sequence of activity from the septal to the temporal pole in each theta cycle in order to properly interpret hippocampal messages. Such space-time relativity is not special for the hippocampal system but for the entire brain. Importantly, the brain rarely gets confused by this relationship (Chapter 3). The confusion typically arises in the

---

49. Lubenov and Siapas (2009); Patel et al. (2012). In the presence of traveling theta waves, at every instant, a segment of physical space is topographically mapped along the septotemporal axis, and every position is represented not as a point in time but as duration in hippocampal volume. Traveling brain waves have been known since the earliest simultaneous recordings and analyses of electroencephalograms (EEG; Hughes, 1995; Ermentrout and Kleinfeld, 2001).

mind of the experimenter because of the different interpretations of sequential patterns by the brain and the human interpreter (Chapter 10).

A potential brain solution to the segmental readout problem is that the neocortical readers learn to regard the spatial travel of hippocampal output as a syntactical unit, a sentence, so that they can predict and take into account events happening over a time segment. This requirement may explain why the frequency of the hippocampal theta oscillation covaries precisely across the entire hippocampal-entorhinal-prefrontal system.

Overall, in this chapter, I showed that sequences of neuronal patterns are not always imposed on brain circuits in an outside-in manner by the sensory inputs. Instead, internally organized processes can sustain self-organized and coordinated neuronal activity even without external inputs. If so, these mechanisms should be also at work when the brain disengages its operation from outside cues, as happens during sleep. To see support for this claim and find out why sleep is beneficial for cognitive performance, please proceed to Chapter 8.

SUMMARY

Self-generated, sequentially evolving activity is the default state of affairs in most neuronal circuits. All it takes is the presence of two competing mechanisms, such as excitation and inhibition, and an adaptation component. With these ingredients, the activity moves perpetually, and its trajectory depends only on the initial conditions. Large recurrent networks can generate an enormous number of trajectories. No experience is needed. Each trajectory is available to be matched by experience to simulate something useful for the downstream reader mechanisms.

Sender and reader mechanisms in the brain are often intermixed, and the direction of information can be biased by oscillations. Readers typically initiate the transfer of information, coordinating the onset of messages from multiple senders via slow oscillations. The messages then arrive in the form of faster oscillatory packages at the perturbation (i.e., sensitive) phase of the reader-generated slow oscillation.

Reading out and properly interpreting neuronal messages take time, similar to words and sentences in language. Unless we wait until the end of the sentence, we may misinterpret its content. Similarly, messages in the brain are distributed over neuronal space and coordinated over time in the form of traveling waves. Thus, it is critical for the reader mechanism to be "aware" of the traveling nature of activity. For example, place cells that fire at the same time in different hippocampal segments may be interpreted to represent the same spatial position by a simple reader. In contrast, a more sophisticated reader mechanism, which takes into consideration the traveling nature of hippocampal content, may correctly infer that perfectly synchronous spiking from different hippocampal segments correspond to past, current, and future positions of travel.

Birdsong and grooming are examples of neuronal trajectories that can be interpreted by relatively simple reader mechanisms due to their limited variability. In contrast, large recurrent circuits can generate numerous trajectories, each of which can become a meaningful pattern when matched to experiences. The richness of the information depends not on the sequence pattern generator but on the reader mechanisms. This may explain why the neocortex (reader) is so much larger than the sender (hippocampus).

8

Internally Organized Activity During Offline Brain States

Alice remarked. "I can't remember things before they happen."

"It's a poor sort of memory that only works backwards," the Queen remarked.

—Lewis Carroll[1]

To think is to forget a difference, to generalize, to abstract . . . to sleep is to be abstracted from the world.

—Jorge Luis Borges[2]

When social animals are gathered together in groups, they become qualitatively different creatures from what they were when alone or in pairs. Single locusts are quiet, meditative, sessile things, but when locusts are added to other locusts, they become excited, change color, undergo spectacular endocrine revisions, and intensify their activity until, when there are enough of them packed shoulder to shoulder, they vibrate and hum with the energy of a jet airliner and take off.

—Lewis Thomas[3]

---

1. *Through the Looking-Glass*, by Lewis Carroll, chapter 5. [NeuroNote: Lewis Carroll had relatively frequent migraines and at least two documented epileptic seizures. His own experience with seizures may have informed Alice's experiences in *Alice in Wonderland* (Woolf, 2010)].

2. The quote is from Borges' "Funes, the Memorious" (1994). [NeuroNote: The piece was written, according to Borges, as a means of staving off his insomnia (Borges and Dembo, 1970).]

3. Thomas (1972).

ewis Thomas's beautiful metaphor of population cooperativity could equally well describe a peculiar hippocampal population pattern, called sharp wave ripples—just substitute neurons for locusts. The sharp wave ripple is a randomly emerging local field potential event occurring when many neurons emerge from their sessile state and fire together shoulder to shoulder. I have been enthralled by their beauty and power from the first moment I heard their buzzing sound in my postdoctoral years. I felt as if I had been listening to a group of orchestra musicians idly tuning their instruments, and the next moment they united in the thrilling harmonies of Beethoven's *Fifth Symphony*. I still think it is the most beautiful pattern the brain produces. Sharp wave ripples represent among the most synchronous population patterns in the mammalian brain, more synchronous than the responses evoked by sensory stimulation of any strength. Yet they are self-organized and spontaneously emitted by hippocampal circuits. The brain must have a good reason to go to the trouble of producing such unusual population activity, which is totally unexpected from the outside-in view of the brain. My first thought was that the sharp wave is an abnormal epileptic activity, perhaps triggered by mechanical trauma inflicted by the recording electrodes I placed into the brain. After numerous control experiments, it became clear that, instead, they represented a potentially important physiological pattern, one whose mysteries waited to be uncovered.

There is no trigger for the occurrence of sharp wave ripples. They are not caused by anything. Instead, they are released, so to speak, when subcortical neurotransmitters reduce their grip on hippocampal networks, as routinely happens during nonaroused or idle waking states, such as sitting still, drinking, eating, grooming, and non-rapid eye movement (REM) sleep.[4] Sharp wave ripples are produced by tens of thousands of neurons in the hippocampus, subicular complex, and entorhinal cortex, firing over just 30–100 ms at times when the brain is disengaged from the environment.[5] Any more synchrony than that could trigger an epileptic seizure. This neuronal cooperation is what fascinated me most. Synchrony can bring about enormous power surges at no cost. If a symphony is played one instrument at a time, the cooperative action of many instruments can never be identified. Similarly, when we listen to the spike

---

4. Sharp wave ripples are the main population pattern seen in the isolated and transplanted hippocampus (Buzsáki et al., 1987). Conversely, when septo-hippocampal cholinergic neurons are activated by optogenetic methods, the synaptically released acetylcholine prevents the occurrence of sharp wave ripples (Vandecasteele et al., 2014).

5. Occasionally, multiple ripples occur in a sequence or fuse together and reflect exceptionally long neuronal assembly sequences (up to several hundred milliseconds). These long sequences most often occur in novel environments and memory-demanding conditions.

sounds of individual neurons (over a speaker that allows researchers to monitor the signals coming from their electrodes), we cannot detect a sharp wave ripple, no matter how long we listen to neurons one by one. But when many neurons emit action potentials together, their collective action sounds like a jet airliner. I wondered what the function of this collective action might be.

In the 1980s, many laboratories were searching for natural conditions that could mimic artificially induced synaptic changes in neurons in a dish, used as a model for memory.[6] And there it was. The sharp wave ripple shared numerous characteristics of the electric pulse patterns that brought about long-term changes in synaptic plasticity, such as duration, high frequency, and the super powerful synchrony required for a teaching signal.[7] Sharp wave ripples are found in every mammalian brain investigated to date in the same form and shape—and may serve identical functions (Figure 8.1). They form discrete quanta of spike sequence packages. As early as my doctoral dissertation, I made the bold claim that hippocampal sharp wave patterns represent the sought-after biological marker for information compression and plasticity. Then I spent the next three decades trying to understand their secrets. During these years, the views about the sharp wave ripple have enriched considerably. This peculiar and unique brain pattern is viewed today as a subconscious mechanism to explore the organism's options, searching for stored items of the past in the disengaged brain in order to extrapolate and predict possible future outcomes. It embodies a brain mechanism that compresses the discrete concepts of past and future into a continuous stream.

EXCITATORY GAIN DURING SHARP WAVE RIPPLES FACILITATES SPIKE TRANSMISSION

The sharp wave ripple is a cascade of two events, the sharp wave and the ripple, which occurs relatively simultaneously but at different locations (Figure 8.2). Sharp waves are large-amplitude, negative-polarity deflections in the apical

---

6. *Synaptic plasticity* is often used synonymously with *memory*. However, alternation of circuits could be also brought about by changing the intrinsic properties of neurons and axon calibers without affecting synaptic number or strength.

7. At an early meeting on the physiology of the hippocampus, Tim Bliss, the discover of long-term potentiation, encouraged me after my talk in front of a skeptical audience: "My point about sharp wave was that it is a naturally occurring phenomenon which satisfies the two conditions, high frequency and high strength" (p. 395 in Buzsáki and Vanderwolf, 1985). I have recently written a monograph on sharp wave ripples for my specialist neuroscientist colleagues. Mechanisms, implications for memory, planning, disease, historical details, and the pertinent references can be found there (Buzsáki, 2015).

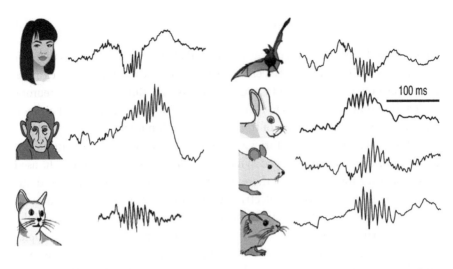

Figure 8.1. Preservation of sharp wave ripples in mammals. Illustrative traces recorded from various species.
Reproduced from Buzsaki et al. (2013).

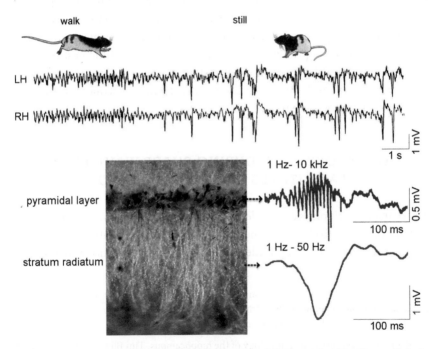

Figure 8.2. Behavior-dependence of hippocampal local field activity. *Top*: Extracellularly recorded events from symmetric locations in the dendritic layers of the left (LH) and right (RH) dorsal hippocampus during locomotion–immobility transition. Note regular theta waves during locomotion and large amplitude, bilaterally synchronous negative waves (sharp waves, SPW) during immobility. *Bottom*: A single sharp wave recorded from the dendritic layer (*red*) and simultaneously occurring ripple event recorded from the pyramidal layer.

dendritic layer of CA1 pyramidal neurons, where most of the incoming synaptic contacts are made by the axons of their CA3 partners. The sharp wave is a telltale of strong firing coordination among many neurons in the upstream CA2 and CA3 regions. Measuring the sharp wave amplitude is, therefore, a convenient way to quantify the magnitude of this synchrony. The self-organized population activity underlying sharp waves emerges from the strong excitatory influence of the extensive recurrent collaterals of the CA3 pyramidal neurons.[8] In turn, the strong depolarization of CA1 pyramidal neurons and interneurons brought about by the synchronous activity of the CA3 neurons produces a tug of war between excitatory and inhibitory neurons, resulting in a short-lived "ripple," a fast oscillatory synchronization (140–200/s) of these two competing populations (Figure 8.2).[9] The pyramidal neurons of the CA1 region provide the only hippocampal output to cortical targets, so this activity must have functional significance.[10] We just have to figure out what it is.

The most interesting part is the neuronal dynamic that evolves around these events. Nearly all neuron types participate in the ripple events. Some inhibitory interneuron types remain neutral or decrease their activity after a transient increase, but the majority robustly increase their firing. The overall result is that while both inhibition and excitation increase during sharp wave ripples, excitation makes a two- to threefold gain over inhibition.[11] This is the highest gain

---

8. Prior to the full organization of the large recurrent excitatory CA3 system, sharp wave bursts are already present in the postnatal brain. However, nearly every event is driven by some kind of movement from the body (Leinekugel et al., 2002). The proprioceptive reafferentation drives neocortical–entorhinal circuits which, in turn, trigger hippocampal sharp wave bursts. By the second week of life, the hippocampus disengages from its dependence on body-derived signals and generates self-organized sharp waves (Valeeva et al., 2018). Ripples in CA1 emerge by the third week of postnatal life, coinciding with the sequential firing of hippocampal neurons (Buhl and Buzsáki 2005).

9. Buzsáki et al. (1992). Ripple patterns occur in many physical systems when a large number of random, independent events are combined. The most famous such event is the Big Bang, where many random events are assumed to have generated stochastic gravitational waves. These gravitational waves, also called ripples, are approximately 250 Hz and last for 100 ms, reflecting two massive bodies spiraling into each other (Cho, 2016).

10. This is not quite correct, though, as a small fraction of hippocampal inhibitory interneurons also send their axons to several other extrahippocampal regions (Jinno et al., 2007). We called this special type of cell a "long-range" interneuron (Buzsáki et al., 2004) in reference to the "short-cut" connections of small-world network architecture (Watts and Strogatz, 1998).

11. Read more about gain in Chapter 11. The large gain may derive, paradoxically, from the highly reliable spike transfer between pyramidal cells and interneurons (Csicsvari et al., 1998; English et al., 2017). Because pyramidal neurons fire highly synchronously in a given ripple wave and because even a single pyramidal neuron spike can discharge fast spiking interneurons, spiking of many additional pyramidal cells within the spike refractory period of the interneuron

of excitation in the hippocampal system, making the sharp wave event especially suitable to transmit information from the hippocampus to the neocortex (see Chapter 11). But what kind of neuronal information is transmitted? After all, sharp wave ripples are self-organized events that emerge in the "offline" or idling states of the brain. It seems odd for the brain to be so active at times when no perceptual inputs bombard its sensors.

ORGANIZED SPIKE SEQUENCES OF SHARP WAVE RIPPLES

What makes hippocampal sharp wave ripples so special, in addition to their slender oscillatory patterns and strong excitatory gain, is the way that neurons emit their spikes relative to each other. They can play palindrome, a sequence of letters that spell the same words whether read forward or backward. "Was it a car or a cat I saw?" This palindromic role makes sharp wave ripples suspicious of some coding mechanism because similar tricks have been exploited in molecular evolution by retroviruses. A retrovirus can enter the cell nucleus and patch its own DNA into the cell's DNA, adding new information to an existing knowledge base, which then becomes permanent. A palindromic repeat of DNA coding would look like GTTCCTAATGTA–ATGTAATCCTTG.[12] Sharp wave ripples can also do such things, and the ability to switch the direction of the trajectory must also have special importance for neuronal computation.

Recall from Chapter 7 that, during exploration, neuronal sequences occur at both the theta time scale and the time scale of behavior, and that these sequences are specific to the animal's path. For example, in Figure 8.3 place cells 1–13 are active in different segments of the maze track. At any given time on the track, only a small fraction of the neurons fire together within the theta waves. However, at the beginning and the end of the track, where the rat is sitting immobile, virtually all neurons fire together. When we zoom in and examine the firing patterns of these same neurons at an expanded time scale, we find that they also show a sequence. The sequences occur either in the same order as the sequential firing on the track or in the reversed order, but at an accelerated pace in both cases, and they coincide with fast ripple oscillations in

---

(1–2 ms) cannot recruit more spikes and more inhibition. This refractoriness may be the key mechanisms for the large excitatory gain during sharp wave ripples (Csicsvari et al., 1999).

12. Work on the special coding abilities of retroviruses paved the way to today's CRISPR magic of molecular biology (Quammen, 2108). CRISPR, which stands for *clustered regularly interspaced short palindromic repeats*, sounds like spike patterns of sharp wave ripples.

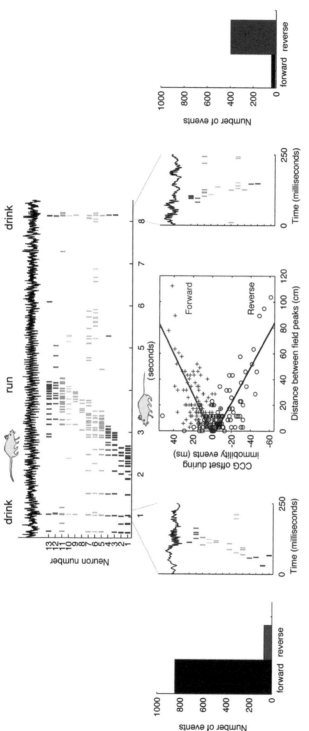

Figure 8.3. Forward and reverse replay of place cell sequences. *Top*: Spike trains of thirteen neurons before, during, and after a single lap (*top continuous tace*: CA1 local field potential). The animal runs from the left to the right ends of the track and collects water reward at both ends. The insets below magnify 250-msec sections of the spike train, depicting synchronous forward and reverse replay, respectively. Bar graphs (bottom left and right) show that 95% of forward events took place before the run, whereas 85% of reverse replay events occurred after the run. *Middle*: Temporal offsets (CCG) between spikes of neuron pairs during the synchronous forward (+) and reverse (o) replay events are correlated with distance representations between place field peaks on the track. From Diba and Buzsaki (2007).

the field recordings. It is as if the brain played a movie back for us on fast forward or reverse that recapitulated the motion of the rat in the maze. Without further analyses, we can only wonder what functions these accelerated replay events may serve.

The majority of fast-forward events occur at the start of the track, and their incidence steadily rises until the run begins as if the hippocampus rehearses or plans the future travel path. In contrast, reversed replay events dominate at the other end of the track, after the run as if the brain recapitulated the path in a reversed order. The preponderance of reverse trajectories can be manipulated by the magnitude of the reward the animal obtains at the end of the journey.[13] In short, the sharp wave ripple is a compression of event sequences that have either already occurred or will evolve in the future. Beyond faithfully preserving sequential order, both forward and reversed events during sharp wave ripples also contain information about the distances between place fields on the track (Figure 8.3). Recall that this distance-to-time conversion is reminiscent of the theta time scale compression of place cell distances discussed in Chapter 7, with two notable differences. First, the time compression factor is typically larger during sharp wave ripples than during theta oscillations.[14] Second, during theta oscillations, only forward sequences are observed, whereas the neuronal movie of the animal's journey in the maze can be seen in both forward and reverse order during sharp wave ripples. Importantly, the fast replay events during ripples are not assisted by external signals or body-derived cues. Compressed

---

13. Typically, many more neurons participate in sharp wave ripples than the number of place cells in a maze. This makes it complicated to compare real-time sequences with replay sequences. Foster and Wilson (2006) introduced a novel and effective method by regarding place cell sequences in the maze as a template and searching for a similar rank order of spikes during sharp wave ripples. They observed only reversed replay events on the track, in support of a previous prediction (Buzsáki, 1989), and contrasted them with the assumption that only forward sequences occur during sleep. According to Foster and Wilson, reversed "replay is suggestive of a role in the evaluation of event sequences in the manner of reinforcement learning models." I expressed a similar view: during sharp wave ripples, "recurrent collateral excitation in the CA3 region spreads hierarchically: the most excitable cells fire first followed by less excitable ones, that is in the reversed order to that in which they were potentiated during exploration" (Buzsáki, 1989). Reverse replay can be enhanced by the magnitude of the reward (Ambrose et al., 2016). Compressed replay of learned information has also been observed in humans (Kurth-Nelson et al., 2016).

14. Distance compression is approximately 30% stronger during sharp wave ripples compared to theta waves, commensurate with the stronger network drive during ripples (Diba and Buzsáki, 2007; Drieu et al., 2018). The speed of replay during sharp wave ripples corresponds to 8 m/s, which is 10–20 times faster than the average speed at which the rat runs or walks through those place fields (Davidson et al., 2009). These time scales are determined by internal brain states rather than by sensory inputs.

replay is like the scaled version of a song or speech; it is identified as the same sequential pattern, just played faster.

Although sharp wave ripples are self-organized events, their spike content is related to the firing patterns evolving in real time in the behaving animal. We can conclude that forward events before the run may serve to plan the upcoming locomotion trajectories. Conversely, reversed replay events at the end of the run may recapitulate the trajectory in the maze, so the animal remembers that choice. Thus, the forward and reversed sequences of spontaneously induced sharp wave ripples may serve both prospective and retrospective roles (i.e., prediction and postdiction), just as the Queen noted in Carroll's masterpiece. Based on the discussion in Chapter 5, we can speculate that internalization of travel trajectories can be used both to repeat snippets of past experience and to imagine or plan actions. After all, the only biological utility of memory is to help predict the future.

CONSOLIDATION OF LEARNED EXPERIENCE DURING NON-REM SLEEP

Manipulations that affect synaptic strength distributions in hippocampal networks can have a profound impact on the neuronal membership of spontaneous network events such as sharp wave ripples. Artificially altering synaptic strengths between CA3–CA3 or CA3–CA1 neurons can alter the waveform of sharp wave ripples, make new place fields, or make place fields disappear. These findings suggest that the neuronal composition of sharp wave ripples can be biased by the recent past of the hippocampal network. Once a new sharp wave–associated neuronal sequence emerges, the sequence repeats spontaneously many times after the initial condition that induced it. In this way, firing patterns of the entorhinal inputs can modify intrahippocampal synaptic connections, and these modifications are reflected in spontaneous, self-organized population events.[15] Let's pause here briefly to appreciate what good service the sequence of neuronal events can do for the hippocampus and its targets: the synaptic-cellular modification of neurons brought about during learning can be read out later from the spontaneously emerging neuronal activity. This sequential

---

15. Manipulations that affect synaptic strength distributions in hippocampal networks have a profound impact on the neuronal membership of spontaneous network events. Electric stimulation of different sets of afferent entorhinal axons can induce a unique spatial-temporal constellation of evoked responses in the hippocampus. After tetanic stimulation (trains that induce long-term potentiation of synapses), the modified connections are reflected in the spontaneous, self-organized population events, so that the spontaneously emerging events resemble the evoked responses (Buzsáki, 1989).

mechanism may be regarded as the physiological foundation for a two-stage model of memory formation.

The hypothesized consolidation model works in two steps. During learning, associated with the theta rhythm, afferent activity from the neocortex–entorhinal cortex pathway brings about a transient change of synaptic strengths in the CA3 hippocampal regions and within CA3–CA1 connections, where the learned information is temporarily held. This first stage is followed by a consolidation phase during which the same neurons and synapses that were active in the learning process are repeatedly reactivated during sharp wave ripples. Because hundreds to thousands of sharp wave ripples occur during non-REM sleep, I hypothesized that they are critical for consolidating labile memories into a more permanent form. Sharp wave ripples repeat snippets of the previously learned information over and over during sleep, after the experience. This protracted process fits nicely with research on episodic memories. They are not instantaneous veridical pictures of a situation, like a picture taken by a digital camera. Instead, episodic memories resemble Polaroid pictures, in which an initially faint scene gradually becomes crispier. During the process, the picture is vulnerable to modification, as are our memories.

The first convincing evidence to show replay of learned experience came from a groundbreaking study by Matt Wilson and Bruce McNaughton at the University of Arizona at Tucson.[16] They examined the co-occurrence of spikes of hippocampal CA1 pyramidal neuron pairs in 100-ms windows, which they called "coactivations," during sleep both before and after behavioral experience on a familiar open-field foraging task. As expected, the activity of neurons with overlapping place fields showed a highly positive correlation during exploration, whereas pairs with nonoverlapping fields showed no reliable temporal correlations. Critically, neuron pairs with overlapping place fields and strong correlations during foraging continued to fire together more strongly during sleep afterward, although they had been less correlated during sleep before foraging. Numerous experiments in many laboratories have supported and

---

16. Wilson and McNaughton (1994); Kudrimoti et al. (1999); Skaggs and McNaughton (1996); Hirase et al. (2001). By and large, these physiological findings support previous ideas, based on experiments with hippocampus-damaged patients, that the initially labile memory trace becomes consolidated over time. However, while the psychological experiments promoted the view that the initially hippocampus-dependent memories are transferred to the neocortex and become independent of the hippocampus (Scoville and Milner, 1957; Squire, 1992a, 1992b; Squire et al. (1975); Ferbinteanu et al., 2006; Squire and Alvarez, 1995), an alternative view is that consolidation corresponds to preservation of a crude "index" of neocortical memories in the hippocampus (Teyler and DiScenna, 1986; Nadel and Moscovich, 1997). In this way, hippocampal sequences can sweep through representations of detailed neocortical semantic information and concatenate them into a virtual offline episode (Buzsáki, 2015).

expanded these findings. One important extension showed that those neuronal trajectories (similar to those seen in Figure 8.3) reactivated many times during the sharp wave ripples of non-REM sleep are an accelerated version of place cell sequences during exploration.[17] Thus, replay of learned information has been demonstrated.[18]

A Revised Version of the Two-Stage Model

These early speculations and experiments supported the idea that, during learning, large-scale synaptic modifications occur in the hippocampus, and the transiently modified synapses get stronger or at least persist (i.e., consolidate) during offline states of the brain. In turn, learning-induced synaptic changes are believed to entirely determine how neurons organize themselves into sequences during sharp wave ripples. In a sense, this description of events can be interpreted to mean that novel experience builds up a new neuronal trajectory from scratch, in line with the outside-in view. Yet the same findings and the experiments we are going to discuss herein are also compatible with the alternative, inside-out view that we discussed in Chapter 7: namely, that

---

17. Nádasdy et al. (1999); Lee and Wilson (2002) introduced a behavioral event-based template-matching method where the sequences of smoothed place fields of hippocampal pyramidal cells during exploration defined the template (a "neuronal word"). Higher order relationships are important for making "associative inference." For example, if A is related to B, and B is related to C, then A is indirectly related to C (in a higher ordered manner, mediated by the hidden variable B). Such inference depends on the integrity of the hippocampus (Dusek and Eichenbaum, 1997; Schacter and Addis, 2007; Schacter et al., 2007). In all these early experiments, reactivation of waking sequences was in the forward direction, largely because these were the only templates examined. When researchers looked for reverse sequences, they were also found during sleep (Wikenheiser and Redish, 2013), thus demonstrating that reverse replay is not specific to sharp wave ripples during waking states (Foster and Wilson, 2006).

18. These ideas did not remain unchallenged. Lubenov and Siapas (2008) suggested that because replaying waking experience involves a small fraction of neurons, neuronal firing during sharp wave ripples reflects a transition state between randomness and synchrony. They suggest that the synchronous bursts can lead to long-term depression of synapses and the selective erasure of information from hippocampal circuits as memories are transferred to neocortical areas. However, memories do not cease to exist in the hippocampus (Nadel and Moscovitch, 1997). The potential role of sharp wave ripples in sleep homeostasis is discussed in Grosmark et al. (2012). Foster (2017) questions the role of waking sharp wave ripples in memory and, instead, emphasizes their prospective role. However, this view leaves the coexistence of both forward and reverse replays during both sleep and waking unexplained. Norimoto et al. (2018) suggested that sharp wave ripples protect experience-active neurons while downscale memory-irrelevant neuronal activity.

the hippocampus can generate myriads of sequences internally even without experience and that various aspects of learning can be linked to some of these preexisting sequence patterns.

According to this revised view, the first stage of the two-stage model is a selection process in which one or a few of the existing sequences or their concatenated variants are matched to the experience. Because every novel situation has elements of familiarity (Chapters 12 and 13), the brain can select a neuronal trajectory that best describes the current situation. The brain always guesses; that is, it attempts to deal with any event even in the most unexpected situation by matching it with a particular existing neuronal pattern. If needed, new neurons can be added (or deleted) to this selected backbone to refine the sequence adapted to the novel aspects of the situation. In turn, these new neurons are incorporated (or eliminated) in the selected sequence during sharp wave ripples (stage 2). While the selection-incorporation version of the two-stage model appears to differ from the previous consolidation version only in small details, it is more economical. Only small changes are expected to occur in the hippocampal and neocortical networks because a large part of learning is a selection process—matching of the experience to preexisting patterns (Chapter 7) rather than a new construction.[19] To develop this argument further, we need to learn important details about the statistics of neuronal dynamics, which we discuss in Chapter 12, and we revisit the selection hypothesis in Chapter 13.

Furthermore, the experiments on sharp wave ripples discussed thus far only demonstrate that reactivated neuronal patterns of waking experience during sharp wave ripples may do something useful in the hippocampus, but not beyond. If the occurrence of sharp wave ripple–associated replay is irregular, rather than predictable, then how can the neocortex read the messages reliably without being on the alert all the time? As discussed earlier, the neocortex reboots itself from complete silence of neuronal activity ("down" states) hundreds to thousands of times during non-REM sleep. If replay from the hippocampus occurs during neocortical silence, it might fall on deaf ears unless hippocampal and cortical activities are somehow coordinated. Fortunately, they are.

---

19. Admittedly, these dense paragraphs are based largely on speculation, derived mainly from theoretical considerations and, to a smaller extent, on recent experiments (Luczak et al., 2009; Dragoi and Tonegawa, 2011, 2013a, 2013b; Mizuseki and Buzsáki, 2013; Buzsáki and Mizuseki, 2014; Grosmark and Buzsáki, 2016; Liu et al., 2018). I expand on these ideas in Chapter 13.

Coupling Sharp Wave Ripples with Neocortical Events

The two major neocortical events during non-REM sleep, slow oscillations and sleep spindles, have a nuanced relationship with hippocampal sharp wave ripples. The direction of communication may depend on sleep state, target structure, and previous experience. The strongly synchronous hippocampal output during sharp wave ripples can latch prefrontal circuits to the down state.[20] In the return direction, when neocortical spindles successfully invade entorhinal–hippocampal networks, they can time sharp wave ripples. These triggered events are likely important because the punctate hippocampal output can further enhance the activity of the same still-active neurons that initiated the neocortical–hippocampal dialogue.[21] Although the hippocampus receives disparate inputs during successive cortical events, this process allows it to find or index the active cortical patches. Thus, the interplay of hippocampal sharp wave ripples and neocortical ripples creates temporal windows that facilitate the exchange of information.

These correlational experiments have been supplemented by experimental manipulation of sharp wave ripple content. In one clever experiment, the spiking activity of place cells during sleep was coupled with rewarding brain stimulation. Stimulation of many sites in the brain feels good and could be used experimentally as a convenient substitute for reward. Stimulation of the chosen rewarding brain site was triggered by the spontaneous emission of each spike of a place cell during sleep ripples. When the mouse awoke, it made a beeline for the location represented by that place cell as if searching for the reward. The stimulation formed a new relationship between reward

---

20. Peyrache et al., 2011. Interictal spikes in the epileptic hippocampus may represent super-strong, pathological versions of sharp wave ripples and consistently induce down–up transitions and spindles in the prefrontal cortex even in the waking brain. The abnormal epileptic spikes broadcast "nonsense" messages to the neocortex and induce spindles at times when the prefrontal areas are not in the "ready" state. Thus, all three suspected patterns of memory consolidation—hippocampal ripples, neocortical slow oscillations, and spindles (Diekelmann and Born, 2010)—are aberrant in the epileptic brain and can be responsible for memory/cognitive impairment in chronic epileptic patients (Gelinas et al., 2016).

21. Siapas and Wilson (1998); Sirota et al. (2003); Isomura et al. (2006); Mölle et al. (2009); Johnson et al. (2010); Sullivan et al. (2014). A few experiments examined joint replay of waking experience in the hippocampus and neocortical structures (Ji and Wilson, 2007; Peyrache et al., 2011; O'Neill et al., 2017; Khodagholy et al., 2017). Replay in the neocortex also occurs in brief bursts, lasting 100 ms or so, typically coincident with the up-state of slow oscillation (Takehara-Nishiuchi and McNaughton, 2008; Johnson et al., 2010; Rothschild et al., 2017). Most cortical assembly reactivation events occur in concert with hippocampal sharp wave ripples.

and spatial location.[22] To complement the role of replay in memory consolidation, another experiment demonstrated that selectively abolishing sharp wave ripples during sleep following learning in a maze impairs spatial memory performance.[23]

The two-stage model, using sharp wave ripples as a key mechanism for replay, explains why episodic information can be remembered even after a single experience. Instead of repeating the learning process behaviorally, the endogenously organized sharp wave ripples, perhaps in partnership with thalamocortical spindles and slow oscillations, do the heavy lifting. They repeat various aspects of recent experience many times at a compressed time scale and in a strongly synchronized manner so that the neocortical targets take notice of the powerful hippocampal outputs.[24] In short, sharp wave ripples may allow the brain to solidify recently learned material. But they do more than that.

CONSTRUCTIVE FUNCTIONS OF SHARP WAVE RIPPLES

The conclusion that sharp wave ripples strengthen neuronal patterns that occurred before "now" is based on the conceptual distinction of the past and present. However, recent thinking questions the validity of this traditional distinction (see Chapter 10). In line with the idea about the contiguity of postdiction and prediction, experiments paint a richer picture for sharp wave ripples, especially the waking variant, beyond consolidating memories. One interesting

---

22. de Lavilléon et al. (2015).

23. Girardeau et al. (2009). The results of Ego-Stengel and Wilson's (2010) studies in a "wagon wheel" maze further support the contribution of the sharp wave ripples of post-learning sleep to spatial memory. Jadhav et al. (2012) interfered with sharp wave ripples in the waking, rather than sleeping, rat. Truncation of waking sharp wave ripples during task performance increased choice errors in spatial working memory but not in reference components. In addition, selective truncation of sharp waves that occurred in the vicinity of reward destabilized the spatial map (Roux et al., 2017). Finally, increasing the coupling between sharp wave ripples and slow oscillations by electrically inducing down states of cortical activity or spindles by optogenetic means increased memory performance (Maingret et al., 2016; Latchoumane et al., 2017).

24. Phrased differently, a single experience (incidental learning) can be remembered because "the SPW event would in essence compress time and allow temporally distinct neuronal representations to be combined into a coherent whole" and sharp wave ripples repeatedly replay or "reactivate the same subset of neurons in CA3 and CA1 precisely determined by the recent past of the neural network" (Buzsáki et al., 1994; p. 168). Buzsáki (1989, 1996, 1998, 2015); Lee and Wilson (2002); O'Neill et al. (2006, 2008). Dupret et al. (2010); van de Ven et al. (2016).

study used a long maze that was twisted around to fit the recording room. The duration of replay sequences appeared to depict behavioral trajectories at a constant speed; that is, with duration proportional to distance depicted. This explains why the researchers found up to 700-ms sequences, concatenated from multiple ripple events, that wended their way around the bendy track without jumping between physically close (but uncrossable) corridors. For comparison, smaller mazes typically evoke 50–100 ms events. However, most ripple events were shorter and jumped between discontinuous corridors along the length of the maze track. In other words, replay events do not necessarily represent temporally compressed veridical episodes of travel. Some sequences instead represent snippets and fragments of traversable distances, as if evaluating alternative paths while others play out sequences corresponding to trajectories the animal has never traversed, starting and ending at positions different from that of the animal.[25] Thus, similar to the egocentric-allocentric divisions of navigation, simulated navigation of fictive paths during sharp wave ripples can be either egocentric (moving to or away from self) or allocentric (ordering events at other places). These experiments, together with the forward replay of upcoming place cell sequences, suggest a prospective or constructive role for sharp wave ripples.

In most early experiments, replay sequences were compared to a template based on the ordered sequence of place cells on the maze tracks. In essence, it is a targeted search for a needle in a haystack. However, this method comes with a cost: it throws out the many other neuronal sequences that occur during sharp wave ripples and so cannot examine their potential role. In a T maze (which has a central arm and two choice arms), only two templates are examined during sharp wave ripples: left and right turns and their corresponding place cell sequences. In such a simple memory task, neuronal sequences related to sharp wave ripples may correspond to either left or right turns, as if the hippocampus contemplates the proper choice.[26] However, during natural foraging, many choices are available at each position of travel. Forward replay may serve to evaluate alternative scenarios and calculate optimal choices.

---

25. Davidson et al. (2009); Wu and Foster (2014); Foster (2017); Karlsson and Frank (2009); Gupta et al. (2010); Liu et al. (2018). Theta sequences can also serve such a "contemplating" function, especially at maze choice points, where sequentially active neuronal assemblies can "look ahead" toward either the right or left arms of a T maze (Johnson and Redish, 2007; Redish, 2016).

26. Singer et al. (2013). In this task, the neuronal trajectories reactivated during sharp wave ripples preceding correct trials were biased toward representing sequences that proceeded away from the animal's current location.

A laboratory equivalent of foraging is a search for reward in a large open arena with numerous wells. In this task, the rat is trained to find food hidden in a random well and then return to home base for another reward. If the home base is varied from one day to the next, the rat learns to locate the new home base in just a few trials, after which it returns to it from any food well in a few seconds. The main finding of such an experiment was that before the animal returned to the home base from any location in the arena, its sharp wave ripple–associated sequences often closely matched the place cell sequence between the current location and the home base. The forward neuronal sequences in this two-dimensional exploration task better matched the rat's future path than its past path.

These findings, therefore, demonstrate that previously acquired memories can be flexibly manipulated internally to evaluate alternative future paths and plan optimal behavioral trajectories to a remembered goal, even when those trajectories have never been taken before.[27] Other experiments suggest that neuronal sequences are often "stitched together" from fragmented representations involving multiple concatenated sharp wave ripples containing both forward and reversed events.

Before closing the discussion on the constructive aspects of sharp wave ripples, it is important to emphasize that replay does faithfully mimic the true sequential order of the experienced past. Instead, recent experience is embedded into an existing network dynamic. A large part of the dynamic is preformed and guided by the brain's genetically determined construction and existing knowledge base (Chapter 12). Thus, replay sequences reflect more the brain's "beliefs" than mirroring the order of world events. We can speculate that this constructive role of sharp wave ripples can assist in weaving together paths and events that relate to each other even if they have not been experienced physically.

---

27. Pfeiffer and Foster (2013); Papale et al. (2016); Pfeiffer (2017); Pezullo et al. (2017); Liu et al. (2018); Xu et al. (2018). These experiments confirmed previous theoretical predictions (Schmajuk and Thieme, 1992). The hippocampus graph model of Muller et al. (1996) suggested that distances in a two-dimensional map are reflected by the reciprocal value of synaptic strengths between place cells in the CA3 recurrent system. Once the main landmarks of an environment are mapped onto hippocampal assembly representations, all possible combinations (e.g., the route from any location to the home base) can be computed even if those routes were never traversed by the animal. It has been postulated that sharp wave ripples are exploited for this purpose (Muller et al., 1996; Samsonovich and Ascoli, 2005), and possibly the same mechanisms can be used to generate novel solutions to non-navigational problems (Buzsáki and Moser, 2013).

Are Animals Aware of the Content of Sharp Wave Ripples?

Single sharp wave ripples may be too short to induce a conscious experience, which requires as long as half a second.[28] Yet stitching together neuronal trajectories from sharp wave ripples can be conceived of as a mechanism for preconscious contemplation. When you let your mind wander off the leash for a while, that may be a better way to reach the correct solution than trying to concentrate. We all have such experiences: a while after we give up on an effortful search for a name, it may just pop up. Sequences predicting future actions or solutions can be constructed by chaining together clusters of sharp wave ripples so that spiking activity in each sharp wave ripple in the cluster corresponds to different potential segments of a potential route.

Overall, the selection of possible forward paths during sharp wave ripples can be conceived as an internalized vicarious trial-and-error process (Chapter 5) that flexibly "imagines" real or fictive alternatives to select an optimal path or construct novel inferences without the need for movement-based exploration.[29] Because neuronal assembly sequences in the hippocampus serve not only spatial but numerous nonspatial functions, the compressing-mixing functions of sharp wave ripples can also support covert simulation of many real-world scenarios). Recently acquired and preexisting knowledge can be combined during sharp wave ripple replay to influence choices, to plan actions, and, potentially, to facilitate generalization, abstraction, and creative thought. In support of this outlined scenario, failure to replay the appropriate trajectory sequence before a choice can predict behavioral errors.[30]

---

28. Libet (2005) called it "mind time." This time may be needed to mobilize large resources in multiple brain structures and allow them to communicate effectively.

29. Schmajuk and Thieme (1992), following Tolman (1932), discussed the idea that head-scanning movements at maze choice points may correspond to "vicarious trial-and-error," contemplative behavior to select optimal travel paths (Tolman, 1932). Artificial intelligence researchers have also been inspired by the two-state model of memory. The celebrated deep Q-network (DQN) can outperform humans in numerous video games by learning to transform a vector of image pixels into a policy for selecting actions (e.g., joystick movements). The network stores a subset of the training data in an instance-based way and then "replays" it offline over and over, learning anew from successes and failures that occurred in the past. This replay is critical to maximizing data efficiency (Hassabis et al., 2017).

30. Ólafsdóttir et al. (2017). These authors also demonstrated that forward and reverse replays of maze trajectories in sharp wave ripples occur within 5 s before the animal starts the journey or arrives at the goal location. If the animal pauses for longer times at these locations, then the hippocampus disengages from the current situation and the subsequently emerging sharp wave ripples replay more remote experiences, mixed with the current one. In these late-occurring events, the deep-layer neurons of the entorhinal cortex are also recruited. In

To translate plans into actions, the hippocampus needs to shift from a sharp wave state to a theta oscillation state, during which the subconsciously "primed" circuits can be deployed more efficiently, with goal-contingent valuation of the simulated outcomes occurring in areas downstream of the hippocampal system, likely including the lateral septum–hypothalamus action path or the prefrontal cortex, orbitofrontal cortex, and basal ganglia.[31]

In the preceding four chapters and especially in Chapter 5, I argued that multiple episodes of experience or perhaps multiple sharp wave ripples are needed to generate abstract, semantic information because only multiple experience allows the spatial-temporal context of episodes to be stripped off, leaving behind explicit information. In the next chapter, I explain that acquisition of semantic knowledge can be accelerated by externalizing brain functions and that such externalized functions in the form of artifacts, language, and other forms of communication. Such externalization has been key to the exponential growth of humankind's knowledge.

SUMMARY

The sharp wave ripple in the hippocampus is the most synchronous population event in the mammalian brain. Ripple events can play palindrome, so that place cell sequences prior to choosing a particular path are replayed in the same order as during crossing the path, only much faster. At the end of the travel path, the same sequence is replayed but now backwards, as if the brain recapitulated a virtual reversed run. Thus, ripple sequence events represent mental travel both into the future and the past.

There are at least two functions of the forward and reversed cell assembly sequences. First, such sequences reach back to the past to replay snippets of waking experience at times when the brain is disengaged from the outside world. This process may consolidate episodic memories and stitch together discontiguous experiences, thereby giving rise to creative thoughts. Second,

---

contrast, superficial-layer entorhinal neurons, the inputs to the hippocampus, appear to replay sequences largely independent from those in the hippocampus (O'Neill et al., 2017).

31. Priming is an example of what is often referred to as implicit memory (Graf and Schacter, 1985) or "level of processing" (Craik and Tulving, 1975). Redish (2016) discusses extensively how the hippocampal system can cooperate with the networks of the basal ganglia to perform planned actions. Tingley and Buzsáki (2018) demonstrated how the spatial map of the hippocampus is converted into a firing rate-independent, theta oscillation-dependent phase code in the lateral septum, an important conduit from the hippocampal system to the motivational and movement controlling circuits of the brain.

the forward sequences of neuronal assemblies can be viewed as internalized vicarious trial-and-error mechanisms, which can assist with subconscious optimization of future plans. Because the same neuronal substrate can perform both retrospective and prospective operations, the traditional separation of postdiction (i.e., memory) from prediction (i.e., planning) needs to be readdressed (see Chapter 10).

Studying the mechanisms and behavioral relevance of sharp wave ripples raises a deeper problem, which is the "causal" direction (Chapter 2) of the relationship between behavior and replay sequences. Is every internally generated sequence imposed on hippocampal circuits by experience (i.e., from outside in)? In that case, the number of sequences should scale with the amount of experience. Alternatively, the extensive CA3 recurrent system can presumably self-generate very large numbers of sequences independent of any prior experience. Under this second scenario, the role of experience is to pick and choose among preexisting sequences, concatenate them, and modify them a bit, but without the danger of destabilizing the balanced dynamic of the brain.

9

Enhancing Brain Performance by Externalizing Thought

Ars longa vita brevis.

—Attributed to Hippocrates

Science works for evil when its effect is to provide toys for the rich, and works for good when its effect is to provide necessities for the poor.

—Freeman Dyson[1]

Technology is a mode of revealing. [It] makes the demand on us to think in another way.

—Martin Heidegger[2]

In the textile industry of nineteenth-century Britain, the invention of the "knitting frame" and "shearing frame" made the cloth manufacturing process simpler so that workers with limited training could be employed. At the same time, these innovations threatened the weavers and artisans in Nottinghamshire, Yorkshire, and Lancashire who had spent years learning their craft. These highly skilled people saw the new machines as the cause of their decreasing income and lost jobs, so they picked up sledgehammers, invaded the mills during the night, and wrecked the machines. Ned Ludd, a young apprentice, was rumored to be the first attacker, and his followers called themselves Luddites. Between 1811 and 1816, Luddites raided and burned down dozens of mills and damaged machinery in hundreds of them. They also lobbied the

---

1. Dyson (1997).
2. Heidegger (1977).

British Parliament to restrict automated looms and knitting frames. However, the government sided with the mill owners, gunned down the rebels, and hanged the leaders or transported them to the colonies.[3] The movement was crushed, but the word "Luddite" survived. Today, it is a blanket term used to describe technophobes who feel that industrialization, computerization, or new technologies in general do not always bring happiness. The Luddite movement can be viewed as a leaderless resistance to frightening technologies and consumerism that have damaged the livelihoods of part of the population.

Although the Luddites wrecked machines, likely their real goal was to dissuade the owners, the aggressive new class of manufacturers who drove the Industrial Revolution, from using cheap labor. The Luddites clearly knew who owned the technology that concerned them. Today, we are no longer sure who controls technology and who drives the need for constant innovation. An even more complex question is how much benefit we gain from it.[4] Since President Franklin Roosevelt signed the Wages and Hours Bill in 1938, the five-day, forty-hour workweek has been the norm for many workers in the United States, followed by most other countries.[5] Can we reduce our workload further because technological inventions allow it to happen and devote more time for recreation? Industries and offices can be shut off, and even season-dependent agricultural work can be operated by shifting groups of people. Therefore, in principle, we can determine the number of hours we work.

If bad decisions are human-made, it should be possible to unmake them. Several grassroots movements have tried to reduce the workload of humankind. Reasonable calculations show that current technological developments would allow people in industrialized nations to work only three or four days a week while maintaining the living standards of the 1960s. Such a reduction in

---

3. Sale (1995); Binfield (2004).

4. An irrational hostility to technology fueled the bombing acts of Theodore Kaczynski, the Harvard-educated mathematician, Assistant Professor at the University of California, Berkeley, and the infamous Unabomber. In his manifesto (*Industrial Society and Its Future*), published by the *New York Times* as part of his ransom demand, he said: "Industrial Revolution and its consequences have been a disaster for the human race." [NeuroNote: A court-appointed psychiatrist declared that he suffered from paranoid schizophrenia, but Kaczynski dismissed this evaluation as "political diagnosis."]

5. The seven-day grouping is an arbitrary epoch introduced by the Babylonians who believed there were seven planets in the solar system. To mark the end of the week, a worship of the gods was introduced. The Sabbath is on Friday, Saturday, or Sunday in Muslim, Jewish, and Christian religious traditions, respectively, and reflects the belief that God rests on the seventh day. The brain does not "feel" one week of time, and there is no clear indicator in the outside world to mark seven-day epochs. An exception may be in today's urban areas, where air pollution varies significantly between work days and week days (Ward et al., 2015).

time spent working could address problems with unemployment, high carbon emissions, social inequalities, family care, and personal happiness for many. Yet, instead of reducing workloads, more of the world's population works long hours, and the full-time employment of women has increased over the past several decades.[6]

How did technology mysteriously descend on us? Who is to blame? It seems that nobody is in charge, and no one can stop it. Instead, all of us contribute to the chaotic system and impose more work on ourselves. We are driven by the short-term benefits of technology and economic envy of others instead of exploiting these innovations for our long-term well-being. When all of us are involved in a distributed process, nobody can make a decision on our behalf.[7]

In this chapter, I attempt to tackle this complex issue by examining how the externalization of brain function affects our brains and the brains of others. Some of the text may appear off-stream from the main topic of this book. However, without discussing these issues, it is not possible to understand why the relationship between episodic and semantic memories appear so different in humans than in other animals.[8] I conclude that human-made instruments and other artifacts have become an extension of the action–perception loop, as well as the media through which abstract ideas emerge and spread quickly. Through this process, humans use their brains in a way that no other animal can approach. Without taking this externalization process into consideration, we cannot understand how complex thoughts and ideas are created by human

---

6. I am not competent to discuss the sociological, economic, or political implications of technology. Instead, I focus on the impact of externalized innovations on the brain in this chapter.

7. Naturally, humans complained about technological carnage after the two world wars. Similar hatred of technology was aroused in my generation, who witnessed the horror of napalm bombs in Vietnam. The hippie movement was largely fueled by the evil aspects of science. There is a catch-22 with the use of technology. A minority of people profit from it tremendously, while the majority benefits much less or not at all. Even if technology improves the quality of life for all, control of technical innovations comes with a disproportional benefit for those who are in control of the society. This economic scissor always makes the majority *feel* like exploited losers even when they benefit from it.

8. In Chapter 5, we discussed the idea that the evolutionary roots of episodic and semantic memory systems are the dead reckoning and landmark-based forms of navigation and that semantic knowledge requires multiple episodic events, through which process the spatiotemporal aspects of the events are stripped off (Buzsáki, 2005; Buzsáki and Moser, 2013). In contrast, Mishkin et al. (1998), studying the effects of early hippocampal injury in humans, conclude just the opposite. In their view, semantic information is primary and needed for constructing hippocampus-dependent episodes. The difference in emphasis may be the "shortcut" mechanisms in humans afforded by the ability to rapidly acquire semantic information through externalization of mental representation.

brains, which are not much more complicated structurally than the similarly large brains of some other species.

THOUGHTS ARE ACTIONS: EXTERNALIZATION OF MENTAL OPERATIONS

Brain outputs come in various flavors. The first one that comes to mind is muscular action to move the body and its sensors. A second output acts via the autonomic nervous system to affect our internal organs and glands. A third output is the release of hormones from the pituitary gland to affect growth, blood pressure, sexual hormones, thyroid hormones, metabolism, temperature, childbirth, lactation, and water/salt concentration in the kidneys. The fourth large category of brain output is thoughts, even if we do not intuitively think of thoughts as actions. A *thought*, which can be conceptualized as a buffer for a deferred action (Chapter 5), is useful only when it benefits the brain's owner.[9]

Imagination and thought rely largely on the hippocampus and prefrontal cortex. The prefrontal cortex can be considered an internalized offshoot of the motor cortex with a similar neural architecture. The motor and higher order prefrontal cortical areas (divided into medial, orbital, and insular areas) receive remarkably similar inputs and send similar outputs to their targets. The main difference is that while the primary motor cortex sends direct projections to the spinal cord to control the skeletal muscles, the prefrontal cortex instead targets autonomic and limbic sites, including the basal ganglia, basolateral amygdala, thalamus, hippocampus, and lateral hypothalamus.[10] These projections can be viewed as corollary circuits through which prefrontal areas inform other higher order cortical areas and motivation or action-preparing structures about pending action plans, similar to the way that motor cortex informs sensory areas about ongoing actions. By analogy, we can conclude that prefrontal cortex projections compare pending or potential actions and their expected consequences against recalled information and desired goals. Instead of

---

9. Buddhist teaching emphasizes that physical and mental actions (thoughts) are related. "Karma" literally means action or doing. I thank Liset Menendez de la Prida for pointing out this relationship.

10. Other targets with notable projections from the medial prefrontal cortex in the rat are the mediolateral septum, dorsolateral periaqueductal gray, ventral tegmental area, parabrachial nucleus, nucleus tractus solitarius, rostral/caudal ventrolateral medulla, and even the thoracic spinal cord, with some notable differences across infralimbic (area 25), prelimbic (area 32), and dorsal anterior cingulate (area 24b) areas. Many of these connections are reciprocal (Gabbott et al., 2005).

directly evaluating actions, this hypothetical higher order corollary mechanism compares and evaluates the internal workings of the brain to form multiple-step action plans before any overt movement.

In humans and other great apes, several prefrontal areas contain large neurons in cortical layer 5; these are called *spindle cells*. These large cells are the first cousins of the similarly large layer-5 Betz cells of the primary motor cortex, which project to the spinal cord, another indication of the general architectural design of prefrontal cortical areas. These special-looking spindle cells were first observed in hominids, prompting several investigators to attribute to them unique functions in cognition and emotion control.[11] However, they have since been described in cetaceans, at higher densities than in humans, in elephants, and, to some extent, in rhesus monkeys and even raccoons. These neurons have the strongly myelinated, long axons needed for sending action potentials quickly across long distances in large animals. These anatomical considerations reveal that motor and prefrontal areas of the cortex share many anatomical features. The main functional difference is that while activity in the motor cortex leads to immediate action, activity in the prefrontal cortex may only simulate action, which we call *plans* and *imagination*.

Are Humans Superior?

Many investigators have sought to explain the "superior" qualities of *Homo sapiens* by some anatomical feature, such as brain size, neuronal numbers and densities, axonal connectivity, special cell types, or axon conduction speed. They found none. Our cousin, the Neanderthal, with whom we overlapped 50,000 years ago, had a bigger brain than modern humans do.[12] They surely had some special abilities compared to us, but what exactly they did better with so many neurons remains unexplained.

Modern humans who walked the planet 50,000–200,000 years ago had the same vocal cords, hands, and brains as we do today.[13] Imagine that a child born 50,000 years ago could be swapped with a newborn of today. Would the child

---

11. Spindle cells are also known as *von Economo neurons*, named after Constantin von Economo (von Economo and Koskinas, 1929). Allman et al. (2002) studied these neurons extensively in several species.

12. Henneberg and Steyn (1993); Ruff et al. (1997); Bailey and Geary (2009). An excellent recent account of evolutionary comparison of the human brain with that of other animals is by Herculano-Houzel (2016).

13. The newly found skull at Jebel Irhoud, Morocco, was dated at 300,000 years old. The remains of the skulls of five individuals had modern facial traits, although the brain cases were elongated, with a somewhat smaller brain volume and not round like those in humans living today. Jebel Irhoud people are considered protomodern humans who controlled fire and hammered

born in New York City and teleported back in time to a loving family become a constructive member of a hunter-gatherer community? Conversely, would the teleported child from tens of thousands years ago, adopted by an average family in the United States, have the same chance of getting into college as my daughters? My answers are a resounding "yes." I assume that our ancestors had the same cognitive and communication capacity as we do, and I would also venture that they had pretty much the same brain wiring and dynamics.[14] If you side with me on this issue, then we also likely agree that something besides anatomical differences in the brain must explain why *Homo sapiens* managed to become the ruler of the world. In my view, that something else is that modern humans learned to externalize their minds by first making tools, and then inventing speech, writing, and art with ever-increasing sophistication. Along the way, they redesigned themselves into a cooperative, knowledge-sharing species.[15]

EXTERNALIZATION: A QUICK OVERVIEW

Homo habilis and *H. erectus* were already making bifacial hand axes to scrape meat from bones and butcher large animals 1.5 million years ago. Their technology remained relatively the same, with minor modifications, for a million years.[16] When the climate got warmer, our ancestors migrated to mid-latitude Eurasia, where food availability was more seasonal than in Africa. These more dispersed food resources required increasingly sophisticated navigation,

---

stones to use as tools (Hublin et al., 2017). At about the same time, Middle Stone Age blades and spears with higher precision and sophistication were manufactured by protohumans in today's Kenya. These were more abstract than earlier hand axes and meat scrapers that still preserved the shape of the original lump of rock. The leftover stone dust from these new tools was used paint the body of the creator, as a symbol of individuality or group identity (Brooks et al., 2018).

14. In fact, we do not have to go back in time literally. There are still a few uncontacted peoples left on our planet who live like their ancestors did 50,000 years ago. The North Sentinelese people who live in the Andaman archipelago between India and Malaysia do not know how to make fire and still use spears and arrows to hunt.

15. Boyd et al. (2011).

16. From flaking stones to mastering sharp bifacial hand axes requires multiple steps and premeditation. Tools exhibit a mental template imposed on unorganized, dead material. The large numbers of artifacts found in the Olduvai Gorge, West Africa, show a continuum of refinement (Toth, 1985). Capuchin monkeys use a rock and a log as a hammer and anvil to crack palm nuts for food in the cerrado of Brazil. Some monkeys bang rocks together and break off stone flakes. Big brains are thus not required for intentionality (Proffitt et al., 2016).

memory, and planning.[17] The hunter-gatherers slowly transformed into farmers, established permanent domestic spaces, and invented more elaborate artifacts, including weapons, paintings, sculptures, and musical instruments. Many of these artifacts no longer mimicked or represented anything real in the world. They had to be imagined before their maker could start working. Artifacts are externalized versions of a thought, a reflection of contemplation, and a way to communicate personal knowledge to others even after the creator has vanished. Artifacts are semantic entities, which can be labeled and remembered as separate from other things. The root of this mirroring between action and perception may be built upon mechanisms analogous to active sensing, with its corollary circuits (Chapter 3).[18] However, the feedback in this extended loop is not a specialized circuit within the brain but a sequence between action-produced artifacts and their reflection back to the brain.[19]

These externalized mental products come to exist outside their creator's brain as permanent social memory, enriching the collective mind. The creative action of tool-making might be considered a prelude to verbal language. Conceiving a multiple-component tool, such as a rock tied to a rod to make a hammer, requires a hierarchical algorithm, a natural precursor for language syntax. The manual dexterity required for tool-making is also necessary for subsequent writing ability.[20] Artifacts and, eventually, words can readily communicate semantic information from one brain to many without laborious episodic explorations by each individual (Chapter 5). Instead, the grounding of meaning is simply achieved by guidance or approval from others. Externalized information can be named, and, therefore, it rapidly spreads semantic knowledge. This ability comes with a cost, though. We accept the definition of events and phenomena too often without personal experience, accumulating and

---

17. Some view that such ecological factors comprised a critical evolutionary pressure and selection on human brain size (e.g., Parker and Gibson, 1977; Gonzáles-Forero and Gardner, 2018).

18. A good part of archeology is "neuroarcheology," a fossil record of thought (Malafouris, 2009; Renfrew et al., 2009; Hoffecker, 2011). Human-created objects not only document the migration of early people but also are a rich source of the intentions, lifestyles, and cognitive abilities of their masters. However, tools carry meaning only to those who share the "cipher" and can envision their use.

19. This loop is analogous to the way we do experiments in the lab. A thought is acted out by producing an experiment (e.g., making a lesion in the brain) and then watching and interpreting the consequences of our perturbation.

20. Greenfield (1991) hypothesized that the same neuronal substrate can serve the syntax necessary for both multiple-step object-making and construction of sentences. Although there is an analog of Broca's area in great apes, it can only support the complexity of both tool construction and symbol manipulation at the level of a two-year-old human (Holloway, 1969).

using a huge vocabulary in which we do not understand the true meaning of many words (Chapter 1).

Urban Revolution

Once hunter-gatherers started farming and domesticating plants and animals about 12,000 years ago, group knowledge began to increase dramatically. Agriculture required division of labor, specialization, hierarchy, and trade, associated with an acceleration of demographic growth. Tens of thousands of people lived together in the Sumerian cities of Uruk and Shurupak 6,000 years ago. The pressure on dwelling solutions led to the introduction of geometric designs that were entirely the product of the externalized mind, such rectangular and square house floors. As population density grew and cities were formed, the collective sharing of external thought expanded explosively. A city of 10,000 people working as a cooperating group can generate much more accumulated knowledge and much more rapidly than a hundred small communities of 100 people each.[21] This stage of *H. sapiens* history, known as the *Urban Revolution*, brought us writing and mathematics.[22] Kings, pharaohs, and their bureaucrats emerged, and the texts they created gained authority. The earliest pictographic or ideographic scripts gradually transformed into Sumerian wedge or cuneiform writing over a thousand years and then to the current forms of writing. Pictographs worked mainly as a mnemonic device to help with spoken language and had to be interpreted rather then read. They are a bit like the private notes and symbols I scribble in the margins of the books I read. Even I have a hard time interpreting them properly without consulting the related text. Yet pictographs are very useful for storing thoughts outside the brain. Such externalized thoughts, collectively agreed upon by a small group of powerful people, became the rule of the land for generations. Written texts served to remind members of the society that they must obey certain laws and that breaking them would result in punishment. The rulers invented money

---

21. Not only humans but the cognitive abilities of other animals also benefit from the size of a social group, including non-human primates (Humphrey, 1976) and Australian magpies (Ashton et al., 2018).

22. Because of the relatively simultaneous emergence of writing and math in Mesopotamia, several scholars have speculated that they required the same kind of cognitive processes and perhaps brain regions. However, this cannot be fully correct. The Inca people in Mesoamerica built a civilization comparable to that of Egypt, built pyramids with the help of sophisticated math and geometry, and had a method of record keeping with knotted cords but did not invent writing (Ascher and Ascher, 1981).

that allowed exchanges across variable services and products of labor. This externalized metric of value enabled further stratification of work and allowed even more people to cooperate and work together.

Cognitive Revolution

The ability to read the unique minds of other individuals through externalized objects, symbols, texts, and institutions gave rise to the cognitive revolution phase of human evolution. This sort of information reflects back to the brain to profoundly affect its circuits. In a modern experiment, illiterate women in India learned how to read and write their mother tongue, Hindi, while their brains were repeatedly imaged. After just 6 months of literacy training, several areas of their brains reconfigured, including the thalamus and even the brainstem.[23]

Fast forward to Johannes Gutenberg's invention, which enabled the production of printed books in the fifteenth century[24], and then jump to the rotary presses of the nineteenth century, which introduced printing on an industrial scale. Written sentences, mathematical equations, and pictures are externalized products of the collective cognitive powers of the human mind, and they have an unlimited lifetime. Books allow people to check and compare the thoughts of others whose ideas were developed under different circumstances. Unlike a busy college professor, a book can provide all the time that the reader needs. The reader can consult its contents over and over again. Only the possession of the cipher is needed to convert the abstract patterns of letters and words into organized neuronal activity to create novel thoughts in the brain. Externalization of brain function enables both broadcasting and copying of complex ideas, intentions, emotions, and hopes from brain to brain without the need for immortality of the ideas' creator. By virtue of the hierarchical, generative, and

---

23. Previous studies focused on changes mainly in the early visual system (Carreiras et al., 2009; Dehaene et al., 2010). In the study of Skeide et al. (2017), the effects of literacy learning also extended to V3 and V4, the thalamic pulvinar, and the superior colliculus of the brainstem. The coupling of blood oxygen level-dependent (BOLD) signals between these subcortical structures and visual areas V1 to V4 increased after the acquisition of basic literacy skills. The Devanagari script of Hindi is an alpha-syllabic writing system that represents sound simultaneously at the syllable and phoneme level. The involvement of the superior colliculus might be a consequence of the fine-tuning of the initiation of saccadic eye movements necessary for fixating the eyes to relevant parts of the printed text (Dorris et al., 1997). The importance of the action system is also illustrated by the finding that visuospatial motor skills predict reading outcome in preliterate children (Carroll et al., 2016).

24. It is hard to be first in anything. Books had been printed in China and Korea for centuries before Gutenberg.

recursive nature of language syntax and its virtually limitless scope of expression, people can find their way to God or forbidden thoughts, come up with new hypotheses, and publish earth-shaking discoveries.

EXTERNALIZATION AND INTERNALIZATION ARE COMPLEMENTARY PROCESSES

As discussed in Chapter 5, disengagement of the brain from its environment and its ability to manipulate models of the external world are prerequisites for cognition. This operation is facilitated by externalization: internally generated thoughts are tested against action-produced artifacts, which reflect back on the original thought as a validating, grounding process. This is a protracted action–perception mirroring loop. We test the validity of a thought by acting in certain ways and, as a result, seeing things differently. Externalization is an elaborate test of internal thoughts. With the help of artifacts, such as books, thoughts can be communicated to many other brains. This practice can give rise to agreed abstraction or semantic knowledge within the community. Human-made objects are therefore a medium through which semantic knowledge and abstract ideas can rapidly spread.[25] No thought or concept possesses validity unless it is communicated to other brains. Meaning is created when a group of people weaves together a common story: "my thoughts are not so crazy." Thus, a concept becomes a concept only by externalized grounding, and the process of grounding involves interactions with other brains. This extended view of cognition implies that cognition can be understood only as a social phenomenon that transcends the brain of an individual.[26]

Inventing Numbers

How is a communicable abstract idea created? Nobody knows for sure how it happened first, but we can make up a plausible scenario. Let's suppose that a smart individual wanted to throw a party for the village. As this person sacrificed one sheep after another, he scratched a line on a bone for each sheep. Tally marks were used for counting before the construction of cities in Mesopotamia.

---

25. The meaning of the Latin verb *abstracte* is to "remove, detach, or pull away." A cave painting or other nonpractical artifact fuses the essential features of multiple concrete situations by detaching the conditions in which the concrete events occurred (Chapter 5).

26. The term "con-sciousness" literally means shared knowledge and is meant in this way in many languages.

After this process happened many times, the idea may have struck him that five vertical tally marks resembled five fingers. With such an insight, the next time no groove marks were needed; he just counted each sheep with a finger. He or his descendants who learned this approach might have called five sheep a "handful." At this stage, there was still a match between symbols and the physical things they represented. Yet this was already a major abstraction that likely took several hundred years.

From here, archeology can assist us in reconstructing the evolution of externalized thought. For hundreds of years, arbitrary number systems were used to count different groups of things, such as animals and bags of grains, while other systems were applied to count other things.[27] People in different cities also used different tally mark systems. The next level of abstraction was the realization that numbers apply equally to all things. Once this generalization was made, tally marks could be manipulated without knowing what they referred to. Through such mental manipulation, the marks were separated from the circumstances that created them, and they became explicit knowledge. Numbers thus lost their dependence on their physical correspondences. Recall that this mental transformation is similar to the process of making maps from exploration or creating semantic knowledge from multiple personal episodes (Chapter 5). Thus, externalizing ideas in the form of tools and objects could have helped to generate novel thoughts, which, in turn, could be internalized into other minds. The initial process may have taken a thousand years, but it was accelerated rapidly by the need for many individuals to do the same activities again and again in increasingly large cities. No genetic change in brain hardware was needed to gain this ability.

We can witness the evolution of abstraction at an accelerated pace when observing our children. First-graders can count five apples, five pencils, and five other things by touching and holding them. Then they learn that "5" corresponds to the word "five" but often ask: "Five what?" as if looking for the grounding role of the objects. Adding concrete objects is a relatively easy task when kids can manipulate them. In the internalization phase, when counting imaginary apples, they touch their fingers prior to gradually treating "5" as an object and eventually as an abstract number. The entire process is speeded up not (only) because of brain maturation but because learning numbers is supervised by experienced adults whose brains have already abstracted the concept of numbers.

---

27. Clegg (2016).

Many mathematicians and physicist believe that numbers and math are the fundamental organizers of the universe, so that we could simply infer them from our mental imagination, independent of the outside world.[28] In this chapter, I have tried to make the case that the process acts in the opposite direction. Without externalization and human-made artifacts, it is unlikely that the concept of numbers and arithmetic would ever have emerged.

CLOCKS FOR SYNCHRONIZING HUMAN COOPERATION

Perhaps no externalized artifact has had a larger impact on our ideas than the clock. Vague concepts of space and time predated measuring instruments, mainly to explain the vastness of the surrounding world and our place in it.

Seasonal changes have predictable effects on the accessibility of various food resources and on agriculture. In addition, daily circadian changes influence the activities of all living creatures. Early humans routinely used celestial "time keepers," such as the relative position of the sun and the moon, but activities can be better allocated within these natural time windows if shorter, discrete intervals can be determined. From ancient Egypt came the idea that the day from sunrise to sunset and the night from sunset to sunrise should be divided into twelve hours. There are different explanations for the origin of the number twelve. There are roughly twelve moon cycles in a year, each cycle corresponding to one complete orbit of the earth around the sun, and twelve zodiacal signs. Also, there are three phalanges in each of the four fingers. When numbers emerged by making correspondences between things in the world and references in our body, marking the phalanges was likely a good method for counting twelve with one hand. The Babylonians used a duodecimal (twelve) and a sexagesimal (sixty) system. Why they chose sixty is not clear except that it is a nicely "divisible" number (by 1, 2, 3, 4, 5, 6, 10, 12, 15, 20). Furthermore, ancient astronomers believed that the Heavens comprised sixty concentric translucent spheres, the Earth being at the center. In geometry,

---

28. According to Roger Penrose (2004), numbers can "seemingly be conjectured up, and certainly assessed by, the mere exercise of our mental imaginations, without any reference to the details of the nature of the physical universe." Max Tegmark (2014) argues "that our physical world not only is described by mathematics, but that it is mathematics. . . . Everything in our world is purely mathematical—including you." But there is no alternative method to quantify the world out there. Thus, math as a tool cannot be falsified. Mathematics starts out with *axioms*, fundamental assumptions without a need of proof. Math often makes predictions that the real world cannot absorb.

a sexagesimal system works better than decimal numbers. Modern clocks are still based on the 12/60 method.[29]

Time is intimately related to movement and geometry. However, dividing the interval between sunrise and sunset to calculate time is problematic because its length varies with seasons and locations. Noon time and the length of the hour, measured in the old way, were different in different regions of the vast Roman empire, where sundials were ubiquitous. The ancient Greeks realized this and defined the hour based on equinox days, when day and night are of equal length. This celestial duration served to calibrate the shadow-casting sundial. Yet this approach did not solve the time compression/expansion issue at different longitudes.

Concrete Time and Duration

Water clocks were invented at about the same time as sundials but for a different purpose. They served to measure the duration of certain sacred or bureaucratic activities as opposed to the time of the day. In Greece, water clocks (called *klepsydra*) were used in courts to limit the time allowed for presenting complaints.[30] They served as a calibration device to settle the dispute over whether days and nights are of different lengths in different seasons. Much later, when the geometrical relationship between the height of the water in the tank and volume was understood, water clocks were used to calibrate the hour (*horos*).

In medieval Europe, churches were built with a bell tower to solicit the faithful to attend Mass.[31] The monasteries became the first institutions to use mechanical clocks, which helped to organize their complex bureaucratic

---

29. Dohrn-van Rossum (1996). Of course, water clocks do not measure time flow per se but the volume of water drained from the container, which is then converted to time units calibrated by other measures.

30. Lamps fueled by oil were used in magical rituals. The volume of oil placed in the lamp determined the duration of the event. This ritual still exists in churches where the duration of the offered prayer is timed by a burning candle. Lamps, water clocks, and similar instruments measured duration only and were only much later synchronized to some celestial event to tell the time of day. Demonsthenes demanded "Stop the water!" when his court proceedings were interrupted by depositions (Hannah, 2009a, 2009b). In fact, the word "water" (*hidor*) was synonymous with time. Of course, water clocks do not measure time flow per se but the volume of water drained from the container, which is then converted to time units calibrated by other measures. Thus, the measurable volume grounds the abstract duration.

31. In 1456, the Hungarian-Serbian Army, led by János Hunyadi, defeated the Ottoman Turks besieging Belgrade (then Nándorfehérvár). The victory was important because it prevented the expansion of the Ottoman Empire for seventy years. In recognizing the importance of the victory, Pope Callixtus III ordered the bells of all churches to be rung every day at noon as a call to

Figure 9.1. Conceptualization of change or lack of it. *Left*: Maya calendar. Life is a circle, nothing ever changes, everything repeats periodically. This view is shared by Hinduism, Buddhism, and Egyptian cultures. *Right*: Time flows, nothing is ever the same (*panta rhei*), a view adopted by Aztec, Christian, Jewish, and Islamic religions. Upstream is past; downstream is future. Relative position determines time flow.

apparatus. In the Muslim world, *müezzins* recited the call to prayer from the minarets at approximately the same local times of each day. Mainly as a result of such practices, the hour acquired a practical meaning both as duration and time of the day. In turn, it took centuries and numerous steps of innovation[32] before mechanical clocks became accurate enough to divide hours into minutes and later into seconds. Today's atomic clocks are based on the duration of a particular number of energy transitions of the cesium atom. However, the discrepancy between instrument-kept time and time based on astrological movements persists even today. Leap seconds need to be inserted every now and then to compensate for the mismatch.

Internalization of Externalized Time

The increasing prevalence and precision of measuring instruments changed the human concept of time. Egyptian and other early societies did not see time as an arrow. Instead, change was envisioned as cosmic cycles (Figure 9.1). The wheeling of the sun, moon, and stars through the heavens meant time. Although they

---

pray for the souls of the defenders of the Belgrade battle. Of course, noon meant different times in different countries in old Europe. The noon bell ritual persists to this day.

32. The story of John Harrison, the English clockmaker who invented a marine chronometer and solved the problem of calculating longitude at sea for dead-reckoning navigation, has been told many times. My favorite account is by Dava Sobel (1995).

surely realized that humans and animals were born and died, this applied only to individuals. Species were eternal. This view was in harmony with the dominant ideology of preservation of the existing system—for almost 3,000 years in Egypt. Instead of historical explanations, powerful societies created a myth whose content equally applied to the past, present, and future. Past events, such as creation, were imagined as the actions of gods, but the distinction between past and future was inconsequential. In ancient Greece, Plato insisted that nothing is new, the soul already knows the truth, and everything is just reminiscence (*anamnesis*). "Nothing is new under the sun" (Ecclesiastes 1:9).

The concept of linearly progressive time is rooted in the Judeo-Christian view and the encouragement of manual labor and practical innovation, especially after the thirteenth century, as agricultural productivity and population density were rising in Europe.[33] The technology that was key to the emerging Industrial Age was the mechanical clock. The church embraced the clock, which allowed timing of prayers. By installing bells in virtually every community, these human-made devices created an "arbitrary temporal matrix within which everyone functioned. They generated a component of the daily landscape of the mind."[34] The clocks and other instruments with their arbitrary units became metaphors for the universe and contributed to the mechanistic worldview of the Middle Ages. This new interpretation of the world was radically different from, say, the world of hunter-gatherer societies, which was filled with timeless ghosts and deities.[35] The linear concept of time gradually but fundamentally transformed the European mind, although it took centuries. The habit of thinking within the framework of time's arrow for interpreting history is a relatively recent phenomenon.[36]

Let us pause here for a second and summarize that the concept and meaning of time evolved considerably over the history of humankind. The externalization

---

33. Those who believed otherwise were punished. The atheist Giulio Cesare Vanini wrote in 1616: "human history repeats itself. Nothing exists today that did not exist long ago; what has been, shall be." He was executed for these views by the church (as quoted in Wootton, 2015). Yet Christianity retained elements of recurrence. The liturgy is organized around the endless cycle of Christ's life.

34. Hoffecker (2011).

35. This connection was first articulated most clearly by Gordon Childe (1956). [NeuroNote: Childe suffered from social anxiety.] While rods were used routinely for measuring distances while building pyramids and water clocks were used to divide the day into hours by the aristocrats of Egyptian society, it was not until the Middle Ages that everyday people began to have access to them. There existed a fully articulated philosophy of the universe, based on purely mechanical lines, which served as a starting point for all physical sciences (Mumford, 1934).

36. Collinwood (1946); Lévi-Strauss (1963); Hannah (2009a).

of brain function facilitated the emergence of concepts and the spread of abstract ideas. The more people interacted with measuring instruments, the more real their abstraction-based units appeared to be. By measuring distance and duration, we formed a more concrete concept of space and time.[37] That is to say, emerging societal knowledge can shape concept formation in individuals. In support of this hypothesis, children below the age of nine intuitively grasp the notion of speed but not time. When one toy train is moved faster than another but for the same duration, children most often respond that the faster train traveled for longer time. The idea of velocity is natural at a young age, whereas the concepts of space and time are acquired from adults.[38]

Humankind has spent considerable effort in perfecting time-keeping instruments. The clock has become the ultimate externalized synchronizer of human thought, with a tremendous impact on our concepts of the world and also on our activities (see Chapter 10). Without precise clocks, we could not conceive of travel, communication, and cooperation at today's global scale.

Paradoxically, the idea of clocks as servants of humans changed radically starting with the Industrial Revolution. Clocks became tools to measure productivity and regulate labor, and through this process time got linked to value. The abstract concept of time has become concrete money. However, the worker's sense of time and the time marked by the manager's clocks always compete forcefully. The tyranny of clocks is something we do not fully understand. Clock- and calendar-mandated demands, from morning alarms to grant deadlines, place extraordinary pressure on our daily lives. Over the past century, time has been perhaps the most important factor in advancing technology while also being the number one burden on human endeavors. This is strange because we know that rats, monkeys, humans, and likely other species perform worse under time pressure than without it.

THE INFORMATION REVOLUTION PHASE

Externalized thought in the form of objects, books, and communication media allows the spread of an individual's knowledge to many people. The process is

---

37. "By the use of a clock the time concept becomes objective" (Einstein and Infeld, 1938/1966).

38. "Time appears as a coordination of velocities" (Piaget, 1957). Jean Piaget concluded that temporal concepts are understood much later in development than are spatial concepts, whereas the idea of velocity is present from early on: "time relationships constructed by young children are so largely based on what they hear from adults and not on their own experiences" (Piaget, 1946).

creasing division of labor. As a systems neuroscientist, I am not an expert in molecular issues, neuroimmunology, or artificial intelligence, and my knowledge of the numerous diseases of the nervous system is rudimentary. It is hard to be a polymath these days, even within the confines of a single discipline. This situation, however, does not mean we should not try. The knowledge is there, waiting to be absorbed.

Perhaps some of my thinking is misguided. We find ourselves simultaneously saddened and enthusiastic when we learn about a new robot that outperforms a human in balancing skills, memory, or other cognitive games. However, this is not a fair comparison. The robot or supercomputer was produced by hundreds to thousands of interactive human brains. No wonder it can beat a single *Homo sapiens*. We often mix up the comparison between human versus machine knowledge and the knowledge of humankind versus "machinekind." But the argument can go the other way as well. Our knowledge is rich not only because of the performance of the human brain but also because there are 7.4 billions brains on this planet, and a good fraction of them can communicate effectively, facilitated by the Internet. It does not seem realistic to believe that energy-hungry robots can build billions of other robots, each comparable with the human brain.

This is not to say that AI will not undergo an unprecedented and breathtakingly spectacular revolution. There is virtually no remaining frontier of brain-based intelligence that has not been challenged by AI in one way or another. Just a decade ago, it seemed unrealistic that face or scene recognition could be achieved by a machine. Today, they far surpass the best human experts. Their "reasoning" ability is demonstrated by beating every human in chess and other games.

Imagination and creativity? Hm. What are these faculties? Writing this book required reading tons of works by others, extracting and combining existing information, and creating a novel story by my brain. Can AI accomplish this? Nothing says it cannot. AI can already do such things by writing interesting stories, transcribing text to speech and vice versa, translating from one language to another better than most of us, composing new music that many of us can enjoy, and creating esthetically pleasing paintings. Many of these accomplishments have been suggested to be inspired by the brain. Yet this inspiration is not much more than analogy or a metaphor since the hardware realization of the machine algorithms are so different in brains and computers. If the hardware base is so different, then the algorithmic solutions are likely to be very different as well. And one more thing. The brain has both intelligence and emotions. Emotions do not rise in the brain alone but also require the rest of the body as well. There is no such thing (yet) as artificial emotion (AE), even if it is fun to imagine otherwise. For this simple reason, the machine and algorithms

reciprocal because the knowledge of many brains can also be communicated to a single individual. Books, computers, and Internet communication externalize brain functions at a grand scale and provide virtually unlimited storage space for accumulated human knowledge. However, like books in a library, this knowledge is only as useful as its accessibility. When I was a student, I looked up references related to my work and ordered the articles through interlibrary loan. When the book or journal arrived a few months later, I had to request permission in writing to make a print copy for myself. This process is much easier today, but our self-imposed demands have increased proportionally since rapid access is available to everyone.

In 1992, it took me forever to filter traces and compute the Fourier transform of a single channel electroencephalogram (EEG) on my Macintosh Classic II with its 80 Mb hard drive and 10 Mb of RAM. Today, Google's DeepMind can teach itself how to beat humans at Go and Atari games.[39] The FaceNet system can recognize a particular face among a million others. Similar systems can locate street scenes anywhere in the world and do a better job than groups of well-traveled humans. IBM's Watson and its brothers are trained to outperform doctors with decades of experience in diagnosing diseases.[40] By 2020, 80% of the adult population worldwide will own a smartphone that provides access to nearly everything humankind knows. Self-driving vehicles are around the corner. Wearable sensors monitor the activities of more than 50 million registered users, and new classes emerge daily. Monitoring the pattern of strikes on the keyboard or tracking eye movements, pupil size, and heart rate during reading of tens of millions of users provides an unprecedented amount of analyzable and quantifiable data. While you are being interviewed for a job or reading this page, sensors can collect information about your reactions, health states, and even desires, generating unprecedented amounts of biometric data that will be eventually owned by unknown groups of people. Artificial intelligence (AI) technologies will transform nearly every aspect of life from agriculture and medicine to transportation and finance. Tomorrow's AI machines will learn on their own, rivaling human performance in surprising and unforeseen domains and complementing the human brain.

39. See Silver et al. (2016). An excellent review about deep learning and neuroscience-inspired artificial intelligence is by Hassabis et al. (2017). A longer treatment of the topic is by Sejnowski (2018).

40. The cash flow of Apple, Microsoft, and Facebook exceeds by tenfold the annual federal budget for biomedical research in the United States (Insel, 2017). These rapid changes bring about a massive transformation of the workforce. Today, only 2% of Americans work in farms and 20% in industry, while the remaining part of the population provides services, with an increasing share of those working with "information" (Harari, 2017).

The Internet is becoming a world communication language, the new *agora* for exchanging externalized knowledge at a global scale. Words, expressions, and idioms can already be translated into virtually any language instantaneously, and the translation programs are getting better every day. We can internalize new knowledge at an unprecedented speed, test "what will happen if" scenarios without leaving our chairs, and, in the process, make new discoveries. Yet such globalization comes with homogenization. We are guided to follow the same trends and brands, whether we live in the United States or the Philippines. For a small, short-term benefit, we reveal our personal lives voluntarily on the Internet, encouraged by the many companies that benefit from learning about our needs, goals, and income.

Neurotechnology: The Next Revolution

As a culmination of the externalization of brain function, science has become perhaps the most powerful driving force of societal change. Perhaps the next new era in human evolution will be neurotechnology, the development of noninvasive techniques for monitoring and manipulating brain circuits with high spatial and temporal precision. Getting information about neuronal activity indirectly from the brain at relatively slow speeds is already possible with functional imaging. High-density recording of electrical and magnetic signals generated by large aggregates of superficial cortical neurons at the speed of neuronal communication is also routine. There are still a number of hurdles blocking the next step, recording from representatively large numbers of neurons with single action potential resolution and manipulating critical numbers of neurons one by one or in small groups. But, once implemented, such externalizing brain activity at high speed will allow the design of feedback actuator mechanisms to reshape the firing patterns of neuronal assemblies brought about by disease or substitute for the actions of missing neurons in brain repair.

Signals representing anything from brain states to the intention of one brain can be routed to many other brains, bypassing the current "slow" brain-to-brain communication via language. In speech, only one speaker speaks at a time, and others respond one by one. In direct brain-to-brain communication, we could download thoughts, intentions, and emotions from several brains and route them to many other brains, allowing for collective thinking and consensus manipulation. Messages sent and received on Twitter and Instagram are already precursors of semi-online multibrain communication. New senses for magnetic fields, infrared and ultraviolet light, or radioactivity, could be incorporated into the brain. While such "telepathic" communication sounds futuristic at the moment,[41] the laws of physics do not warn us that such tech[nology is impos]sible. Rather, the big question is who needs such technology a[nd what] it offers. The development of these techniques is likely to be diff[erent] on whether they are driven by scientific curiosity, therapeutic [goals,] or commercial interests.[42]

BRAIN RESET

When the Ancient Library of Alexandria burned down, an enor[mous amount] of hard-earned knowledge vanished forever. Imagine waking up i[n a world] to learn that all knowledge that existed in digital form has disap[peared. This] loss would reduce humankind's externalized knowledge to the lev[el of the pre-]Sputnik era. Almost no one could build a computer or design a s[oftware] from scratch. Today's technologies require the cooperative effo[rt of] highly trained individuals, all of whom gain their smartness from [the ac]cessibility of externalized human brain products.

So who is smarter? An average hunter-gatherer who depends on [a va]riety of skills to survive in the African savannah, or an average New Y[orker who] has a cell phone, can call a cab, finds a Starbucks at every corner, [and orders] takeout food after spending eight hours in an air-conditioned work[place with] easy access to the vast accumulated knowledge of humankind create[s the illu]sion that our brain's performance increased over the past 50,000 year[s. Yet the] stand-alone brain is just the same hardware filled with different kinds [of know]ledge, updated to current needs. When we search for the superiori[ty of the] brain of *H. sapiens* over other animals, we need to understand not only [the] number, connection density, and clustering rules, but also how huma[ns] externalize information to learn from the experience of others. Exte[rnalized] knowledge of a community of minds allows for efficient internalization[, a pre]requisite for redesigning ourselves.

There is a price to pay for these gains. While the knowledge of h[uman]kind increases exponentially due to the multiplicative interactions of [many] brains, the relative share of the individual decreases dramatically, leading [to]

---

41. Entertaining accounts of such futuristic perspectives are by Harari (2015) and Te[gmark] (2017).

42. The driving force of new technologies has recently shifted from scientific curiosity to [com]mercial interest. Especially in the case of applied and translational science, the decision-ma[king] committees are increasingly composed of managers and business executives who tend to [sup]port proposals for technologies that rich customers like themselves can buy (Dyson, 1997).

do not have a first-person perspective on anything. A robot has no agenda. Current AI devices are autistic in the sense that they cannot read intentions or deduce our beliefs from our actions. Even if future-generation machines can read our minds and respond accordingly, they do not take such messages "personally." They remain obedient partners, like well-trained seeing-eye dogs.

Therefore, I am not scared that self-organized AI machines will turn against us, as is often depicted in science fiction movies.[43] For robots, happiness, loneliness, and pain are absurd concepts. They have no desire to clone themselves, work together with other robots and with us, or develop agendas against humans. In the unlikely case event that they do, we can just pull the plug, and they will become lifeless and useless objects. Machines are not dangerous. We humans are.

A key aspect of externalizing brain function is the creation of measuring instruments and conceptualization of their abstract units. These instruments, such as clocks, accelerated the formation and communication of novel concepts and affected how we view the world and ourselves. Two concepts appear fundamental to our current world view: space and time, whose relationship to brain activity we discuss next.

SUMMARY

The brain areas in charge of generating plans and thoughts share many similarities with the motor cortex in terms of cellular architecture and input–output connectivity. The main difference is that prefrontal cortex does not directly innervate motor circuits. Instead prefrontal cortical areas can be designated collectively as an internalized action system, and thus plans and thoughts can be conceived of as internalized neuronal patterns that serve as a buffer for delayed overt action. Thoughts are only useful if they lead to action, even if that action is delayed by days or years. These same brain areas and mechanisms are also responsible for externalizing thought in the form of artifacts, language, art, and literature. In turn, externalized objects, as the tangible products of abstract thought, can have a profound impact on the creator's mind and on the minds of others. Thus, externalized brain function facilitates the communication of explicit knowledge, hard-earned by a few, to all members of the community, enabling the quick and efficient spread of semantic knowledge.

---

43. Futurist forecasters always make the mistake of including timelines in their prophecies. Arthur C. Clark (1987) predicted thirty years ago that, in 2017, we would ride orbiting space stations and reach old age with the strength and energy of a teenager, but he failed to predict the Internet or the collapse of the Soviet Union.

This effective externalization of brain function—not special wiring or other qualities of our nervous system—explains why we became the rulers—and potentially the destroyers—of our planet. The cognitive evolution of the human race started with tool-making and language and transitioned through an Urban Revolution to our present Information Age. In the process, the individual's relative share of humankind's collective knowledge has decreased dramatically, driving the need for a increasing cooperation for survival. Without extensive cooperation and externalized brain products, our knowledge would be set back to the level of our hunter-gatherer ancestors.

10

Space and Time in the Brain

An unmistakable difference exists between spatial and temporal concepts.
—Henrik Lorentz[1]

[T]ime is an abstraction, at which we arrive by means of changing of things.
—Ernst Mach[2]

Space and duration are one.
—Edgar Allen Poe[3]

After illustrating and discussing the advantages of the inside-out framework, let's return to the Table of Contents of William James's book (Chapter 1). I hope that I have convinced at least some readers that searching for the brain mechanisms of human-invented terms is not the best way to proceed in neuroscience. But which terms should we abandon or redefine? There are two prominent concepts in James's Table of Contents—space and time—which seem to be non-negotiable, not only in neuroscience but also in everyday life. We can live without sight, sound, smell, or touch, but everything we do seems to occur in space and time. These concepts are built into our language and thinking. We feel ourselves to be distinct from others largely because we cherish our own episodic memories that occurred in a specific place at a specific time. An implicit prediction of this framework is that experiences

---

1. As cited in Canales (2015).
2. Blackmore (1972).
3. Horgan (2015).

are broken down into components of what, where, and when, which together can recreate the original episode. Thus, if we want to know ourselves, we must understand not only how the brain stores the "what" but also how the place information and time stamps associated with those experiences are processed, stored, and recalled.

Yet there is something fishy about this postulated triad. In this chapter, I invite you to think about space and time in an open way. You may or may not agree with me, but at least you will experience a different perspective. And if I manage to convince you that time and space are human-invented concepts, perhaps we can begin to think about the rest of the items in James's list similarly: they are constructs of the human brain rather than brain "re-presentations" of real things out there.

> **Partial list of the Table of Contents of *Principles of Psychology* (William James, 1890)**
> Chapter XI—Attention
> Chapter XV—**Perception of time**
> Chapter XVI—Memory
> Chapter XVIII—Imagination
> Chapter XIX—The perception of "things"
> Chapter XX—**The perception of space**
> Chapter XXI—The perception of reality

INTERCHANGEABLE SPACE AND TIME IN LANGUAGE

Discussions of space and time often cite the philosopher Immanuel Kant, who argued that these concepts are a priori categories that cannot be studied directly. "Space is nothing but the form of all appearances of outer sense . . . [that] can be given prior to all actual perceptions, and so exist in the mind *a priori*, and . . . can contain, prior to all experience, principles which determine the relations of these objects."[4] Space and time are thus the axioms of the universe, orthogonal to each other and independent from everything else. For episodic memory, such separation also makes a lot of economic sense. If the brain had to store a list of every individual experience of our lifetime (i.e., every combination of *what*, *where* and *when*), the list would be extraordinarily long. Encoding it would require an extraordinarily large storage capacity, and recalling so many

---

4. Kant (*Critique of Pure Reason*, p. 71). For Kant, time and space are not concepts but unique preexisting "schemes": only one time and one space in the entire universe. Concepts do not preexist. They are created by some humans and explained to others.

memories would be complicated. Borges's character Funes the Memorius has an impeccable memory and can recall every single moment of his previous day's activity, but it takes him another full day to do that.[5] An alternative solution is to store the what, where, and when components separately and recreate the original episode by re-embedding the what into the where and when. Viewed from this perspective, neuroscience of episodic memory needs to focus on space and time mechanisms in the brain. But do we really know what we are looking for?

The notion of universal space and time in most cultures is used to contrast the vastness and complexities of the universe against the shortness of life. In our everyday conversations, these dimensions are often used interchangeably. In numerous languages, distance and duration are synonymous: "My lab is an hour from home," or "This was a short meeting." The longitudes of the earth are known as *time zones* because high noon occurs at different times in different countries and on different continents.[6] A unit of distance is defined by time: a light-second is the distance traveled by light in 1 s in a vacuum. Today, much of our knowledge of where we are depends on global positioning system (GPS) data, which have no meter metric at all. GPS computes position by determining the time interval that the signal takes to reach the receiver from its various satellites.[7]

Moreover, not all humans share the need for Kant's essential categories. Some cultures have no concept of the passage of time. In fact, almost half of the world's languages do not have grammatical tense (e.g., Mandarin). In all languages examined by linguists, most temporal words have a spatial sense as their primary meaning. The Amondawa, an Amazonian tribe known to the rest of the world only since 1986, lack linguistic structures that relate to time altogether. Not only do they have no word for the abstract concept of time, but they also have no words for the more concrete month or year. The Amondawa do not refer to their ages or celebrate birthdates. Instead, they change their names

---

5. Borges (1994).

6. Before the cesium atomic clock existed, the official time for both civil and astronomic use was calibrated by the orbital position of the sun and the moon based on a dynamical theory of their motions, called *ephemeris time*.

7. Satellites of the GPS constellation orbit at an altitude of about 20,000 km from the ground with an orbital period of approximately 12 hours. Local time on each satellite is computed by an atomic clock that has 1 ns accuracy. According to general relativity, a clock aboard a satellite should tick faster than a clock on Earth (corresponding to 45 μs advancement per day). If this time advance was not taken into account, the accumulated error would be almost 10 km each day, making the GPS practically useless.

each time they achieve a different status within their community.[8] Similarly, for the Aborigines of inner Australia in the Great Victoria Desert, the concept of Creation exists in the past, present, and future all at once. As they do not believe in tomorrow, they keep no possessions. If food is left from one day to the next, they simply abandon it, even though food is scarce in the desert. Today is today, not a continuation of yesterday. They pass songs or "songlines" from generation to generation to tell of the travels of the creator across the land. The songlines describe the Creator's path, including the location of mountains, waterholes, landmarks, and boundaries, and thus function as a verbal map for navigation.[9] When members of another Australian aboriginal tribe were asked to arrange a set of pictures that illustrated temporal progression (e.g., pictures of a boy, man, old man), they did not arrange the pictures from left to right, as we normally would. Instead, they arranged them from east to west. That is, when they were facing south, the cards were placed from left to right. In contrast, when they faced north, the cards were arranged from right to left. When they faced east, the cards came toward the body and so on.[10] Thus, even without the concept of time, these people can understand ordering, sequences of events, or arrangement, but they do not have a notion of time as independent of all other things or as a context that events occur within.

The indigenous Wintu people, who have lived in the Sacramento Valley, California, do not distinguish personal space from allocentric space. They do not refer to the left or right related to their body scheme but to the river side or mountain side, depending on whether they travel north or south, the two key directions in the valley.[11] In contrast, Finnish people, who live in a flat land with thousands of lakes, have separate words for the cardinal directions (S, E, N, and W = *etelä, itä, pohjoinen,* and *länsi,* respectively) and the intercardinal directions (SE, NE, NW, and SW = *kaakko, koillinen, luode, lounas,* respectively). Far from Finland, members of a small Kuuk Thaayorre aboriginal community on the western edge of Cape York, in northern Australia, also use cardinal directions

---

8. Of course, this only means that the Amondawa people live in a world of events with sequences, rather than seeing events as being embedded in continuous time (Sinha et al., 2011).

9. Songlines are essentially oral maps of the vast desert, enabling the transmission of navigational skills in cultures that do not have a written language (Wositsky and Harney, 1999).

10. Boroditsky (https://www.edge.org/conversation/lera_boroditsky-how-does-our-language-shape-the-way-we-think).

11. In Wintu culture, the self is continuous with nature; therefore the self never vanishes (Lee, 1950). There are, in fact, mechanisms in the brain that can code for such information. Neurons in the post-subiculum of mice can integrate head direction information with the presence of a wall to the left or right. All we have to decide is what to do with the time that is given us. Thus, the boundary between egocentric and allocentric information is blurred (Peyrache et al., 2017).

even when they talk about personal (egocentric) space: "There is a spider on your east ear." So if you are one of the Kuuk Thaayorre clan members, you need to know which way you are facing to identify the whereabouts of the spider.[12]

These cultural differences suggest that space and time may not be inseparable ingredients of neuronal computation, but instead mental constructions represented only by trained brains. Our concept of memory requires space and time because our thoughts need to be placed into an imagined coordinate system. But these coordinates may exist only in our conceptual world.

If space and time are immaterial, how can they "cause" any change in the brain, which does not have specialized sensors to perceive them? Finding a reasonable answer to such a difficult question first requires a quick tour of space and time as seen by physicists.

SPACE AND TIME IN PHYSICS

Conceptual space and time are dimensionless and unmeasurable so they cannot be studied directly. Modern science transformed these concepts by replacing the abstract philosophical versions with their measurable variants: distance and duration.[13]

Defining Space and Time

The new tools introduced to calculate distance and duration allowed our predecessors to address important questions, such as "Where am I?" [14] and

---

12. Levinson (2003).

13. Measuring instruments as the yardstick of truth have only been adopted recently. Galileo palpated his pulse and heartbeat to determine a pendulum beat in the tower of Pisa, not the other way around (Canales, 2015)! According to Richard Feynman and colleagues (1963) "What really matters anyway is not how we define time, but how we measure it." However, if we accept that distance and duration can only be interpreted with the help of human-made rods and clocks, we deny that non-human animals can feel and use these dimensions.

14. Mishkin, Ungerleider, and colleagues (1983) distinguished two hierarchically organized and functionally specialized processing pathways: a "ventral stream" from the visual cortex to the inferotemporal cortex for object vision (what) and a "dorsal stream" from the visual cortex to the posterior parietal region for spatial vision (where). Goodale and Milner (1990) pointed out inadequacies in such pathway–function assignment and suggested that instead the ventral stream plays a role in the perceptual identification of objects, while the dorsal stream mediates the required sensorimotor transformations for visually guided actions directed at

"What time is it?" However, the relationship between distance and position, or duration and clock time, is inherently circular. Position is the end point of a displacement, relative to an arbitrarily defined first position. A similar circularity also applies to the instant or "now," which is defined as the end of a duration relative to an arbitrarily defined beginning.[15] Therefore, the question should be rephrased to "What agreed date/time is it?" because time depends on who is measuring it. As I write this paragraph, it is February 17, 2017, by the international calendar, which corresponds to 21 Jumada I 1438 in the Islamic calendar, 22 Shevat 5777 in the Jewish calendar, and 30 Bahman 1395 in the Persian calendar. The calendars show duration from an arbitrary beginning, and thus the definition of "now" depends on the observer's viewpoint.

If space and time require different instruments to measure their quantities, they may have different qualities. In classical physics, the time axis is added to the three dimensions of space. Distance and duration cross each other only at a single point of time and position. This inert and empty theater of space and time allows movement of things within. Special relativity continued in this tradition and always gave the exact location and speed of a particle.[16]

However, even before Newton, physicists recognized that space and time are related to each other through speed, which is the ratio of distance and duration.[17]

---

such objects. Thus, the function of the dorsal stream is not to extract the *where* but to support object manipulation (*how*).

15. In *Being and Time* (1927/2002), Heidegger wonders "How is this mode of temporalizing of temporality to be interpreted? Is there a way leading from primordial *time* to the meaning of *being*? Does *time* itself reveal itself as the horizon of *being*?" He suggests that the Being (*Dasein* or existence) is not a question about *what*, but about *who*. This existentialist question is reminiscent of the constant inquiry about the observer's role in interpreting the world.

16. The assumption of an exact location of a particle location violates Heisenberg's uncertainty principle. In quantum physics, the instantaneous velocity of a particle is independent of its exact position. Both can never be recorded simultaneously due to the particle-wave equivalence. Any measurement affects either the particle or the wave state. Because waves spread out in space (and time), the phase is meaningless without knowing the frequency. On the other hand, frequency can be determined only over a space (time) span. This uncertainty is also the fundamental source of the debate about the noise versus rhythmic nature of the local field potential or electroencephalogram (EEG) because frequency and time are independent from each other. The presence or absence of an oscillation can be quantitatively determined only over a time span.

17. The "Mertonian mean speed theorem," named after fourteenth-century scholars at Merton College, Oxford, is credited for the modern formula of velocity as the ratio of displacement and duration. However, recently discovered clay tablets of cuneiform writings have indisputably proved that Babylonian astronomers already used geometrical methods as abstract mathematical space and calculated Jupiter's position from the area under a time-velocity graph (Ossendrijver, 2016). To be precise, speed is not distance divided by duration, but by some

Knowledge of any two parameters is sufficient to fully describe the third. As discussed in Chapter 2, when two things are invariably correlated, we often assume that one causes the other or that both have a common cause. Does distance cause duration or vice versa? Or do they have a joint explanation? The independence of space and time in classical physics has been questioned by many. If time is a medium through which things travel, time does not exist without them. To define time, some particles must move across space. Conversely, no event can happen in no time or as Hermann Minkowski, Einstein's teacher, phrased it: "Nobody has ever noticed a place except at a time, or a time except at a place." [18] Time is chronicled by matter.

Time Is an Arrow

In classical thermodynamics, heat flows from hot bodies to cold ones. Every moment is different because collisions of the molecules produce more collisions and therefore friction and heat. The second law of thermodynamics states that the universe moves to increasingly higher disorder, quantified by the term "entropy,"[19] and never backward. According to Arthur Eddington, this one-way flow of entropy is the reason that time moves forward, and it explains the "arrow of time."[20] If time is an arrow, it is a vector, and as such needs to be characterized by two measures: duration and direction. One should notice, though, that increased entropy is only a metaphor and not equal to time or duration. Importantly, using the logic of "entropy explains time," time should move backward in biological systems since in self-organized system entropy decreases.

---

tangible and measurable change, such as another convenient distance measure or the number of swings of a pendulum. Only when clock time measurements are available does the distance divided by duration formula become valid for speed calculation.

18. As cited in (Barbour, 1999).

19. Entropy—*energy trope* or transformation (*en-trope*)—is the log of the number of quantum states accessible to a system.

20. "The great thing about time is that it goes on" (Eddington, 1928). Eddington is credited for the expression "time arrow." Heraclitus's *panta rhei* refers to a similar idea. However, a recent experiment shows that heat can spontaneously flow from a cold quantum particle to a hotter one under certain conditions (Micadei et al., 2017). For a recent discussion on the direction of time, see Zeh (2002). The directed notion of time makes it a vector, in contrast to the scalar duration. For a discussion and critique on the relationship between entropy and time arrow, see Mackey (1992) and Muller (2016).

The Big Bang theory also supports the time-arrow idea. The Big Bang is often considered as the beginning of everything—the point from which the universe expanded in space and moved forward in time. The theory received a major boost in 2016, when astronomers identified gravitational waves likely originating from a pair of neutron stars or super-massive black holes.[21] However, preexisting matter did not explode within preexisting space. Instead, the Big Bang created everything, including what we call space and time. Thus, there may be no need to consider these terms separately because expansion of space is inherently linked to dilation of time. In contrast, another theory, known as *inflation*, suggests that an exponential process of increases in both space and mass preceded the Big Bang, even if its duration was super short. The Big Bang time may apply only to our part of the universe.[22] Yet another altenative is the universe undergoes an infinite number of Big Bangs with recurring cyclic expansion and contraction of space and time. This view of uncaused eternity is of course less attractive to scholars in search for causal explanations of things with clearcut beginnings.

Spacetime in Physics

In the nineteenth century, the discovery of the constant speed of light and the recognition that light disobeys the principle of causality[23] paved the way to theories of relativity, which rest on an equivalence between distance and duration and the symmetry of time.[24] The classic model was abandoned in favor of a spacetime model of general relativity, where the time axis is a fourth

---

21. Steinhardt and Turok (2002). A concise account of gravitational field is by Cho (2016).

22. Guth (1997); Tegmark (2014).

23. The experiment by Michelson and Morley (1887), and confirmed by many, refuted "ether" as a medium through which light and electromagnetic fields travel. They showed that the speed of light (~300,000 km/s in a vacuum) is the same whether it is measured in the direction of the Earth's motion or at a right angle to it. Light thus has special qualities; its speed is not accelerated or decelerated by other things moving.

24. Hermann Minkowski formulated relativity theory, in which space and time were coordinates in an abstract four-dimensional "spacetime"; the time coordinate was imaginary. In Minkowski's spacetime, relative times and spaces appear as projections relative to an observer. "Henceforth space by itself, and time by itself, are doomed to fade away into mere shadows, and only a kind of union of the two will preserve an independent reality" (Minkowski, 1909). Prior to Minkowski, spacetime unity was already postulated in a literary form by H. G. Wells in his *The Time Machine* novel: "There are really four dimensions, three of which we call the three planes of Space, and a fourth, Time."

dimension. Einstein postulated that the time of an event affects its location, and vice versa,[25] so that distance and duration cannot be measured independently from each other.

All major theories of twentieth-century physics have been based on the spacetime continuum model. There is no change in spacetime; it does not have a past, present, or future. Spacetime of the universe is like a movie that contains the entire story. But we are looking at the universe through a narrow slit, giving the illusion that what we are watching through the slit is the now. What we recall from memory is the past, and what we attempt to deduce from the past is the future (see Figure 2.1). Thus, from the human observer's point of view, the world is in perpetual motion, while from the view of a very distant observer, nothing changes. Conversely, the Earth is not moving from my current perspective, but it does when I view it from a spaceship. In general relativity, time travel is symmetric. Einstein famously declared that "the separation between the past, present, and future, holds nothing more than a persistent, stubborn illusion, however strong it may be."[26]

Time is reversible: it can be compressed or expanded. It can be perfectly described by clocks without a conscious human brain to sustain it.[27] There is a catch here, though. Clocks neither make time nor represent it. Time does not mean anything to a clock. Ticking has no abstract meaning without an observer.

If time always moves in one direction, but a particle can return to its original place, displacement has no temporal correspondence. That is, no event can happen in zero time. A photon can never rest because at rest it would have zero energy, and so it would not exist. In relativity theory, past and future are completely symmetric, and thus time is scalar. The concept of "now" is irrelevant. In fact, the theory would still hold if the planets revolved in the opposite direction or if the universe went backward. Many interesting things can happen when time moves backward. A positron becomes an electron moving backward in time, as postulated by Richard Feynman, and its positive charge is therefore only an illusion due to negative time.[28]

25. Einstein (1989, 1997). [NeuroNote: Einstein's son Eduard was schizophrenic.] Many excellent books are available about the relationship between space and time and the implications of the spacetime concept (Canales, 2015; Muller, 2016; Rovelli, 2016; Weatherall, 2016; Carroll, 2000).

26. While Einstein put this statement in writing, it was not in a scientific publication but in a personal letter sent to Michele Besso's sister after Besso's death. In the theory of relativity, light is the grounding element of truth, to which everything relates secondarily.

27. I assume that Einstein made this statement out of frustration when he realized that his audience did not comprehend his new spacetime formulation. In the everyday, practical world, clock time tells you what you need for your regular business.

28. In general relativity theory not only the flow of time is an illusion, but change itself as well. Several excellent books discuss these interesting concepts, including the idea that the concept

Is Time Redundant?

While the theory of relativity and quantum mechanics are both internally consistent and beautiful, they contradict each other.[29] The heart of the dispute is whether the world is continuous or consists of discrete quanta. Several attempts have been made to combine these vast fields of physics, most recently in the theory of loop quantum gravity.[30] Here the grains of space make up discontinuous space. This resonates well with the comment of French philosopher Henri Bergson, "When the mathematician calculates the future state of a system at the end of time *t*, there is nothing to prevent him from supposing that the universe vanishes from this moment till that, and suddenly reappears."[31] For Bergson, time is not something out there, separate from those who perceive it. However, the grains of space are not simply quanta, and they are not in space, as the quanta themselves make up space by their sheer interactions. The theory that describes the interaction of space and matter does not contain time, even though everything is in continual motion.

The apparent contradiction between physics and the feeling of the flow of time has upset many people, including the German philosopher Martin Heidegger, who concluded that physics cannot reach the most fundamental aspects of reality. Equations are simply symbols, but reality may be different. "Once time has been defined as clock time, then there is no hope of ever arriving at its original meaning again."[32] But the problem is perhaps different from what Heidegger

---

of time is not needed to describe the physics of the world (Toulmin and Goodfield (1982); Barbour, 1999; Carroll, 2000; Greene, 2011; Tegmark, 2014; Rovelli, 2016). In contrast, Lee Smolin (2013) forcefully argues that the most essential feature of the universe is time, and physics took the wrong path when it fused it with space. Smolin believes that the laws of physics constantly evolve in time. Gravity may act differently at different points in time, so the past, present, and future can be still distinguished. We may still be surprised about what happens in the next moment.

29. To know that duration is *time*, intervals need to be calibrated against something else. To calibrate the duration between the swings of a pendulum, another instrument is needed (e.g., a faster pendulum or clock with ticks), and so on. Even the fastest atomic clocks still measure time by quanta. According to quantum theory, spacetime is discrete. With clocks, we do not measure time per se but instead count quanta and multiply them by duration. But duration is measured by counting faster quanta, and so it goes *ad infinitum*.

30. Smolin (2013); Rovelli (2016).

31. Bergson (1922/1999); Canales (2015).

32. Heidegger (1927/2002). [NeuroNote: Heidegger had a major "nervous breakdown" after World War II, possibly due to his guilt for aligning with the Nazi regime combined with his existential uncertainty.]

thought, and physics alone may not be able to provide answers to the space and time issue. Instead, it is up to neuroscience to address them.

SPACE AND TIME IN THE BRAIN: REPRESENTATION OR CONSTRUCTION?

So far, these discussions within physics and between physics and philosophy have had little impact on neuroscience. Instead, Newtonian space and time are present in every nook and cranny of neuroscience. [33] Research on space and time in the brain has generated two independent fields with separate literatures while largely ignoring the possibility of space-time equivalence. Neuroscience researchers continue to perform experiments in the framework of classical physics using practical rod and clock measurements. There is no problem with such an approach, as long as we are aware that our measurements are based on human-designed instruments. But it is entirely different to assume that correlations with such metrics can illuminate the concepts of space and time.

The Grounding Problem of Space and Time

An alternative to taking physics seriously is to question whether the laws of physics apply to "psychological" time. After all, the math that describes those laws is also a product of human thought.[34] Math is an axiomatic system, always starting with some assumptions to build internally coherent and beautiful theories. Perhaps physics is just a type of science that has laws that are different from biology. After all, increasing entropy is the rule of the inorganic world, whereas decreasing local entropy is the norm in biology. Prediction or even determinism rules the nonliving world, whereas prediction becomes an increasingly more complex problem in biology.

---

33. [NeuroNote: Isaac Newton had what today we would call bipolar personality with grandiose delusions at the end of his life. He thought he was appointed by God to describe truth to the world.

34. This argument can be fueled further by Kurt Gödel's theorem, which traumatized mathematicians and scientists alike: all mathematical theories are incomplete or, in its milder form, any theorem or set of definitions is incomplete. Mathematical truth cannot be challenged through the use of logic. Gödel's argument can be also extended to neuroscience. No branch of science is superior to another. [NeuroNote: Gödel suffered from bipolar disorder, as did the physicist Ludwig Boltzmann, the mathematician Georg Cantor, and many other truly great minds of science.]

A related argument against "physicalism," one often used by philosophers, is that the world does not come with built-in units of distance and duration. Human-invented instruments do not create space or time, nor do they measure them. A rod is just a rod. A clock just ticks, and its ticking moments must be compared to some other change, such as the distance traveled by light in a given time unit. This comparison or calibration needs be made by a third party—a human. Because humans define the units of rods and clocks, this process inevitably defines our notion of space and time.

The problem of the human observer and the need for grounding are analogous to the problems in neuroscience that we discussed in Chapters 1 and 3.[35]

There appears to be a vast gap between the classical or macroscopic world that we live in and the microscopic world ruled by relativity and quantum physics. We can ignore this gap, or we can try to examine the sources of the apparent contractions and the boundaries between the macroscopic and microscopic.[36]

Whatever path we take, we need to face at least two issues. First, we must recognize the difference between the spatial and temporal properties of the *process of representing* something and the *representations* of space and time. Even if neuronal activity is reliably correlated with the spatiotemporal sequence of events (measured against a rod or timer), such correlations do not mean that neuronal activity computes distance or duration per se.[37] In other words, we must not conflate the description of events with their subjective interpretation (Chapter 2). Second, we need to clarify whether the *time* we talk about in neuroscience experiments corresponds to the time of the universe (physics), the "lived time" of philosophy or "practical time" as measured by the clock. The same scrutiny should apply to space as well. The essence of the problem is that the human brain interprets both the meaning of its responses, as measured against instruments, and the meaning of the units measured by instruments without independent grounding. Measuring relative space and time via rods and clocks is essential for relating and contrasting experiments across laboratories and has yielded enormous progress in neuroscience. Yet it does not get to the heart of the problem: What do space and time mean for the brain?

---

35. Before precise instruments to measure time were available, people thought that perception and memory were instantaneous. Indeed, without a comparison to something else, no counterargument could have been made because it feels like no delays occur.

36. The bestseller by Rosenblum and Kuttner (2008) discusses many aspects of quantum physics and its implications for a variety of complex issues, including human consciousness. Unfortunately, nearly everything I have read on the complex issues of physics eventually looks for solutions in God, consciousness, and the like, rather than examining the sources of inconsistencies between internally consistent theories.

37. Dennett and Kinsbourne (1992).

SPACE IN THE WORLD VERSUS SPACE IN THE BRAIN

The brain mechanisms for measuring space and time are inevitably inferential because, while we can sense sight, sound, odor, touch, or our movements through specialized receptors, we have no sensors for space and time. In other words, with the outside-in approach, we cannot track down its existence. Even so, neuroscientists initially defined space from an egocentric perspective that reflects how our sensors feel the world out there. This approach led to the view that many different realms of space exist depending on the perceptual system in use. When the body and head are fixed, but the eyes can move, the space is the visual field itself, called *visuo-ocular motor space*. When head movement is allowed, we encounter the *visuo-cephalo motor space*. Locomotion leads to *visuo-locomotor space*. Even visual-ocular motor space can be divided into *foveal* and *peripheral fields*. Similar divisions were applied to auditory, olfactory, and somatosensory space. An object in the hand sits in *palm space*. A candy in your mouth is in *oral space*. The things within range of an arm movement correspond to *reach space*. *Instrumental space* can be defined as things we perceive through instruments. When riding in a car, the bump we feel when driving through a pothole is referenced to outside the car. We can expand our visual space by orders of magnitude by looking through a telescope.

Although the multitude of space terms in these early experiments may appear bizarre, pioneers of egocentric space exploration provided experimental support for such ideas and found brain-damaged patients with selective deficits in several of these categories. When children were asked to identify mirror-image objects by using both hands without vision, performance was worse when their arms and wrists were immobilized so that only the fingers could move, compared to another condition when the arms and wrists were free to move but the fingers could not perform multidigital touch. The interpretation is that identifying asymmetry or symmetry is easier when the perception can be grounded to another axis; in this example, the hand position in reaching space relative to the postural body space. Similarly, congenitally blind children learning bimanual Braille reading are more confused by patterns that are mirror images of the other Braille patterns than are sighted children of the same age. Possibly, visual information aids the placement of palpated objects in both tactile space and body-centered space.[38]

---

38. From numerous experiments, Jacques Paillard concluded that mental "maps" of the external world are constructed by the body's own movements (Paillard, 1991). The experiments using blind children were performed by Martinez (1971). This framework was perhaps driven by Nicolai Bernstein's model suggesting that movement is not driven by neuronal representation of individual of joints and muscles but instead guided by space representation (Bernstein, 1947/1967).

Patients with posterior parietal cortex damage illuminate the difference between object and space perception. These patients have difficulty in following a moving object, pointing or reaching to a visible target, learning routes, and recognizing spatial relationships, but they have relatively few deficits in navigation. The personal space effects are more dramatic when the damage affects the right hemisphere than the left hemisphere. These individuals neglect their contralateral body space, shaving only one side of their face or finishing food only from one side of the plate. If asked to copy a picture, such as a clock, they often draw only one half. When a well-studied patient in Milan, Italy, was asked to imagine himself facing the Piazza Del Duomo and describe the scene, he correctly identified buildings on his right but could not recollect things on the left. When he was asked to imagine standing at the opposite end of the Piazza, he listed the buildings and structures on the other, previously neglected, side, which was now to his right.[39] Such "hemi-neglect" patients can perceive and recall objects per se but are unable to access them from imagery or describe them in proper geometrical relationship. An image on the retina is not sufficient to perceive space. The brain must also know where the eyes and head are pointing.[40]

Physiological experiments in monkeys corroborate these observations in humans. The posterior parietal cortex and its partners form a body-centered coordinate system that encodes spatial relationships between objects and the body.[41] Population firing patterns of neurons in the parietal area combine inputs from the environment and various parts of the body. However, this area does not contain a topographic map of either the body or the environment.

---

39. Holmes (1918) described deficits in spatial orientation in soldiers after World War I with penetrating missile wounds in the parietal lobe. The paper by Bisiach and Luzzatti (1978) was the first to describe that patients with right posterior parietal damage not only ignore things to their left in their vision but also in their imagery. See also van den Bos and Jeannerod (2002).

40. Organizing hand trajectories and other multijoint movements is a super-complex task, yet we do it without effort. *Trajectory* refers to the configuration of hand in space from its initial to final position. Giving distance and duration commands to multiple muscles on the fly would require extraordinary computations. Instead, it seems that a higher order plan of geometric simulation guides movement. Signatures written on a very small piece of paper or on a large board by the same individual are scaled versions of each other. This fact illustrates that trajectory planning and execution are size invariant, which would be difficult to achieve with distance–duration transformations only and without a normalization process (Flash and Hogan, 1985).

41. Numerous psychophysical observations and clinical data with parietal cortex–damaged hemineglect patients had been difficult to reconcile under the assumption that a fixed coordinate frame is used for object localization. However, when positions of objects are represented in multiple reference frames simultaneously in a model of parietal neurons, the inconsistencies diminish (Pouget and Sejnowski, 1997a, 1997b).

Instead, the main job of this network is to integrate such information from the many inputs it receives, several of which do have topographic maps.[42] Such brain areas include the premotor cortex and the thalamic nucleus putamen, where neurons represent visual space near the face in "head-centered" coordinates and the space near the arm in "arm-centered" coordinates.

The parietal cortex also has prominent connections with the dorsolateral prefrontal cortex. The so-called memory fields in the prefrontal region are topographically organized and respond when the eyes move to a remembered position in two-dimensional space. This idea is beautifully illustrated by the induction of a "memory hole" in a circumscribed area of the remembered visual field by a small cortical lesion. Objects presented in the visual field without parietal-prefrontal cortical representation are seen, but they cannot guide the eye to focus on them. Overall, these and many other experiments[43] show how sensory inputs are converted not just into two-dimensional scenes and isolated objects, but also into relationships between the objects and the body. Such mappings can be acquired only via active movements to ground the relationships.[44]

A general principle that emerges from physiological experiments is that the brain does not encode any general and uniform space somewhere within its networks. Instead of a single master coordinate system, most behaviors require multiple coordinate systems and spatial structures represented relationally in different brain regions in varying formats.[45] The next issue to address is how

---

42. Multimodal signals are combined in the lateral intraparietal cortex (LIP) and area 7a of the posterior parietal cortex. The integration of visual, somatosensory, auditory, and vestibular signals can represent the locations of stimuli with respect to the observer and within the environment. The dorsal medial superior temporal area (MSTd) combines visual motion signals with eye movement and vestibular signals. This integration is believed to be critical for specifying the path on which the observer is moving (Andersen, 1997). Many LIP neurons fire in anticipation of eye movements and are part of the corollary loop of movement planning (Chapter 3).

43. These short paragraphs did not do justice to this very large area of active research. For reviews, see Goldman-Rakic et al. (1990) and Gross and Graziano (1995).

44. Transformation may be symmetric when properties remain unchanged, such mirror symmetry, rotational symmetry (rotating the coordinated), and translational symmetry (shifting the coordinates), or asymmetric ("symmetry breaking") when the transformation produces new qualities; for example, the transformation between sensory cues and an abstract feature, such as a name associated with a face. The "Jennifer Aniston" or "grandmother cells" in the hippocampus (Quian Quiroga et al., 2005) have nothing to do with the intrinsic properties of those neurons (they are just regular pyramidal cells), but their property arises from their complex relationship to other neurons. For a general model of transformations, see Ballard (2015). Properties thus reflect mainly relationships.

45. Graziano et al. (1994).

eye- and body-centered (egocentric) spatial representation is transformed into a world-centered (allocentric) coordinate system.

Navigation in Space and the Hippocampal System

Another major target of the posterior parietal region is the parahippocampal cortex, which projects to the hippocampus by way of the entorhinal cortex. The computations of hippocampal circuits primarily relate to spatial navigation. Hippocampal and entorhinal principal neurons have place fields and grid fields, which explicitly define the x-y coordinates of a two-dimensional environment to generate a map (see Chapters 5 and 7). The positions defined by these cells collectively provide an allocentric, nonvectorial representation of space. This spatial map theory, proposed by John O'Keefe and Lynn Nadel, was inspired by Kantian philosophy.[46] In this model, space is not an independent entity out there; the brain constructs it from parts and relationships. How does the brain do this, given that there are no dedicated sensors for space per se? Distance and duration are not derived from first principles. Instead vision, hearing, olfaction, and proprioception are used to deduce the location of and distances to objects. As the animal moves through the environment, the map is formed by multiple mechanisms, including counting the number of steps, self-motion–dependent optic and somatosensory signals, and vestibular acceleration signals.[47]

---

46. Kant (1871). In *The Hippocampus as a Cognitive Map*, O'Keefe and Nadel (1978) discuss Kant's view of space and how Kant's view relates to allocentric definition of space (pp. 23–24).

47. Various combinations of activity in the otolith organs and semicircular canals uniquely encode every translation and rotation of the body (Angelaki and Cullen, 2008). This information is complemented by signals from the retina. A subset of ganglion cells, the direction-sensitive ganglion cells, is specialized for detecting image motion, called *optic flow*. These neurons mainly encode two motion axes of high behavioral preference: the body axis and the gravitational axis. Translation movement ("direction of heading") induces optic flow that diverges from a point in space around the animal and follows lines of longitude in global visual space. Rotary optic flow follows lines of latitude, circulating around a point in visual space. Subtypes of direction-sensitive ganglion cells align their preferences with one of four cardinal optic flow fields and encode the gravitational and body axes, which are decoded by the brain into translational or rotational components. Binocular ensembles form a panoramic view, and their combinatorial activity signals when the animal rises, falls, advances, or retreats (Sabbah et al., 2017). This system is hard-wired and works hand in hand with the vestibular system. Motion-induced somatosensory stimulation, such as whisker activation by walls, also produces valuable signals, called *haptic flow*. The vestibular signals compensate for eye and head movements to stabilize the retinal image. Discrepancy between visual and vestibular signals induces motion sickness in a car, airplane, train, roller coaster, or even a virtual reality simulation.

TIME IN THE WORLD AND IN THE BRAIN

Our travels and memory recall feel as though they involve time.[48] Is time really critical for brain computations? Many experiments have explored the brain mechanisms that may support timing. We often argue that timing is critical to survival in the real world. A well-timed extension of the paw may mean lunch for a predator, and a quick jump at the right moment may save the life of its prey.[49] Estimation of elapsed time (duration) is essential for anticipating events, expecting reward, preparing decisions, planning actions, and structuring working memory. Time researchers often make a distinction between perceptual-motor timing at the subsecond scale and cognitively mediated interval detection at the suprasecond scale. They associate these scales with mechanisms supported by the cerebellum for short durations and networks in the basal ganglia and in the prefrontal, motor, and parietal cortical regions for longer durations.

Three procedures are used to investigate the passage of time: estimating an event duration, producing intervals, and reproducing intervals, as in a syncopation task. In animal experiments, the experimenter programs the apparatus so that reward is available at, say, 10-s intervals. The participant then reports the learned rule by strategically concentrating responses near the end of the target interval. Many animals are good at this game, giving the impression that their nervous systems have specialized time-keeping mechanisms. After sufficient training, the mean response intervals will correspond to the target intervals. The dispersion of the responses around the mean is proportional to the criterion duration in all species investigated: short interval targets evoke small errors, whereas longer target intervals evoke proportionally larger errors. The

---

48. I expect that the immediate reaction of the reader is to say that "integration occurs over time." Yet, no matter how natural this link feels, integration is a sequential operation and is just the same without time. Mathematical integration does not need a temporal factor. Physical instantiation of integration needs succession but no time. Integration can occur quickly or slowly, but this speed parameter is not time per se either and can be substituted by the engineering term "gain" or rate of change that can accelerate or decelerate sequences (Chapter 11). Several chapters of Hoerl and McCormack (2001) contain interesting discussions about the relationship between time and memory.

49. Gallistel and Gibbon (2000) suggest that computation of time by some hypothetical brain circuit represents an elementary aspect of cognition. Alternatively, brain rhythms can serve the same purpose, temporally coordinating neuronal events across structures, without dedicating timekeeping mechanisms (Buzsáki, 2006). However, all neurons that take part in brain oscillations compute other functions as well. Here it is important to reiterate the broad definition of computation, which is calculating and describing the relations between algorithmic steps.

lawful relationship between the mean and the distribution of error responses, quantified by standard deviation, is termed the *scalar property* of interval timing mechanisms. What brain mechanisms are in charge to calculate intervals?

Timekeepers in the Brain?

With analogy to ticking clocks and hourglass sand timers, two different brain mechanisms have been assumed to track time: *neuronal clocks* and *ramping timekeepers*. A candidate mechanism for neural clocks is brain rhythms, which can produce relatively discrete "ticks" with a frequency span over several orders of magnitude (Figure 10.1). An example ramp mechanism is the accumulation or integration of spikes of many individual neurons. Integrating spikes over time is analogous to sand flowing through the hourglass. The integrator can be reset at any point (i.e., at preselected threshold), a command is forwarded to a response system, and the integration can start all over again. Thus, while the

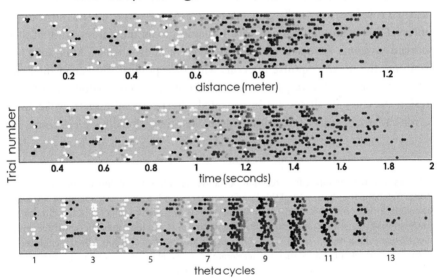

Figure 10.1. Clock time versus rate-of-change in the brain. Place cell assemblies in the hippocampus are organized by theta oscillations ("brain time"). Each raw of dots is a trial of the spiking activity of ten place cells (indicated by different shades of gray). *Top panel*: Trials are shown against distance of the maze. *Middle panel*: Trials are shown against elapsed time from start. *Bottom panel*: Trials are shown as phase-locked activity of neurons in successive theta cycles.
Courtesy of Eva Pastalkova and Carina Curto.

integrator model can have many real intervals as measured by the accumulated sand, the clock has discrete ticks and only virtual intervals between the ticks. With some variation, all timekeepers follow these basic ideas. Duration estimation based on counting spikes deteriorates over time so that the estimated error at any time point is proportional to the elapsed duration since the beginning of the integration (Chapter 12).[50]

Where are the timers in the brain? This question has been asked repeatedly, and two solutions have been offered. An early idea was a central clock that distributes ticks to various brain regions.[51] However, no such brain command system has been found for intervals of milliseconds to minutes.[52] A more contemporary idea is that time is created in each brain system[53] according to local needs. Thus, time is everywhere but calibrated locally and coordinated across networks when needed. As discussed in Chapter 7, any network that can support the self-sustained, sequential activation of neuronal assemblies can potentially track the passage of time. Yet, the local timer model has its own problems. Physiological, pharmacological, and lesion experiments inevitably face the

---

50. I am aware that these short sentences reflect only a telegraphic summary of the large literature on the psychometric and neuroscience of timing behavior. Luckily, several excellent review papers are available for the interested reader: for example, Church (1984), Michon (1985); Gibbon et al. (1997); Buonomano and Karmarkar (2002); Mauk and Buonomano (2004); Ivry and Spencer (2004); Nobre and O'Reilly (2004); Buhusi and Meck (2005); Staddon (2005); Radua et al. (2014); Mita et al. (2009); Shankar and Howard (2012); Howard et al. (2014); Buonomano (2017).

51. Church (1984). A recent update to these internal clock ideas is that the striatum provides the beats and distributes them to the rest of the brain as needed (Matell and Meck, 2004). However, it is unlikely that the brain has a universal master clock. As in physics, time is a ubiquitous, localized phenomenon for each structure and function in the brain, referenced to some other measure (such as clock time). Time is a relation of events.

52. Neuronal mechanisms that track time at minutes-to-hours scales are scarce. It is believed that such long timing relies on external mechanisms, such as hunger, sleepiness, light, and temperature.

53. In the human EEG literature, the best-known correlate of such time accumulation mechanisms is the *contingent negative variation* (CNV; Walter et al., 1964) and its *Bereitschaftspotential* kin (Kornhuber and Deecke, 1965) that builds up with maximal amplitude above the supplementary motor area seconds before a voluntary action. Because these electrical patterns begin to emerge before the subject is aware of moving, they triggered an intense discussion about free will (see, e.g., Libet, 1985). Error correction in the hippocampus is achieved by the distance–time compression mechanisms of theta oscillations. In each theta cycle, segments of navigation distance and segments of memory and planning are replayed repeatedly and in an overlapping manner. Approximately seven items are repeatedly replayed in multiple theta cycles, and each subsequent cycle gains and loses only one new and one old item, respectively (Lisman and Idiart, 1995; Dragoi and Buzsáki, 2006; Diba and Buzsáki, 2008). This is an effective error-correcting safety mechanism.

problem of dissociating the temporal sequences of representations from the representation of time or duration per se. If behavioral sequences and time are so intricately intertwined that one can perfectly predict the other, why do behavioral sequences need a time label? Let us examine how behaviors are modified under external clock control.

Behavioral Mediation of Time

In interval-timing experiments, animals are always doing something. They figure out ways to pass time by generating stereotypic behaviors. Rodents superstitiously circle around, scratch the wall or the food dispenser, rear repeatedly, or otherwise occupy themselves to avoid a premature press of the lever, especially if that action comes with a penalty. If animals are given the opportunity to run in a wheel, dig in wood shavings, climb ladders, or crawl through tunnels, their timing performance drastically improves compared to an environment where there is not much to do.[54] Humans are not very different. We scratch our heads, bite our nails, or sip coffee while waiting for an idea to pop up in our brains. When in line for a table in a restaurant, we engage in conversations with strangers, check our i-Phones, or impatiently bug the host. Just watch teenagers rhythmically move their legs in the doctor's office while waiting for their appointments. Timekeeping involves doing.

"Time Cells" in Navigation Memory Systems

Without a ruler, distance can be calculated by a time-generating mechanism combined with instantaneous travel velocity, as in the GPS system. All ingredients needed to calculate a route, including speed, duration, and direction, are present in the hippocampal-entorhinal system. Many neurons are modulated by velocity in these structures, where directional input is provided by the head direction system (Chapter 5).

Elapsed duration can be tracked by the accumulation of neuronal spikes in a cell assembly sequence relative to a time-measuring instrument. For example, the passage of time is faithfully recorded by internally evolving neuronal trajectories that maintain information about past memories and planned goals without spatial displacement of the animal (Figure 7.6). The duration estimation error from such cell assembly sequences does not increase proportionally

---

54. Staddon and Simmelhag (1971).

with elapsed time, but stays relatively asymptotic even after delays of tens of seconds.[55] Based on activity from a hundred neurons in the hippocampus, the accumulated error after 15 s can be less than 20%. This timekeeper may not sound very impressive, at least compared to clocks that have been perfected by humans for centuries, but by increasing the number of recorded neurons, the estimation error can be substantially reduced. Moreover, such precision is good enough for the brain because errors of this size are well-tolerated when computation spreads over tens of seconds.

Why would timekeeping be an important function of the hippocampal system? The typical answer is that the hippocampus and its partners are in charge of our episodic memories, which by definition must be placed into space and time. The postulated need for temporal embedding of episodic memories and the ability of cell assemblies to track time effectively prompted Howard Eichenbaum at Boston University to dub hippocampal and entorhinal neurons "time cells."[56] His group designed a series of experiments to dissociate duration coding from distance coding. They trained rats to run on a treadmill for a target duration or a target distance in a hippocampus-dependent memory task. Most hippocampal and entorhinal neurons responded to both time and distance; that is, they fired reliably and repeatedly on subsequent trials at the same distance or at the same time from the beginning of the run. A minority of neurons was relatively selective for time spent on the treadmill, while the activity of an equally small fraction was better correlated with distance. These results led to the hypothesis that the neurons whose activity correlates with duration may be the key missing ingredient of episodic memory.[57] Investigating the relationship between coding for distance and coding for duration is important because such experiments may allow neuroscientists to confront the concepts of space and time in the context of neuronal mechanisms. However, the more we scrutinize time and its alleged brain mechanisms, the more puzzles arise.

It is hard to justify the hypothesis that there are neuronal networks whose sole function is to clock time. Even when overt motor correlates can be factored out, the neuronal assemblies always turn out to compute something else as well, in addition to the alleged time tracking. For example, in monkey parietal cortex, neuronal spikes are assumed to be related to the judgment of time. At the same time, these neurons also compute velocity and acceleration, measure distance,

---

55. Pastalkova et al. (2008); Fujisawa et al. (2008); Itskov et al. (2011).

56. Of course, the proper term should be "duration cells" because cell assemblies track duration rather than absolute time.

57. Eichenbaum (2014); MacDonald et al. (2011); Kraus et al. (2015); Tiganj et al. (2017). However, once velocity is monitored at high resolution, the distance and duration of a run are naturally identical (Redish et al., 2000; Geisler et al., 2007; Rangel et al., 2015).

code for spatial attention, plan movements, and prepare for decisions. Thus, these labels largely reflect the perspectives and questions of the experimenter rather than segregated streams of spike information for downstream readers. Indeed, computational models of decision-making use virtually identical logic to the ramping accumulator-threshold models of time-interval mechanisms.[58] Again and again, we find that duration calculations are linked to something else.

In summary, sequential neuronal activity can track the succession of events, which can be placed on a timeline when compared to units of a clock. However, they do not show that neuronal activity actually computes either clock time or duration. It is not yet clear whether neuronal circuits anywhere in the brain are dedicated to computing time as an independent function or, alternatively, whether circuits compute specific functions that evolve sequentially and therefore correlate with the ticks of an external instrument.

Time Warping

Albert Einstein noted that time flies when we in the company of pleasant people, and it slows when we are bored.[59] Experiments support this intuition. Highly motivated states, uncertainty, novel situations, and focused cognitive activity are associated with underestimation of time. Conversely, fearful or aversive situations, fatigue, and sleepiness are associated with overestimation of time. Brain-arousing drugs can accelerate, while sedating drugs and substances decelerate or distort, our subjective sense of time. Drugs affecting dopaminergic signaling prominently modulate timing behavior, potentially through their impact on the basal ganglia, a structure that is critical in time-interval estimation. Dopamine-producing neurons in the substantia nigra and ventral tegmental area become activated around the time when reward is expected and

---

58. Accumulation of "evidence" and reaching a threshold, determined by comparison to previous knowledge, is the act of decision (Leon and Shadlen, 2003; Janssen and Shadlen, 2005). Accumulation of "something" is calibrated against clock units and can be called time (Machens et al., 2005; Lebedev et al., 2008; Finnerty et al., 2015). An example of such a process is the accumulation of electric charge (evidence) in neurons, brought about by numerous excitatory postsynaptic potentials and culminating with the action potential at a threshold. From an anthropomorphic view, this is a decision made by the neuron. During the delay period of conditioning, increasing population activity can be viewed as a time command (McCormick and Thompson, 1984; Thompson, 2005). Alternatively, one can interpret the same accumulation of spiking activity as preparation for the response.

59. The brain state also affects subjective estimation of distance. When we are engaged in deep thought or a conversation with someone, long walks appear short (Falk and Bindra, 1954). See also Jafarpour and Spiers (2017).

decrease their spiking when the expected reward is omitted or when the expected time of reward delivery varies.[60] Transient selective activation of dopaminergic neurons in mice is sufficient to slow down the judgment of duration, while their inhibition has the opposite effect.[61]

There is a difference between "feeling the moment" and "remembering the moment." Subjective time-bending goes in the opposite direction during feeling versus remembering. When I am giving my 30-minute talk and the moderator politely stops me at 40 minutes, I am in disbelief. Time feels so fast. But when I recall the talk, I wonder why the few things I wanted to convey did not fit into 30 minutes.[62]

Our brain can cheat us not only in judgments of duration but also in judgments of time points. Test this yourself by touching your nose with your toe. They seem to feel the contact at the same time, right? Your eyes saw your nose and toe touching simultaneously, and this event is registered by the visual system. On the other hand, touch information from the toe takes several times longer to reach the brain via the spinal cord than does information from the nose, which is just next door to the brain. Thus, signals from the events do not arrive at the same time in the somatosensory cortex. Yet the brain, the overall observer, learns to compensate for such delay differences and to incorporate them into its computation.[63] Similarly, when we snap our fingers, the somatosensory, visual, and acoustic information blends into a simultaneous event, even though signals of these modalities reach the brain at different times. Experience-driven time-prediction in the brain compensates for such physical difference. However, when the brain is not prepared to compensate, bizarre

---

60. The discovery of the relationship between dopamine signaling and reward expectancy (Hollermann and Schultz, 1998; Schultz, 2015) gave rise to the burgeoning field of reinforcement learning and machine learning (Sutton and Barto, 1998) and recently linked neuroscience with economics and game theory (Glimcher et al., 2008).

61. Honma et al. (2016). Direct optogenetic manipulations of the dopaminergic neurons not only affect the postulated internal clock but also motor behaviors as well, posing the question of whether the delayed or premature responses reflect fundamental motor effects or internal timekeeper mechanisms (Soares et al., 2016).

62. Judging duration and feeling duration are different things (Wearden, 2015).

63. The physicist would say that this is simply an issue of the reference frame. If two events occur at different locations, they can be judged simultaneous in one reference frame or delayed to each other in another reference frame. An observer standing halfway between two distant flashes in a thunderstorm would see the flashes as simultaneous, while another observer standing closer to one than to the other would see the closest first, followed by the more distant. Note that the latter situation is exactly analogous to the relationship between nose, toe, and brain. However, the brain can learn about such delays through action and edit the delays by computation.

temporal illusions (Chapter 3) can occur. For example, participants in an experiment were instructed to make things appear on a computer screen by pressing a key. Unknown to the participants, the experimenter then introduced various delays between the key presses and the screen events. The brain learns to compensate for such delays, and the cause–effect feeling persists. After this adaptation training, however, the normal relationship between key presses and screen events returned to normal, participants reported that it felt as if the screen events were causing the key presses.[64] Their perception of time had reversed.

Subjective time compression happens with every saccadic eye movement, although we do not notice it. Vision is typically thought of as a space-exploring mechanism, but it also affects timing. When the eyeballs are jumping, the duration of a visual stimulus is underestimated by approximately a factor of two. A 100-ms stimulus around the time of a saccade is judged to be the same duration as a 50-ms stimulus presented to a stationary eye. This time compression is specific to the neuronal mechanisms that support saccadic eye movements because it does not occur during blinking or when auditory clicks, instead of flashes, are used for duration judgment. Remarkably, if research participants are asked to judge the temporal order of events, their performance deteriorates in a small window (–70 to –30 ms) before the saccade, and often their order judgments are reversed so that events that occurred second are perceived to occur first.[65] The saccade-induced time compression is tantamount to a slowdown of a hypothetical neural timing mechanism on which judgments of subjective time depend.

Here is another everyday example of a false reconstruction of the order of events. You are driving on a highway, and a deer crosses the road. You slam on the brakes. This is the sequence of events that you believe happened: you noticed a deer and, as a result, pushed the brakes and turned the car. In reality, your motor reaction time is much faster than the time required for awareness. You slammed on the brakes before consciously noticing the deer. Just as in the toe-nose touch situation and the saccade-induced time reversal, the brain

---

64. Eagleman et al. (2005); Yarrow et al. (2001). These observations are the modern-day replication of the findings by William Grey Walter in patients with implanted electrodes in the motor cortex, which he presented to the Ostler Society, Oxford University (1963; see Dennett and Kinsbourne, 1992). In Grey Walter's experiment, the patients advanced slides of a slide projector at will by pressing the controller button of the projector. However, the controller in fact was not connected to the projector. Instead, the CNV signal from the motor cortex of the subject was the trigger. Since the CNV builds up tens to hundred milliseconds before action, the carousal often started to move before the actual button press. The subjects were startled and felt that an agent was acting on their behalf when they were about to push the button.

65. Morrone et al. (2005). These movement-induced, time-warping illusions are reminiscent of the "time traveler" problem of relativity.

retrospectively constructs temporal order and causality. This happens because brain-constructed (subjective) time is relativistic; it can flow forward or backward. Copies of the automatic, short-latency actions from the brainstem are forwarded to cortical areas, where such information is interpreted. This neuronal interpretation gives rise to the internally generated prediction that "I am the agent of this action." In turn, watching yourself produces an illusion of temporal order and cause-and-effect relationship.[66]

To summarize, time-estimation experiments show that subjective time can be warped by many manipulations. Einstein was right: duration is relative. In addition, these observations demonstrate that the brain does not faithfully emulate the passage of physical or clock time. Time mechanisms in the brain run at multiple scales and manifest relatively independently in different structures in the service of their own computations. Most importantly, no experiment shows convincingly that there are dedicated mechanisms and neurons whose sole function is time generation or counting the number of oscillation cycles.[67] While all neuronal computation evolves on a time line, subjective time can move either forward or backward.

SPACE AND TIME IN THE BRAIN

Space and time have deep conceptual similarities and they share relational structure. Spatial schemas can be used equally well for temporal schemas to organize events in sequential order. Distance and duration appear to be represented in the brain by the same structures and likely by the same neurons

---

66. This may explain why we experience the urge to move a finger, for example, several hundred milliseconds after the onset of a readiness potential (CNV) above the supplementary motor area, a brain region believed to be critical in voluntary action. Neuronal "interpretation" needs approximately 500 ms to become available for awareness (Libet et al., 1979). An interesting question is whether we notice the feeling of the urge to move if we are not actively attending to the action. At this moment, you do not feel that you have a shirt on. But after reading this sentence, you might. The internal interpretation can be viewed as "attention." In turn, the act of attention and the underlying neuronal mechanisms can be regarded as the source of conscious experience (Graziano, 2013).

67. Neurons in the hypothalamic suprachiasmatic nucleus of the mammalian brain are portrayed as dedicated timekeepers that oscillate with an internal period of approximately 24.2 hours and maintain a stable phase relationship with solar time. Such a circadian clock is present in virtually all terrestrial and most aquatic animals and enables them to coordinate their behavior with the day–night cycle. Such recurrences and environmental regularities allow organisms to reliably predict the consequences of their actions, whether we call them timing behaviors or not. By this logic, the rotation of the earth is also a timekeeper because it changes cyclically.

and mechanisms. The preceding physiological data show that duration computation relates to distance computation. If these computations in the brain are related, as they are in physics, they should co-vary.

Imagine that our friends come over to watch a movie. After a while, we stop the movie and ask everyone to mark on a rod how much time has passed. Will the markings be similar on an identical length rod? What if everyone gets a rod of different length? My guess is that everyone's marking will reflect approximately the same ratio, independent of the rod length. Furthermore, nobody will argue that, "hey, we are measuring distance not time."[68] Analogous to egocentric and allocentric view of spatial relationships (Chapters 5 and 7), we also conceptualize time by such dual measures: we move through time or it passes us. In the egocentric view, the observer is moving on a timeline ("We are approaching the deadline of grant submission"). The future is in front of us. In contrast, in the allocentric metaphor, the observer is stationary and time is a river which carries things relative to the observer ("The deadline of grant submission is approaching"). When two objects are moving, their temporal relationship is based on their direction of motion (A is ahead of B or vice versa). Because of the duality of the event sequencing nature of conceptual time, we often get confused: "Should I turn my watch forward or back six hours when flying from New York to Europe?"[69]

Feeling duration depends not only on mood but on many other factors. When two objects are presented separately but for the same length of time, the duration of the larger object feels longer. Such subjective time expansion also occurs when a low-probability "oddball" event is embedded in a familiar sequence of stimuli. The novel stimulus feels longer than the familiar stimuli. Similarly, the first stimulus in a repeated train is judged to have longer duration than successive stimuli.[70] Subjective time moves at a different pace than clock time.

An elegant demonstration of the space-time similarity of subjective experience comes from research in architecture. In that experiment, students of architecture were asked to imagine that they were a Lilliputian figure in scale models

---

68. This experiment shows that our guess is a comparison against something that we have previously calibrated (learned). To guess the elapsed time, we had to have experience with many movies to learn that they typically last 100 minutes. The duration guess would be identical if we asked our friends to mark the pitch of sound along a sweep from, say, 1 kHz to 10 kHz. In each case, we compare the ratios of different modalities. Ratio estimation is easy for the brain, and it is independent of the measuring units. Such relations are universal (see Chapter 12).

69. Lakoff and Johnson (1980) point out that we often use metaphors to talk about abstract domains such as space and time and that we move from concrete domains to the more abstract, similar to the episodic-semantic memory transformation. See also Boroditsky (2000).

70. Tse et al. (2004).

that were 1/6, 1/12, or 1/24 of their full-size lounge on campus and to engage in the activities that they usually do there. The miniature lounges were furnished with chipboard furniture and proportionally sized figurines. The key parameter of the experiment was subjective time: the participants were asked to inform the investigator when they felt that 30 minutes had passed. The participants found that subjective time accelerated. Even more surprisingly, the magnitude of the acceleration, measured as the ratio of experiential duration to clock time, was quantitatively predicted from the proportions of the models relative to the full-sized environment. For example, if a participant experienced 30 minutes in a 1/6 scale environment in 10 minutes of clock time, the subjective 30 minutes in a 1/12 model occurred in just 5 minutes. This experiment demonstrates that temporal and spatial scales are related in subjective experience and that spatial scale may be a principal mediator of experiential time. Such space-time relativity implies that the experience of space and time refers to the same thing.[71] Experiencing time is experiencing space.

Space and Time Warping by Action

A number of neuronal mechanisms compensate for the disturbing sense of motion during saccadic eye movements. Under rare conditions, these useful mechanisms produce illusions. Saccades not only affect temporal judgments but also distort space. Objects flashed briefly just before and after the beginning of the saccade appear to be in false positions. They seem to move in the direction of the saccade target, and they are squashed parallel to the path of the saccade. As a result, geometric relationships between objects get distorted temporarily. Such warping of space and time is not a direct consequence of the fast-moving eyes because the distortion starts tens of milliseconds before and lasts until after the saccade. Instead, it reflects the corollary discharge mechanism (Chapter 3) that keeps track of motor commands.

Neuronal recordings also support the equivalence of perceived distance and duration. Neurons in the lateral intraparietal cortical area of the monkey appear to encode temporal duration as well as spatial distance. Approximately

---

71. DeLong (1981) performed these experiments in the School of Architecture, University of Tennessee. Spatial scale was based on the linear dimension, not volume. This suggests that, for humans, two-dimensional scaling is related to duration. The large number of subjects allowed proper quantification of the findings and formulation of the relationship as $E = x$, where E is the experiential duration and x is the reciprocal of the scale of the environment being observed. See also Bonasia et al. (2016).

one-third of the neurons in this brain region change their coding predictively during saccadic eye movements, and their activity is believed to be responsible for the perceived compression of space and time.[72]

In a locomotion experiment, normal participants were asked to reproduce a previously walked straight path while blindfolded. Although the velocity was reproduced accurately, the walking duration was affected to the same degree as the distance, suggesting that brain computation of space and time are closely related. In a more sophisticated version of this experiment, each participant was passively transported forward in a robot from 2 to 10 meters while blindfolded and head-restrained. After the robot came to a complete stop, the participant drove the robot with the joystick in the same direction in an attempt to reproduce the distance previously imposed. Not only the sample passive distances but also their velocity profiles were varied in a random sequence of triangular, square, and trapezoidal patterns, yet the participants reproduced them reliably. The results demonstrate that humans can reproduce a simple trajectory using only vestibular and possibly somatosensory cues and can store dynamic properties of whole-body passive linear motion without seeing the world.[73]

Examining the distance–duration issue makes us wonder about the role of velocity on neuronal firing. When a rat runs through the place field of a place cell at different velocities in the maze, the number of spikes emitted in different segments of the field remain largely the same (Figure 7.5). Moreover, the oscillation frequency of individual place cell spikes varies as a function of the rat's running velocity, and, for this reason, they are called *velocity-controlled oscillators*.[74] From the elapsed duration measured by the time-tracking ability of assembly sequences within the place field and instantaneous velocity calculated

---

72. Not only space and time but also numbers are distorted by saccades (Burr and Morrone, 2010), leading to the suggestion that parietal neurons code for magnitude rather than for time or space (Walsh, 2003). A related view is that several types of cognitive tasks, such as mental arithmetic, spatial and nonspatial working memory, attention control, conceptual reasoning, and timekeeping recruit similar structures in imaging experiments; therefore, the underlying computation may be similar, perhaps described by the term "cognitive effort" (Radua et al., 2014). Consult also Basso et al. (1996); Glasauer et al. (2007); Cohen Kardosh et al. (2012).

73. Berthoz et al. (1995); Perbal et al. (2003); Elsinger et al. (2003). Parkinsonian patients strongly underestimate intervals above 10 s on time-production tasks compared to controls, and the time underestimation correlates with the level of striatal dopamine transporter located on presynaptic nerve endings (Honma et al., 2016). Patients with parietal lesions not only suffer from spatial neglect but also grossly under- or overestimate multisecond intervals.

74. This term is a deliberate reference to the voltage-controlled oscillators in electronics; in these devices, the output frequency is relatively linearly controlled by a voltage input (Geisler et al., 2007; Jeewajee et al., 2008). We used the term "speed" rather than "velocity" because identical observations were made during wheel running in a memory task without translocation.

from the oscillation frequency of the place cells, the travel distance and instantaneous position can be continuously derived. How the brain performs such calculations remains to be learned. But we have some clues.

As in physics, displacement and duration are two sides of the same coin in the brain and are linked by velocity. This relationship explains why most hippocampal neurons report distance and duration equally well. If the animal runs a very long distance or duration, every hippocampal neuron will become a place (distance) cell and will also report duration via its position in the neuronal trajectory sequence.[75] If the same neurons function as both place cells and time cells, then downstream reader mechanisms have no way to tell whether the spikes represent distance or duration. Short of a demonstration of separate downstream reader mechanisms of distance and duration, we can conclude that hippocampal cells are neither place nor time cells.[76]

DISTANCE–DURATION UNITY AND SUCCESSION OF EVENTS

The dashboard of my Lotus Elise has multiple displays, including an odometer, a speedometer, and a clock. The odometer keeps track of the distance the car travels. This measure is essentially the number of strokes of the piston in the engine's cylinder.[77] My lab is approximately 1.5 million strokes from home. Because this number is not a very informative measure beyond the Lotus's world, one can relate the number of strokes to the circumference of the tires. In turn, the circumference can be related to some agreed unit and expressed as inches, feet, yards, meters, miles, light years, or whatever units one chooses and displayed on the odometer. Hence, we derived distance from the number

---

75. The small fraction of neurons that appear to be differentially sensitive to either duration or distance may reflect a wide distribution of velocity-dependence within the neuronal population or a potential error in measuring the instantaneous speed (velocity) of the animal (Kraus et al., 2015).

76. During treadmill running in a spatial memory task, sequential firing of entorhinal cortical neurons can also keep track of elapsed time and distance. Grid cells are more sharply tuned to duration and distance than are non-grid cells. Grid cells typically exhibit multiple firing fields during treadmill running, similar to their periodic firing fields observed in open fields, suggesting a common, rather than separate, mechanism for processing distance and duration (Kraus et al., 2015).

77. This is a super-simplified description of the engine's working, of course. A four-stroke cycle engine completes five strokes in one operating cycle, including intake, compression, ignition, power, and exhaust strokes. The piston movements in the 4, 6, or 8 cylinders need to be coordinated.

of strokes. The speedometer uses a small magnet attached to the car's rotating drive shaft and a coil. Each time the magnet passes the coil (the sensor), it generates a bit of electric current and the magnitude of the electric current is converted to units on the speedometer's display. Older mechanical versions use the same conversion principle. Ultimately, the displayed instantaneous reading on the speedometer is related to the rate of the movement of the piston in the engine's cylinder. The faster the piston moves up and down, the faster the car runs. If we divide the 1.5 million strokes by the average readings of the speedometer, we can tell how long it takes for me to get to work. Relating this ratio to some agreed units, such as ticks of a clock, the amount of sand flowing through a sand clock, heat beats or other relational units, we can call it duration or time. The relationship between the piston movement-generated "time" and time on the dashboard clock may vary depending on the "subjective mood" of the car. Using a higher octane gas makes my car happier (i.e., it runs faster) as is also the case when I press the accelerator harder and supply more gas.

The important wisdom of my car metaphor is that the very same mechanical movement (the piston's movement and its rate of movement) can be interpreted as speed, distance or duration, as long as we relate the mechanical movement to units of human-invented measuring instruments. But the car itself does not make either distance or duration, even if the piston and the other moving parts of the engine have the same mechanism as many mechanical clocks.

Neuroscientists search for the dashboard of the brain with a speedometer, odometer, and a clock. But let's face it: despite their seductive allure, we have no direct experimental evidence for neuronal mechanisms dedicated to selectively and specifically compute either space or time, although we have hard evidence for the existence of a speedometer (the vestibular system, muscle reafferentation, optic flow, etc). There is no such a thing as "perception of time" or "perception of space"—they are simply titles in James's list (Chapter 1).[78] Perhaps they are only grand abstractions of the human mind and we actually do not need them to describe brain operations. We need the space and time metaphors of classical physics so that that our mind can travel freely and confidently within them.

Imagine a collection of adjacent cups into which arbitrary items or events can be placed. A walk from event to event then can be described as a journey through space. If the duration of progression from one cup to the next is

---

78. This turns out to be a 2,000-year-old idea: "time by itself does not exist. . . . It must not be claimed that anyone can sense time apart from the movement of things or their restful immobility" (Lucretius, 2008, Book 1). We assume that time exists, but we never observe it directly. Instead, we deduce it from counting events.

considered, it is a trip through time.[79] If the rates of the transitions from cup to cup are known (velocity), the journey in time is identical to the journey in space because the distances between cups can be deduced. Conversely, if the distances between the cups are combined with velocity information, the journey through space is equivalent to the trip duration. Alternatively, we can just call the journey a trajectory or a sequence with transitions. This simplistic description has the advantage of creating a one-to-one relationship between sequences in the outside world and neuronal trajectories in the brain without the need of referring to units of human-made tools. As discussed in Chapters 7 and 8, brain dynamics generate extraordinarily large numbers of such trajectories, which can be split or concatenated in a number of ways, thus creating versatile opportunities to match learned experience with trajectories. These flexible mechanisms ensure that higher order connections can also be established without any need to refer to time. Time can always be explained in nontemporal physical terms. The measurement of time by precise instruments simply provides a practical imaginary reference so that a succession of events in two or more systems can be compared.[80]

Of course, the reader might say, "OK, change is just another name for time." However, there is a fundamental difference. Change always refers to something ("change of something"). In contrast, time is defined as independent from everything else. Together with space, it is this alleged independence from everything else that gives them such powerful conceptual organizational powers.

Motion, Time, and Space

Can we define progression of events in the brain from first principles? Navigation in real or mental space corresponds to the succession of events. In the most general sense, it is a motion. As defined in physics, motion is relative (between at least two bodies) and always characterized by velocity.

---

79. Theorists may call it a *saddle point attractor-repellor state vector*. Change can be continuous or discrete. A drive on an ideal flat highway is continuous. Most country roads go through hills and valleys, relatively discretizing the drive. In conceptual space, such hills and valleys can be called "attractors" and "repellors," which can change dynamically and determine the rate of change of transitions.

80. Perhaps Karl Lashley (1951) should be credited with this insight. He thought about these issues before dynamical theory gained prominence: "memory appears almost invariably as a temporal sequence, either as a succession of words or of acts. . . . Spatial and temporal order thus appear to be almost completely interchangeable in cerebral action. The translation from the spatial distribution of memory traces to temporal sequence seems to be a fundamental aspect of the problem of serial order."

Although we have no direct sensors for space or time,[81] we do have vestibular sensors to detect changes in velocity (i.e., acceleration) and the head direction system can calculate vectorial displacement. Motion per se can be directly experienced from optic or tactile flow and corollary discharge without the need to sense space or time.[82].

Hippocampus as a Sequence Generator

Navigation in real or mental space is, by its nature, a succession of events. Perhaps that is all the hippocampal system does: encode content-free or content-limited ordinal structure, without the details of particular events. Indeed, a parsimonious interpretation of all the experiments discussed in this chapter is that both the parietal cortex and the hippocampal-entorhinal system are general-purpose sequence generators that continuously tile the gap between events to be linked.[83]

Let me support this heretical statement with an experiment that illustrates the contrast between sequences and apparent space or time representation. A group of patients with damage to the hippocampus and some normal individuals received a guided tour of the campus of the University of California (Figure 10.2). The tour included various events, preplanned by the experimenters, such as discarding a cup, locking a bike, buying a banana, finding a coin, drinking from a water fountain, and so forth. At the end of the trip, the participants returned to the laboratory, where they completed three tests. The first one asked the participants to describe everything they remembered about the walk in six minutes. In the second test, the experimenter provided a prompt for each of the events ("What happened to the bicycle?") and asked the participants to provide

---

81. The circadian clock of the suprachiasmatic nucleus is perhaps closest to a man-made clock in the sense that it oscillates "around the clock." However, oscillations and time are orthogonal dimensions (phase vs. duration). Events in an oscillator always return to the same zero phase, quite different from the infinite time arrow idea. Phase can be reset, advanced, or delayed. Time just moves forward. The circadian clocks in the brain, autonomous nervous system, liver, and gut runs spontaneously and largely independently of each other, but they can be reset and synchronized by light, hunger, and other variables, which have sensors to mediate such inputs.

82. After ambulatory calibration of the distance metric, egocentric distance may be extracted from the spiking of binocular disparity-selective neurons in the primary visual and the posterior parietal cortex (Pouget and Sejnowski, 1994). Optic flow is critical for grid cells and spatial metrics (Chen et al., 2016).

83. Of course, many other brain structures can generate sequences internally. However, the hippocampus, as a giant cortical module, is a special case since its sequences can affect neuronal content in the large cortical mantle.

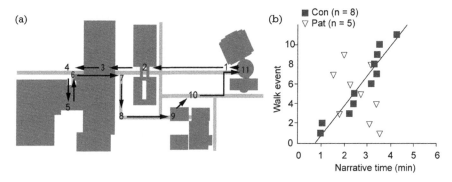

Figure 10.2. The core deficit in patients with hippocampal lesion is their inability to narrate events in the order they occurred. A: Map of eleven events that occurred during a guided campus walk. Sidewalks are light gray. Buildings are dark gray. Arrows indicate the path taken during the walk. B: Events from the walk, described during 6-minute narratives. The control group (*grey squares; Con*) tended to describe the eleven events in the order that they occurred. The order in which the patients (*open triangles; Pat*) described events was unrelated to the order in which the events occurred. Reproduced from Dede et al. (2016).

all the details they could recall in one minute. In the third test, forced-choice questions were asked about the events ("Did we find a quarter or a dime?"). The normal participants also took the same tests again one month later.

As expected, the hippocampus-damaged patients were worse than the normal control participants on most tests. The patients (tested immediately after the tour) performed about as well as the control participants tested one month later on measures like the number of events remembered and duration judgments (except for one participant, who had the greatest brain damage, possibly involving the hippocampus, parahippocampal areas, and frontal cortex). Despite their memory impairment, hippocampal patients recalled many spatial and perceptual details, such as that the bicycle "had a light in the front." The events that were most memorable for the control participants were also most memorable for the patients. Yet there was a striking difference between the two groups. The order in which patients recalled the events was unrelated to the order in which they occurred. In contrast, control participants effectively recalled the order of the events even a month later.[84]

Let me cite another real-life example. A London taxi driver who had sustained bilateral hippocampal damage retained considerable knowledge about city's landmarks and their approximate locations, but he could not compile those

---

84. Dede et al. (2016). For functional magnetic resonance imaging support of sequence coding, see DuBrow and Davachi (2013, 2016).

locations into a novel sequential trajectory. Rats with hippocampal damage showed similarly impaired memory for the order of odor stimuli, in contrast to their preserved capacity to recognize odors that had recently occurred.[85] These experiments support the idea that the hippocampus is a neuronal sequence generator that can create a nominal order of sequential events and reproduce it upon recall.

Whether in reference to space or time, sequences in the hippocampal system may simply point to the items (*what*) stored in the neocortex in the same order as they were experienced during learning. This division of labor is analogous to the role of a librarian (hippocampus, pointing to the items) in a library (neocortex, where semantic knowledge is stored).[86] Thus, the hippocampal system may be responsible for constructing sequences of information chunks, while chunk content is encoded in and retrieved from the neocortex.

The concept of the hippocampus as a sequence generator has the advantages of separately encoding items and events in hypothetical space and time, as discussed at the beginning of this chapter. Instead of storing a huge repertoire of unique states for every sequence encountered, the brain can store sequences (in the hippocampus) separately from the content (neocortex). Instead of storing every possible neuronal sentence, the cortex can separately store all the words, while the hippocampus concatenates their sequential order. This organized access (embedded in neuronal trajectories) to neocortical representations (*what*) then becomes episodic information.[87] Episodic memory is then an ordered sequence of "*whats*."

---

85. Fortin et al. (2002); Maguire et al. (2006).

86. Teyler and DiScenna (1986) used the indexing metaphor for hippocampal computation. The index or table of contents (hippocampus) represents a concise summary of a book. However, instead of a single index that "points" to cortical modules, the "unit" is a sequence of indices (a "multiplexed pointer") such that the hippocampal system is responsible for concatenating neocortical information chunks into sequences for both encoding and retrieval, typically in theta oscillation cycles (Buzsáki and Tingley, 2018).

87. Karl Friston and I gave talks back to back at a meeting in the gorgeous building of the Hungarian Academy of Sciences in January 2016. To the audience, it felt as if we lived in two different worlds (he talked about the Bayesian brain, while I talked about internally generated sequences and log distributions; see Chapter 12). To us, it seemed that our thoughts resonated, and we wanted to discuss the potential links over lunch. After several hours of friendly debate, we left the table with a basically finished paper in our heads about the advantages of sequential order generation. Karl did the heavy lifting of writing most of the manuscript (Friston and Buzsáki, 2016). This lunch discussion was also an opportunity for me to observe how the brain of a genius works.

Memory for the Future

Neuronal trajectories and their ordering-pointing roles can equally well represent the past, present, or future. Predicting the future is possible only after we have experienced the consequences of our actions and stored the outcomes of both successful and failed attempts. We might expect, therefore, that the neuronal mechanisms of postdiction and prediction are not that different. Indeed, imaging experiments in humans show that many structures that were traditionally viewed as part of memory systems are inseparable parts of planning, imagining, and action systems.[88] In laboratory animals, what appears to be a memory of the past in the sequential activity of cell assembly sequences can also reflect planned future action. Perhaps representation via ordered neuronal sequences explains why the brain structures and mechanisms associated with memory and planning show such a strong overlap. The distinction between postdiction and prediction may be analogous to the illusory separation between the past, present, and future in physics.

With or Without Space and Time?

Do not get me wrong. I am not suggesting that neuroscience researchers dispense with the concepts of space and time. The formulation of spacetime by physics did not discredit clocks. Quite to the contrary, physics and all other sciences rely increasingly more on the precision of measuring instruments to calibrate experimental observations. This is not expected to change in neuroscience either. Indeed, in the remainder of this book, I continue to use these terms in the usual manner, with the understanding that they make sense only

---

88. Planning could be referred to as "constructive" memory (Schacter and Addis, 2007) or "memory of the future" (Ingvar, 1985). See also Lundh, 1983; Schacter et al. (2007); Hassabis and Maguire (2007); Buckner and Carroll (2007); Suddendorf and Corballis (2007); Pastalkova et al. (2008); Fujisawa et al. (2008); Lisman and Redish (2009); Buckner (2010); Gershman (2017). These new findings are beginning to bring down the walls that have long stood between concocted words and brain mechanisms. Planning and recalling feel different: one relates to the subjective future, the other to the subjective past. Yet any form of planning requires access to preexisting knowledge (Luria, 1966). Also, a planned action cannot be carried out unless the "plan" is placed into memory until its actions are completed. Bilateral hippocampal damage is associated not only in recalling the past but also with an impairment in thinking about one's personal future (Atance and O'Neill, 2001) and in imagining new fictitious experiences (Hassabis et al., 2007).

in comparison with units of instruments. This should not be taken as an inconsistency but as a way not confusing the average reader. Time and space remain useful concepts for organizing the thought.

If no radical change in practice is recommended, why spend so many pages scrutinizing space and time representations in the brain ? The answer is that the observation that the evolution of neuronal activity is reliably correlated with the temporal succession of events ("temporal sequence of representations") does not mean that neuronal activity computes time ("representation of temporal sequences"). Neither instruments nor brains create space or time.[89] Yet they remain useful concepts because without them complex thinking would be nearly impossible. Yet, it is useful to remember that it is one thing to compare the evolution of brain activity against clocks but quite another to design a research program to search for neuronal "representations" of space, time, or other mental constructs in the brain.

If "my ideas" appear too radical to the reader, here is a reminder that I am not the first weird person to think this way. The Greek philosopher, Philo of Alexandria wrote these lines two millennia ago: "Time is nothing but the sequence of days and nights, and these things are necessarily connected with the motion of the Sun above and below the Earth. But the Sun is part of the heavens, so that Time must be recognized as something posterior to this World. So it would be correct to say not that the World has created Time, but that time owed its existence to the World. For it is the motion of the heavens that determines the nature of Time."

The inside-out approach to space and time may be generalized to other terms listed in William James's list, nearly all of which may exist only in our mind. Our challenge as neuroscientists is how to perform experiments and think about brain mechanisms without resorting to such preconceived ideas.

I left out one thing from the discussion about the relationship between sequences and duration: the rate of succession. Only when sequential order is combined with the rate of change can it replace time. The pace of transition in the brain is supported by ubiquitous brain mechanisms known collectively as "gain control," which I address in the next chapter.

---

89. Bergson (1922/1999); Dennett and Kinsbourne (1992); Ward (2002); Friston and Buzsáki (2016); Shilnikov and Maurer (2016). Ever since Galileo, time has been critical in many equations of classic physics (and in neuroscience; e.g., dv/dt). Yet time is conspicuously missing from equations of contemporary theories. These theories are all about change and relations. Perhaps when a mathematical brain theory eventually emerges, its equations will not use t.

Chapter 10. Space and Time in the Brain

SUMMARY

In this chapter, I used space and time to illustrate that even these seemingly non-negotiable terms can be viewed differently from a brain-centered perspective. In both Newtonian physics and Kant's philosophy, space and time are independent from each other and from everything else. Everything that exists occurs in space, assumed to be a huge container, and on a time line, assumed to be an arrow. Newton could imagine an empty world, but not a world without space and time. Yet space is not a huge theater into which we can place things, except in our minds. We can infer space and time only from things that move.

The science of space and time began with the invention of measuring instruments, which converted these dimensionless concepts into distance and duration with precise units. This process created a special problem for neuroscience. If space and time correspond to their measured variants, we may wonder what space and time mean without such instruments, including for non-human animals that cannot read those instruments.

Contemporary neuroscience still lives within the framework of the classical physics view. Our episodic memories are defined as "what happened to me, where, and when." This definition requires the identification of mechanisms in the brain that store the "what" in independent coordinates of the "where" and "when." This is a typical outside-in approach: assume the concepts and search for their homes in the brain. Accordingly, neuroscience researchers have diligently looked for brain mechanisms that define space and time. Two major hubs have been identified for space: the parietal cortex for egocentric representation and the hippocampal system for allocentric representation. Research on timekeeping has been largely independent from research on space. Initially, it was debated whether the brain has a central clock, similar to that found in computers, or if each structure generates timing for its own needs. The cerebellum and basal ganglia have been favored as central clocks. Recent work emphasizes the parietal cortex and hippocampus-entorhinal system as substrates for time perception and generation. These claims have generated an intense discussion about whether the sole function of some neurons is to deal with space or time or whether they always compute something else, too. Similarly, the main source of controversy about "place cells" versus "time cells" is rooted in the internal contradictions of classical physics. The animal's velocity converts between distance and duration representations, making place cells and time cells equivalent.

The independence of space and time has been debated by linguists and physicists alike. Classical physics revealed that distance and duration are linked through speed: knowledge of any two variables can identify the third, so one is always redundant. Contemporary physics, especially general relativity, produced our current picture of spacetime, the equivalence of space and time, which is different from either space or time. This view teaches us that the world does not *contain* things; it *is* things.

Can we do neuroscience without resorting to these "obsolete" concepts? Alternatively, can we support the philosophical stance that the spacetime of physics and "lived" space and time are qualitatively different? The main problem is that the brain does not have sensors for either space or time—but it has sensors for head direction and velocity. Neither brains nor clocks make time. Similarly, neither brains nor rulers make distance. I suggest that it is safe to continue to do neuroscience as we have done it over the past century, using measuring instruments and relating observations to their units. However, it is not safe to declare that some brain region or mechanism represents space or time or calculates distance or duration.

Everything that we attribute to time in the brain can be accomplished by sequential cell assemblies or neuronal trajectories. Hippocampus-damaged amnesic patients have much less of a problem with estimating and recalling distances and durations than with remembering the sequential order of events.

11

Gain and Abstraction

This idea that there is generality in the specific is of far-reaching importance.

—Douglas R. Hofstadter[1]

An abstraction is taking a point of view or looking at things under a certain aspect or from a particular angle. All sciences are differentiated by their abstraction.

—Fulton J. Sheen[2]

Love is one kind of abstraction. And then there are those nights when I sleep alone, when I curl into a pillow that isn't you, when I hear the tiptoe sounds that aren't yours. It's not as if I can conjure you up completely. I must embrace the idea of you instead.

—David Levithan[3]

Mounted archery was used for hunting and warfare by most nomads of the Eurasian steppe including the Huns, Magyars, and Mongols, as well as by Native Americans on the American plains. Later, as warfare and hunting came to depend on guns, archery evolved into a martial art. Imagine trying to shoot an enemy warrior, a moving target, while your horse speeds along at 50 kilometers an hour. Perhaps nobody knows more about shooting arrows from a galloping horse than Lajos Kassai, known as "The

---

1. Hofstadter (1979).
2. https://www.diannasnape.com/abstraction/.
3. Levithan (2011).

Master" of this ancient art in the remote Hungarian village where my ancestors originated. While on his horse, Kassai appears super-human; he never misses a target.[4] He says masters have an instinct for aiming—after decades of arduous training. Mounted archers rely on a combination of eye–hand coordination, muscle activity patterns, arrow trajectory, and subconscious distance calculations to hit the target. There are just too many things going on to allow deliberate calculation when you aim to hit 10 targets in 20 s. The archer's eyes, head, and torso; the body on the horse; and the horse's body, head, and eyes are all embedded in their own independent coordinate frames. Multiply all this by the same number of degrees of freedom for the enemy warrior and then calculate his angle and direction of movement.

Riding a horse, moving your eyes, or exploring the environment exploits the same principles of coordination. Every coordinate is a different viewpoint (Chapter 10). In every coordinate frame, the information is first generalized and then segregated into categories by some new observer-centered classifier. To achieve effective motor–sensory action, downstream reader networks should extract information according to their points of view.[5] Gain is a mechanism that assists with such coordinated transformation and abstraction problems.

PRINCIPLES AND MECHANISMS OF GAIN

Turning the volume knob on your radio is a gain control mechanism for sound production.[6] Gain modulation requires two sources, an amplifier and a modulator (Figure 11.1). When the sources are multiplied (or divided), such interaction generates an output whose magnitude is larger or smaller (in which case the gain is negative) than expected from adding (or subtracting) the sources.

---

4. You can watch him in action here: https://www.youtube.com/watch?v=a0opKAKbyJw; https://www.youtube.com/watch?v=2yorHswhzrU. Kassai trained Hollywood actor Matt Damon in how to handle the bow in the movie *The Great Wall*.

5. Artificial intelligence algorithms also face the problem of hierarchical abstraction. The best performing deep Q network (DQN; Minh et al., 2017) exploits the local correlations present in images and makes its way up through transformations of viewpoints and scales. The program's explicit goal is to select actions in a fashion that maximizes reward. Thus, grounding is based on (imitated) actions.

6. In electrical engineering, gain is often used synonymously with amplification or boosting the signal. Gain of an amplifier is measured in decibels (dB), a logarithmic unit. A tenfold input–output increase corresponds to a gain of 10 dB; a 100-fold increase represents 20 dB (or 10 to the power of 1 and 2, respectively). The power spectrum of the local field potential (LFP) and electroencephalogram (EEG) is often characterized by such gain factors.

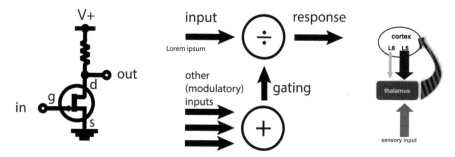

Figure 11.1. *Left*: Gain control in electronic circuits. The magnitude of current flowing from the source (s) to the drain (d) in a semiconductor can be adjusted by applying a small current to its gate (g). Small changes in the input (gate) result in large changes in the output voltage (V_{out}). *Middle*: General implementation of gain. Input–output transfer is modulated by a division or multiplication circuit, modulated by a second, gating input. All subcortical neuromodulators, such as norepinephrine and acetylcholine, achieve their gain control on neurotransmission by this principle. *Right*: Cortical gain control of sensory input transmission in the thalamus. Thalamocortical neurons receive convergent inputs originating from subcortical (gray arrow) and cortical (from layer 5 and layer 6) sources. Layer 6 input rarely discharges thalamocortical neurons, but it can boost their response to sensory inputs, thus providing gain control. Layer 5 inputs can effectively discharge thalamocortical cells and can contribute to the maintenance of persistent thalamocortical activity that can outlast the input. The output from the thalamus is represented by the curved black and gray striped arrow.
Thalamic figure from Groh et al. (2014).

Such nonlinear gain modulation is ubiquitous in the brain, with far-reaching consequences, and is achieved by different mechanisms in individual neurons or synapses or at a larger scale in neuronal networks, indicating that it is a critical operation. Gain control can also determine whether the succession of events in a neuronal trajectory is slow and fast. Thus, gain-controlled rate of change in neuronal sequences can, in principle, replace the concept of time (Chapter 10). Here, I list a few ways in which gain control is implemented in neuronal circuits. If you are not interested in the details, you are welcome to skip the next few subsections.

Neuronal Implementation of Gain Control

The intensity of light hitting the retina varies over nine orders of magnitude (a billion-fold change) from a dark night to a sunny day. Many peripheral mechanisms, including changes in pupil diameter and chemical adaptation of the cones and rods in the retina, effectively reduce the light responses of retinal ganglion cells and the messages sent to the visual system. Still, visual neurons have to deal with a wide dynamic range. The solution is to respond not to the absolute magnitude of the input but to a relative change in intensity, known as *contrast*. By receiving information about both the mean background intensity and about the deviation from this mean, neurons can calculate the ratio or contrast against any light intensity. This process is a type of normalization, to borrow a term from statistics—essentially a division.[7]

Shunting

Multiple mechanisms contribute to achieving such a wide sensitivity range. The first mechanism is gain control in an individual neuron, a change in that neuron's response to a given input based on its other inputs. Each neuron receives hundreds to thousands of synapses from excitatory and inhibitory neurons. Whether it responds with an output action potential depends on the ratio of excitation to inhibition across all its inputs. But it also matters whether that ratio comes from 5 excitatory and 2 inhibitory neurons or from 500 excitatory and 200 inhibitory neurons because neurotransmitters work by opening channels, essentially making temporary holes in the membrane. When many channels are open, the resistance between the extracellular and intracellular environments decreases. In other words, the conductance through the membrane increases as ions can flow freely between the inside and the outside of the neuron. The technical term for this state is "shunting" because the numerous open channels work as a short circuit. This is similar to the situation that occurs when you turn on the oven, toaster, and dishwasher at the same time and the circuit breaker pops because the parallel resistances shunt the circuit. Therefore, even if the membrane voltage remains the same, as happens when the increases in excitation and inhibition are balanced, the inputs have less impact on the neuron when many channels are open because the increased conductance (a short-circuit) reduces the efficacy of the synaptic current. In essence, the efficacy of any synaptic input to the neuron is divided, or normalized, by the combined effects of all other inputs impinging on the neuron at the same time.[8] A change in gain alters the neuron's sensitivity to changes in its inputs

---

7. Normalization is a special case of gain modulation, where the "normalizer" (denominator) is a superset of the signals that determine the responses (Carandini and Heeger, 1994, 2012). In this case, input from one neuron is the factor to be affected by the numerous other inputs to the same neuron.

8. This normalization effect is a consequence of Ohm's law, as the synaptic input current I at a given membrane potential V is scaled by membrane conductance g, as $I = gV$. Multiplication by

without affecting its selectivity or discriminative properties. The gain or efficacy of synapses is amplified or discounted by the background activity of the neuronal population in a multiplicative (divisive), as opposed to additive (subtractive) manner.[9]

Inhibition

A second mechanism of gain control occurs in inhibitory synapses acting through gamma aminobutyric acid ($GABA_A$) receptors.[10] These receptors are permeable to chloride ions, which have an equilibrium potential close to the resting membrane potential of the neuron. As a result, activation of $GABA_A$ receptors barely changes the membrane potential of a relatively quiet neuron, but it shunts the neuron and thus brings about a negative gain (division) on excitatory inputs.[11] This mechanism is important for the sustainability of cell assemblies because they compete with each other through lateral inhibition. Such gain control mechanisms affect the transmission of spikes from neuron to neuron by affecting spike generation.

Short-Term Plasticity

A third family of gain control mechanisms acts on the inputs to the neurons. When many nearby inputs on a dendrite are activated, they amplify each other's

---

a given constant changes the slope, or gain, of the input–output relationship, corresponding to an up- or down-scaling of responsiveness.

9. An increase in neuronal gain corresponds to a multiplication, and a decrease corresponds to a division. A nice summary of neuronal arithmetic is by Silver (2010).

10. Alger and Nicoll (1979); Kaila (1994); Vida et al. (2006).

11. For more details, see http://www.scholarpedia.org/article/Neural_inhibition. Shunting synapses are more effective if the inhibitory synapse is on the path between an excitatory synapse and the spike initiation site. Shunting is spatially localized, whether it is caused by inhibition or excitation because both inhibitory postsynaptic current and excitatory postsynaptic current potential cause leaks in the membrane by opening channels. Gain modulation by shunting becomes more apparent when large numbers of neurons are affected and when there is a relatively balanced increase of excitation and inhibition in short time windows, as typically occurs during gamma frequency oscillations. Balanced background activity produces an approximately multiplicative gain modulation of the response to an excitatory input (Chance et al., 2002).

effects so that their impact is larger than their algebraic sum.[12] Even more selective gain control, acting at single synapses, can be achieved via the short-term plasticity of synaptic strength. The synaptic efficacy of the action potential from a presynaptic neuron is not fixed over time but shows large fluctuations. Some synapses increase ("facilitatory synapses") whereas others decrease ("depressing synapses") their response to the later input spikes in a series.[13] Depression is the predominant form of short-term plasticity in synapses between cortical pyramidal neurons. When the input neuron fires slowly, the magnitude of the evoked postsynaptic responses is similar. However, when the neuron fires quickly, the magnitude of the postsynaptic response diminishes over time. The faster the intervals between input spikes, the stronger the synaptic depression. After the barrage of spikes is over, recovery of the baseline response may take about a second, with a characteristic time constant for each synapse type. Because depressing synapses become increasingly less efficient with increasing rates of input spikes, they can be conceived as low-pass filters of neuronal transmission. Conversely, facilitating synapses are high-pass filters. As a result, the response of the postsynaptic neuron will reflect not the average input rate of the presynaptic neuron but the change in its rate. Short-term synaptic plasticity, therefore, turns neurons into "ratio meters." The nonlinearity of the short-term plasticity transforms the input–output relationship via divisive gain changes.[14] Thus, while the mechanisms are different, gain control operates on both single synapses and single neurons. Gain control through dynamic changes of the synapse can adjust the rate of change across neurons in neuronal sequence, which can be conceptualized as time warping (Chapter 10).

Neuromodulation

In neuronal networks, gain control is achieved by mechanisms that affect large numbers of similar neurons. Recurrent excitation is a frequent source of

12. Rodolfo Llinás was among the first to demonstrate that dendrites are not simply passive integrators but have active voltage-sensitive channels that can powerfully enhance the neuron's computation ability (Llinás, 1988). See also Mel (1999); Magee (2000).

13. Frequency-dependent short-term potentiation (STP) occurs when presynaptic vesicle supplies are quickly depleted or when postsynaptic receptors desensitize. Short-term facilitation can also have both pre- and postsynaptic mechanisms (Thomson and Deuchars, 1994; Markram and Tsodyks, 1996; Zucker and Regehr, 2002). English et al. (2017) described various forms of spike transmission mechanisms between pyramidal cells and interneurons that can perform frequency filtering and amplification at various time scales. These flexible mechanisms can assist in shifting cell assemblies without external influence.

14. Abbott et al. (1997); Rothman et al. (2009).

nonlinear multiplication in neuronal networks. Subcortical neuromodulators, such as acetylcholine, norepinephrine, serotonin, dopamine, and histamine, share one thing in common: they affect the excitability of their target neurons to a large extent by affecting conductance, exerting multiplicative gain control on the output firing rates. In addition to excitability, neuromodulation can affect the balance between synaptic properties and nonlinear membrane properties by targeting ion channels. Several neuromodulators also target presynaptic terminals. Presynaptic receptors can adjust the amount of neurotransmitter to be released into the synaptic cleft, thus affecting synaptic efficacy[15] Because subcortical neuromodulators typically affect many neurons simultaneously, they are critical for gain control in neuronal networks.[16]

Conductance changes, $GABA_A$ receptor-mediated shunting, lateral inhibition, depressing synapses, dendritic boosting, recurrent excitation, and subcortical neuromodulation are ubiquitous features in cortical circuits. Therefore, any of these mechanisms in isolation or combination produce gain control in neuronal networks, influencing their input–output transformation in various ways. Since neuromodulators can also affect the speed of spike transmission across neuronal assemblies, such effects may explain why neuromodulators and drugs that can affect their physiological effects can lead to acceleration or slowing of subjective time.

INPUT MAGNITUDE NORMALIZATION

Perhaps the most obvious example of gain modulation is the response of sensory systems to natural stimuli of differing intensities, such as the adaptation to light intensity changes mentioned earlier. Pyramidal neurons in the primary visual cortex, the first cortical processing station that handles inputs from the thalamus, have a limited dynamic range in their firing rates, which confine their ability to respond differentially to a limited range of contrast. However, neurons that receive thalamic inputs also receive inputs from many other cortical neurons. This arrangement allows the impact of the thalamic input to be

---

15. Ion-channel modulation typically involves binding of a neuromodulator to G-protein-coupled receptors and subsequent activation of second messengers, although ion channels can be modulated through other mechanisms as well. The family of two-pore domain potassium (K^+) channels are responsible for a substantial portion of leak conductance in a variety of neurons, and subcortical neurotransmitters effectively modulate leak conductance (Kaczmarek and Levitan, 1986; Bucher and Marder, 2013). This can explain their strong impact on network dynamics.

16. The role of norepinephrine in gain control is reviewed extensively by Aston-Jones and Cohen (2005).

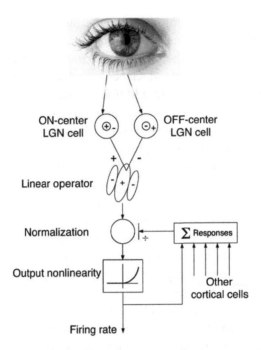

Figure 11.2. Illustration of neuronal normalization. A linear stage combines complementary inputs from the retina by way of the lateral geniculate nucleus (LGN). Neurons in the visual cortex sum responses of "on-center" neurons (i.e., those neurons that respond to the visual input) and subtracts responses of "off-center" neurons (i.e., those neurons whose receptive fields are outside those of the on-center neurons). This is a linear operation. The flanking inhibitory subregions are obtained by the opposite arrangement of excitation and inhibition. A shunting-based normalization mechanism divides the linear stage's response by the pooled activity of a large number of cortical cells. This is a nonlinear operation.
Reproduced from Carandini and Heeger (1994).

reduced proportionally (scaled down) by the conductance increase brought about by feedback excitation from many other neurons (Figure 11.2). It is essentially a division.

A related mechanism is inhibition-mediated competition. A neuron's response in the primary visual cortex to a preferred orientation stimulus can be suppressed by superimposing a nonpreferred orientation. This happens because the activity of neurons with nonpreferred orientation excites inhibitory interneurons, which in turn suppress the neuron's response to its own preferred stimuli.[17] The concentrations of odors and sound intensities also vary

---

17. While there is an agreement about the principle (the need for normalization), a debate persists about whether the division is brought about by conductance change, inhibition, or other

considerably, and a similar mechanism may be responsible for computing intensity change. All it takes is two inputs, a specific one that activates the neuron and another one that produces divisive normalization.[18]

CHANGING VIEWS BY TRANSFORMING COORDINATE SYSTEMS

Now let us examine how gain control mechanisms can be exploited for transforming signals. When you are driving a car, your eyes are fixed on the traffic ahead. If you want to reduce the volume on the radio, you simply reach for the volume knob. In effect, you are multiplexing, attending to the traffic with central vision and to the volume knob with peripheral vision. Because human eyes move a lot in their sockets, the relative positions of the eye, head, and trunk vary from moment to moment. Thus, the retinal output and retinotopic representation of eye-centered space by neurons in the visual system cannot be of much help in the reach movement of the hand. The retinal image must be first translated into head- and body-centered coordinate frames.[19]

Receptive Fields and Gain Fields

Neurons in the visual system have *receptive fields*, meaning that they respond maximally when a stimulus is presented at a particular location (peak response) corresponding to a given spot on the retina (Figure 11.3).

---

mechanisms (Carandini and Heeger, 1994, 2011; Murphy and Miller, 2003). Normalization is believed to be at work in the retina as well, where the output spiking of bipolar and ganglion cells is divided by the activity of other ganglion cells induced by the light contrast in the surrounding regions (Normann and Perlman, 1979).

18. Competitive normalization has been described in the fly antennal lobe during olfactory processing. Here, projection neurons receive two different inputs: excitation from the olfactory receptors and inhibition from other projection neurons (called *feedforward* or *lateral inhibition*). When a single odorant is mixed with many other odorants, the response of a projection neuron to its favored odorant can be decreased by simultaneous activation of many other partner neurons by a mixture of odorants. The divisive normalization scales with the total magnitude of activity of the olfactory receptor neuron population. Increasing the activity of projection neurons makes the responses to the favorite stimuli not only weaker but also more transient (Olsen et al., 2010).

19. The mechanism that compensates for the ever-shifting retinal input relative to the body represents a corollary discharge (see Chapter 3).

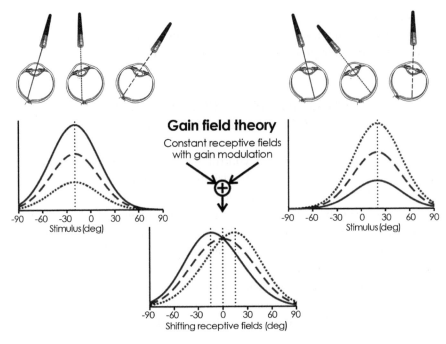

Figure 11.3. Gain field mechanisms in the parietal cortex. The upper panels show the hypothetical receptive fields of two neurons that are gain-modulated by eye position. Three lines in each graph represent visual receptive fields mapped relative to gaze at three different eye positions. The laser light can scan the entire retina. Here only one spot (its maximum receptive field response) is shown. Note that eye position modulates the strength of response of two neurons, but their receptive fields remain the same. However, summation of these two gain-modulated neural responses (*bottom panel*) results in shifting receptive fields in the output as a function of eye position.
From Zipser and Andersen (1988).

The receptive field can be easily mapped when the eyes are voluntarily fixed on a target by projecting a lighted spot successively onto different positions across the visual field. The peak response and the decreasing responses that surround the peak form a cone-shaped response pattern, which is the receptive field.[20]

---

20. The idea is that the visual receptive field is related to the place field of a hippocampal place cell (Chapter 6). The term "receptive field" is central to the study of sensory systems. In addition to position, visual neurons also respond to various features of the light stimulus, such as its velocity, direction, and orientation. The magnitude of responses within the receptive field can be affected by a number of factors, such as brain state. A solid observation in all fields of neuroscience is that brain state changes across sleep and waking to exert a large impact on both the magnitude and often the selectivity of neuronal responses.

Many neurons in visual areas and the posterior parietal cortex have such retinotopic receptive fields. They are typically much larger in the parietal cortex than in the primary visual areas. However, parietal neurons also carry information about eye position, as shown in a landmark experiment by Richard Andersen at the California Institute of Technology. He and his colleagues trained monkeys to fixate on a small cross on the projection screen while a parietal cortical neuron was recorded. After its receptive field had been mapped with a dot stimulus, the fixation point was varied to move the eyes in their sockets. The dot was also moved so that it always stayed in the center of the neuron's receptive field. Thus, the visual input to the preferred retinal spot remained the same throughout the experiment, while the eye position was systematically changed. The firing response of the parietal neuron varied as a function of the eye position. At some positions, the response was large, while at other positions it was smaller or even disappeared. This rate modulation had a spatial pattern, which the authors called the "gain field" (Figure 11.3). The firing rate of the neurons could be approximated by multiplying the firing rates induced by the receptive field profile with the factor due to modulation by eye position—hence the name "gain." This pioneering experiment revealed that the neuronal response to the retinal position of a visual object is multiplicatively scaled by eye position in parietal cortical neurons.[21]

Later experiments showed that visual responses of parietal neurons are also affected by head position and hand position. These eye and hand gain fields are not independent but instead have equal and opposite strengths, so that the eye–hand gain field combines reference frame transformations with movement planning at the same stage of cortical processing. Overall, these mechanisms in the parietal cortex guarantee that the direction of gaze is available for gain modulation regardless of whether the eyes or head are used to guide gaze.[22]

---

21. These observations in area 7a of the monkey parietal cortex by Andersen et al. (1985) have been confirmed and extended to other cortical areas, including the ventral intraparietal area (VIP), by numerous other laboratories. Head position–modulated neurons are found in the lateral bank of the intraparietal sulcus (LIP; Brotchie et al., 1995). Hand-position neurons are found in the medial bank, also called the *parietal reach region* (PRR), because PRR units are spatially selective for visual stimuli when they are used as the goal for reach movements (Chang et al., 2009).

22. In addition to eye, head, and hand positions, other types of signals also produce gain fields, including vergence of the pupils, target distance, chromatic contrast, and the expected probability of the sensory signals themselves, suggesting that gain modulation of neuronal activity is a general computational mechanism.

Retinal View and Movement Coordination

These neuronal mechanisms can explain how we reach for the volume knob on the radio while driving. As discussed in Chapter 10, parietal neurons are sensitive to the direction and speed of moving visual, auditory, and tactile stimuli, so this brain region is a good candidate to support this task. Through its connections to the premotor cortex, the parietal cortex can encode multimodal reference frames and perform coordinate transformations across reference frames. To perform the task correctly, parietal cortex neurons or their targets must calculate the relationship between the current gaze angle and the hand's target location, the knob on the radio.

In general, movement trajectory programming requires transforming information from sensory coordinates into a coordinate system appropriate for the eye, head, arms, or body. Instead of considering the entire population of parietal neurons, let's consider four, for simplicity (Figure 11.4). The goal is to guide the arm to the knob under visual control, regardless of the gaze angle.[23] Because the relationship between the body and the knob can be considered fixed for this simple situation, only the varying relationship between gaze and body must be calculated. As we have seen, parietal neurons "know" the position of the eyes in the orbit, and this information can be added to the object's retinal coordinates. By computing the summed activity of the four parietal neurons, the downstream reader neurons can infer the gaze–body relationship. The response magnitude of these reader neurons shifts as the gaze angle changes. Thus, the spatial coordinates of the arm's target (the knob) can be constantly updated as the gaze direction moves.

Unfortunately, this nice description is based on theoretical modeling rather than real population data and detailed physiological exploration of the transformation. Yet the principles by which gain-modulated neurons can transform coordinate systems illustrate a viable mechanism. Fortunately, some experiments support the existence of the hypothetical reader neurons of the coordinate transformatlion. In both parietal and premotor cortex, some neurons encode the horizontal direction and elevation of object locations independent of gaze direction. These neurons thus represent spatial locations in a head-centered frame of reference.[24]

---

23. For this exercise, we disregard how the movement trajectory is organized. A large field of neuroscience is dealing with that complex issue (Shenoy et al., 2013).

24. Duhamel et al. (1997); Graziano et al. (1997). Fogassi et al. (1992) have shown that neurons in the premotor cortex of the monkey (inferior area 6) provide a stable body-centered frame of reference necessary for programming visually guided movements.

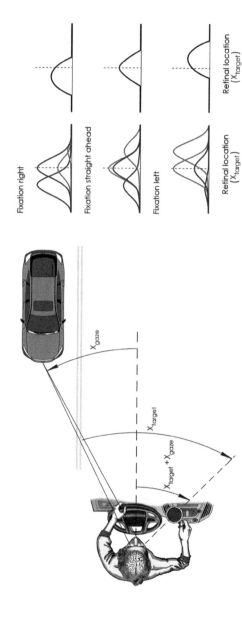

Figure 11.4. Gain fields support coordinate transformations. *Left:* A coordinate transformation performed by the visual system during driving. While driving, you may want to reach the volume control knob of the radio without shifting your gaze. The location of the knob relative to the body is given by the angle between the two dashed lines. For simplicity, assume that initially the hand is close to the body, at the origin of the coordinate system. The reaching movement should be generated in the direction of the knob regardless of where the driver is looking, that is, regardless of the gaze angle x_{gaze}. The location of the target in retinal coordinates (i.e., relative to the fixation point) is x_{target}, but this varies with gaze. However, the location relative to the body is given by $x_{target} + x_{gaze}$, which does not vary with gaze. By this mechanism, a change from retinal, or eye-centered, to body-centered coordinates is performed. *Right:* Combining the activity of several gain-modulated neurons may give rise to responses that shift. The left column shows the responses of four idealized, gain-modulated parietal neurons as functions of stimulus location; the three curves on the right are the summation of the four gain-modulated neurons, which correspond to the drive of a downstream observer unit. Inspired by Selinas and Sejnowski (2001).

Switching coordinate frames is a remarkable feat that involves discarding a lot of information and extracting only the essential features from a new point of view. The structures downstream from the visual-parietal areas need to know only what the eyes are looking at and where to reach, at least in our example. Reaching movements have to be learned by extensive motor practice, which provides the grounding for the transformed neuronal pattern. [25] Without action and feedback from its consequences, the neuronal responses to upstream inputs remain gibberish for the brain.

Abstraction by Different Reference Frames

It is remarkable that the transformation between eye-centered and body-centered coordinates requires only a neuronal gain mechanism. If such a simple mechanism works for one coordinate transformation, perhaps it can be deployed for similar purposes in other circuits. That is, the body-centered information carried by neurons that read the output from the parietal cortex can undergo further transformation at the next level of computation using the same type of gain control mechanism. Along the way, ever more complex features of the retinal information can be extracted.[26] When an object is viewed, approached, and touched from different directions, the multiple experiences may strip off the particular conditions in which the object was

---

25. While there is a backbone of anatomical connections between these layers, most information about the relationship among body parts, their distances from each other (Chapter 3), and their relations to the surrounding world have to be learned and maintained by practice. Experience can ground the sensory information by proprioceptive, corollary discharge signals by way of multiplicative gain modulation. The typical mistake of every novice car driver is steering the wheel in the direction that the head turns. This is the natural action for the brain. Commanding a big extracorporal extension, such as a car, is not rooted in our biology.

26. Salinas and Sejnowski (2001) and Salinas and Abbott (1995) were perhaps the first to hypothesize that gain modulation is a general-purpose coordinate transforming mechanism, extending early ideas by Zipser and Andersen (1988). Their modeling showed that neurons in a recurrently connected network with excitatory connections between similarly tuned neurons and inhibitory connections between differently tuned neurons can perform a product operation on additive synaptic inputs. Impairment of the gain control mechanisms in computer models, using the firing-rate gain-modulation mechanisms described earlier, can account for various deficits observed in hemi-neglect patients (Pouget and Sejnowski, 1997a). Gain control, thus, can be the cipher between different forms of representations.

sensed, leaving only its essential features, regardless of where they appear in the visual field. This is known as *translation invariance*, an abstraction.[27]

GAIN CONTROL BY ATTENTION

The driving scenario can be mimicked in the laboratory by training monkeys to fixate their eyes to a cross (in lieu of traffic) and at the same time to pay attention to a particular location in their visual field where something may appear (in lieu of the volume knob). This is accomplished by extensive training, rewarded with a few drops of juice. The physiological measurement begins after an orientation-selective neuron is found and its receptive field identified in the visual-parietal system. Many neurons in these areas are called *orientation-selective* because they respond maximally to black-and-white stripes moving in a particular direction. By moving the striped gratings in multiple directions, the experimenter can construct a tuning curve. The measurements can be performed under two different conditions. In the first one, the monkey is trained to detect something in the receptive field of the neuron (to pay attention to it, as we do when we guide our hand to the radio knob by peripheral sight while keeping our gaze on the traffic ahead). In the second condition, the animal is asked to detect something in the opposite hemifield of vision (as we do when paying attention to the left mirror by peripheral vision). The two conditions are identical in every physical way because the monkey's gaze is fixed to a cross in the middle of the screen all the time, so only the location of attention is different. The shape of the two tuning curves is similar, as expected, but the magnitude of the response is larger when the monkey pays attention to the neuron's receptive field. The largest difference occurs at the peak of the receptive field; outside the receptive field, the two curves are equal. In short, selective attention multiplies or "gain-modulates" the neuron's response by a constant factor without changing its tuning characteristics. Attention-induced

---

27. Neurons in the inferior temporal (IT) cortex of the monkey (corresponding to the middle and inferior temporal gyri in the human brain) respond equally strongly to objects and, in some cases, to faces wherever they appear in the visual field and against whatever background. Face-selective neurons are more prevalent in the superior temporal sulcus. The same neurons may not respond at all when other objects or faces are shown at these locations (Desimone, 1991; Miyashita, 1993; Logothetis and Sheinberg, 1996). Neurons in the entorhinal cortex of human research participants show highly selective responses to objects and faces, such as the famous actress Jennifer Aniston. These neurons represent highly abstract features because even a cartoon or the written name of the actress was able to activate these cells (Quian Quiroga et al., 2005). This hypothetical coordinate transformation process is reminiscent of the formation of semantic information from episodic experience (Chapter 5).

gain modulation works well at all stages of the visual-posterior parietal cortical system[28] and likely in many other brain areas.

Autonomic Correlates of Attention

Attention is a spooky subjective term and an important item on James's list (Chapter 1). As a covert action, it is not accessible to direct investigation, short of placing numerous electrodes in the brain. But even if we remain motionless and our eyes do not rotate, there may still be some observable changes when we are concentrating on something. As discussed earlier, neuromodulators are important players in gain modulation. Two of these, acetylcholine and norepinephrine, also control the diameter of the pupils. When we are excited, norepinephrine levels are high, neuronal gain increases, and the pupils are constricted.[29] Thus, pupil diameter may be an objective readout of attention.

In an experiment addressing this possibility, human study participants were trained to recognize images with various visual and semantic features. Unknown to the participants, one visual feature and one semantic feature predicted monetary reward. Some participants were biased toward visual and others toward semantic features, which allowed the investigators to focus on the corresponding brain areas in the participants using functional imaging. Pupil response was strongly correlated with the degree to which learning performance followed these individual predispositions for visual or semantic features and also with the strength of the correlated activity across brain regions.[30] Thus, using the pupil diameter or other autonomic responses as a proxy for norepinephrine levels, neural gain can be studied in the intact human brain—at least indirectly. While there are obvious caveats to this approach, the findings nevertheless suggest that attention and gain modulation may have

---

28. Attention-induced gain modulation was discovered by Moran and Desimone (1985), recording from neurons in areas V1, V2, V4, and TEO, and replicated soon by work in area MT (Treue and Martinez-Trujillo, 1999) and in V2 and V4 (McAdams and Maunsell, 1999). The latter studies also showed that attention increases neuronal spiking multiplicatively by applying a fixed response gain factor. The review by Reynolds and Heeger (2009) discusses the many facets and advantages of attention-modulated gain control.

29. Pupil diameter is correlated with tonic levels of spiking activity in locus ceruleus noradrenergic neurons. Phasic activity of locus ceruleus neurons varies inversely with pupil diameter. Pupil diameter has been suggested to track attention and neural gain (Aston-Jones and Cohen, 2005; Gilzenrat et al., 2010; Reimer et al., 2016).

30. Eldar et al. (2013).

overt behavioral correlates.[31] We can speculate that the output commands that control the pupil-affecting neurons send corollary discharges to parietal cortical regions which may be responsible for gain control of visual input.

GAIN CONTROL BY VELOCITY

Gain modulation of firing rates is also a prominent mechanism in the hippocampal-entorhinal system to accomplish running speed-independent place coding (Chapter 7). Each principal cell receives at least two types of inputs. One represents complex features arriving from various neocortical regions (external signals), and another represents information from the velocity system (internal signal). As a result, hippocampal and entorhinal neurons are under dual control. With two types of information at hand, one of which is action-related, the other external input can be grounded.

The firing rates of many place cells are elevated when the animal's locomotion speed increases (Figure 11.5).[32] The middle of the place field shows the strongest effect, and it tapers off gradually toward the animal's entry and exit from the field. Outside the place field, velocity has no or very little impact on firing rate. As in other areas of the brain, the effect of the two inputs is multiplicative. Thus the place fields at different velocities show no change in field size but large changes in firing rates. Although the vestibular system is a likely source of speed modulation, it is not the only one. Firing rates are also modulated by speed during running on a wheel, on a treadmill, or on a ball

---

31. An obvious shortcoming is the lack of a distinction between global and focal attention, although the difference between visual and semantic feature–related changes in the blood oxygen level-dependent (BOLD) signal could be taken as an argument against a generalized arousal effect. Such global changes occur with stress, which is also associated with pupil contraction. Another important caveat is that gain control brought about by subcortical neuromodulators is inevitably slow due to the slow G-protein-coupled receptor responses, whereas gain modulation often occurs at a faster scale by other mechanisms. These fast mechanisms might also be at work during attention.

32. Each place cell has a relatively unique velocity rate profile. Some increase monotonically with rate; others increase and decrease their activity as a function of speed. The same neuron in two different place fields can have different speed rate profiles, indicating that current assembly members are driven by a unique distribution of speed inputs (McNaughton et al., 1983; Czurko et al., 1999). Not all place fields are modulated linearly with speed. Instead, a particular running speed is associated with maximum firing rates. A minority of place cells are negatively modulated by speed.

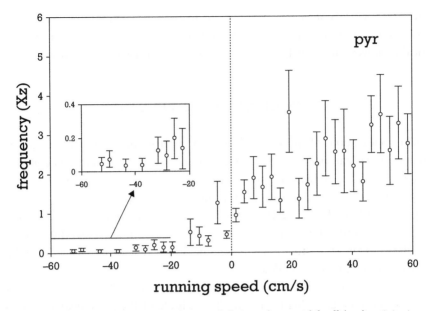

Figure 11.5. Running speed–dependent modulation of pyramidal cell (*pyr*) activity in the hippocampus. Negative and positive speed values indicate running directions (left or right) in a running wheel. Firing rate increases with speed when the rat faces the tuning field of the cell but is suppressed when it faces the opposite direction. The insert emphasizes speed-dependent suppression of firing. This tuning behavior is similar to the behavior of attention modulation in the receptive fields of visual cortical neurons.

or track with the head fixed, perhaps driven by optic flow or feedback from muscular activity.[33]

The velocity input is essential for place cells to represent positions and distances reliably. Suppose that on a slow run the peak firing rate of a place cell is 10 spikes per second. During a faster run, the peak rate may increase to 15 spikes per second. In this case, the neuron fires 10 spikes per second when the animal moves away from the area of peak firing.[34] If neurons downstream from the hippocampus must decode the animal's location from firing rate alone, they will get different answers on different runs because 10 spikes per second corresponds to different positions depending on running speed. Even if the reader mechanisms get inputs from populations of place cells, the ambiguity of

---

33. Neither vestibular input nor optic flow is essential because speed affects the time shift across cell assemblies even in complete darkness while the head of the mouse is fixed (Villette et al., 2015).

34. A related solution is to use the ratios of instantaneous firing rates of place cells with overlapping place fields. In either case, a division is needed.

rate coding remains. To decode the animal's position correctly on the tract, the reader mechanism must divide the arriving pulses by 1.5 during the faster run.

This normalization solution is reminiscent of the mechanisms in intensity-invariant sensory neurons, which use both inhibitory and population feedback responses, as discussed at the beginning of this chapter. Therefore, one possibility is that, during fast runs, the rate increase of a particular place cell is normalized by the velocity-dependent increase of other neurons with overlapping place fields.[35] Another option is divisive inhibition. When the perisomatic inhibition of a place cell is selectively decreased experimentally, its firing rate is enhanced within the place field, with the highest change at the peak of the field. The decreased inhibition outside the place field has little effect. Because interneurons are also modulated by speed, this experimental manipulation shows that changes in inhibition constitute at least one of the factors that induce gain.[36]

Velocity Affects Distance Representation

Velocity affects not only the firing rates of place neurons but their temporal relationships as well. You may recall from Chapter 7 that the cell assembly sequence starts during the late phase of the theta cycle with the representation of places the animal has already passed, and the spikes of subsequent assemblies advance to earlier phases representing upcoming locations. Each theta cycle, therefore, represents a sweep from past to the future or a segment of distance that starts behind and ends ahead of the animal. The forward-looking part of the representation is dubbed "look ahead." Thus, distances are converted into theta time scale durations. This distance-to-duration compression was discussed in Chapter 7.[37] To illustrate its significance, assume for a moment that the velocity signal plays no role in hippocampal computation. In that case, fixed time offsets between assemblies within the theta cycle should be related to variable run

---

35. In Chapter 7, we discussed that hippocampal place cells are velocity-controlled oscillators and that the theta phase of spikes advances proportionally to running velocity so that the theta phase of place cell spikes remains invariant to position. Thus, if the downstream readers have information about the phase of theta oscillation, a phase decoding mechanism is less ambiguous than rate decoding.

36. Royer et al. (2012). Several other experiments have shown that inhibition is, under most conditions, divisive rather than subtractive (Murphy and Miller, 2003; Carandini and Heeger, 2011; Wilson et al., 2012).

37. Skaggs et al. (1996); Tsodyks et al. (1996); Samsonovich and McNaughton (1997); Dragoi and Buzsáki (2006); Diba and Buzsáki (2007); Geisler et al. (2007); Johnson and Redish (2007).

durations between positions *A* and *B*. As a result, the distance representations would vary at different running speeds. The velocity gain is what prevents such distortion. Indeed, experiments show that velocity adjusts the transition times of place cell assemblies within the theta cycle. At faster velocities, the transition from one assembly to the next becomes proportionally faster. Because of the increased gain on assembly transitions, more cell assemblies can fit into a given theta cycle. In other words, larger segments of the environment are represented in each theta cycle during faster runs.

For this reason, distance "representation" (pardon me for using this convenient phrase) in the hippocampus does not vary much with the animal's motor behavior.[38] Thanks to the velocity gain, we can judge the distances between objects irrespective of our speed; for example, when driving behind another car. This mechanism explains how the brain of a mounted archer can calculate the distance between himself and his target. We can conceive the velocity gain mechanism as an extension of the corollary discharge idea, with the goal of informing the brain that the changes in neuronal activity are brought about by the actions of the brain's owner and not by changes in the environment.

FROM VELOCITY TO ATTENTION: INTERNALIZATION OF GAIN CONTROL

How can any of these mechanisms of gain control be mobilized by a subjective mechanism such as attention? During real-world navigation, the rate of change within assembly sequences is controlled by the animal's body velocity or eye movements. As discussed earlier (Chapter 5), many environmentally driven functions become internalized as brains increase in complexity. For example, navigation becomes mental travel and serves seemingly different functions yet uses the same neuronal mechanisms. Self-organized assembly sequences in the hippocampal system and their disengagement from the external world seem perfectly suited for memory, imagination, and planning, but the internalized mechanism that controls gain during these functions is not clear.

---

38. Cell assembly sequence compression is measured as a ratio between distances of places (external measure by the experimenter) and the rate of change between the spikes of the place cells within the theta cycle (an internal mechanism). Maurer et al. (2012) calculated that sequence compression is high at low speed and decreases with the rat's running velocity up to 40 cm/s. A potential explanation for why the compression factor reaches a plateau is because there should be enough time for the hippocampus to compute upcoming locations even at very high speeds.

We can hypothesize that attention is a candidate for this mechanism. In lieu of velocity, attention can adjust the rate of change of assembly sequences. This idea is supported by observations in the visual system. Investigators in several laboratories were surprised to observe that locomotion velocity exerts a robust effect on neuronal firing rate even in the primary visual cortex, showing a general effect of motion on neuronal gain.[39] The mediator of this speed-related gain is a special kind of inhibitory interneuron that inhibits other, dendrite-targeting inhibitory neurons.[40] The overall result is reduced inhibition of pyramidal cells, which increases their activity. These interneurons are excited by subcortical neuromodulators, such as the cholinergic input from the basal forebrain. This input increases its activity not only during locomotion but also when the animal attends to various stimuli. It remains to be explained how attention and cortical circuits can activate cholinergic neurons.[41]

If we tentatively accept that attention, as a multiplicative gain mechanism, can substitute for velocity, then we can explain how internalized functions of place cells, grid cells, and other types of neurons influence the dynamics of cognitive operations without locomotion. Even if the mechanisms are not clear at the moment, the advantage of this idea is that the evolutionary origin and behavioral meaning of attention gain in neuronal networks can be grounded by action-induced feedback to the brain.[42] The vestibular system monitors the speed and acceleration of our own motion and can provide the ground truth about the rate of change of our actions. Although we do not know what neuronal mechanisms step in to substitute for vestibular signals, we do know that such a substitute exists because head-direction neurons maintain their relationships even during sleep, and their speed of change is controlled by internal brain dynamics (Chapter 5).

If division or multiplication underlying gain control and normalization have such far-reaching consequences on behavior and cognition, as we discussed

---

39. This was surprising only within the perceptual framework of vision, where action plays no role (Niell and Stryker, 2010).

40. This special interneuron type is known as the *vasoactive intestine peptide* (VIP)-expressing subgroup, which can be activated by both acetylcholine via nicotinic receptors and serotonin via 5-HTa receptors (Rudy et al., 2011; Fu et al., 2014). This VIP neuron-mediated disinhibitory circuit may constitute the common gain control mechanism activated by both locomotion and attention.

41. Gielow and Zaborszky (2017); Schmitz and Duncan (2018).

42. Krauzlis et al. (2014) also argue that attention is an effect rather than a causal agent. It is a byproduct of circuits centered on the basal ganglia involved in estimating the current state of the animal and its environment. Brain mechanisms responsible for attention existed prior to the emergence of the neocortex.

earlier, one might expect that they exert a profound effect at multiple levels of neuronal organization. And they do. The next chapter discusses how.

SUMMARY

Gain and normalization are simple but fundamental mechanisms that can support numerous functions in the brain. These mechanisms are called by a variety of names, such as coordinate transformation, place anchoring, abstraction, and attention. They are of fundamental importance, as illustrated by the numerous mechanisms for gain control that exist in the brain, including divisive inhibition, short-term plasticity of synapses, and subcortical neuromodulation. Gain control allows inputs from the retina and the positions of the eyes in their sockets, the head, and the hands to affect the magnitude of responses to visual inputs in multiple brain regions, particularly the parietal cortex. Gain control mechanisms can shift coordinate representations, for example, from visual space to head space to hand space or recognize an object as the same when it is viewed from different directions. The mechanisms of translation and object invariance are the neuronal basis of abstraction, a process of ignoring features that are not essential to recognizing entities. Gain control is important in the hippocampal system, allowing judgment of distances independent of locomotion speed. Attention may be viewed as internalized gain control.

12

Everything Is a Relationship

The Nonegalitarian, Log-Scaled Brain

We only use ten percent of our brain[1]

You have no idea, how much poetry there is in the calculation of a table of logarithms.

—KARL FRIEDRICH GAUSS[2]

For things to stay the same, everything must change.

—GIUSEPPE DI LAMPEDUSA[3]

My fellow New Yorkers are tolerant people with a "we have seen it all" attitude. You may wear anything you want or nothing while walking the avenues. You can be a cross-dresser, or even have a Hungarian accent, and the majority of New Yorkers would not even blink. But

---

1. The exact origin of this neuromyth (BrainFacts.org) is not known. "In fact, most of us use only about 10 percent of our brains, if that." wrote the Hungarian-Jewish illusionist, magician Uri Geller in 1996. [NeuroNote: Geller suffered from bulimia and anorexia nervosa. He believed he had paranormal powers given to him by extraterrestrial agents.] The US Defense's Advanced Research Projects Agency (DARPA) wasted tons of dollars to examine the outrageous claims of this parapsychologist (Targ and Puthoff, 1974). The 10% idea may have been derived from a William James quote, "We are making use of only a small part of our possible mental and physical resources" (James, 1907; p. 12). Of course, 10% could refer to brain volume, firing rates, synaptic strengths, etc.

2. http://lymcanada.org/the-poetry-of-logarithms/.

3. di Lampedusa (1958).

a squirrel-sized man walking with a woman the size of a giraffe in the foyer of the Metropolitan Opera would surely raise some eyebrows, not because it is deemed scandalous but because it is so improbable.[4] Such big variation in body size within a species is outside the norm. But in the brain, distributions that range over several orders of magnitude are the norm, whether we look at connectivity or dynamics. Magnitude or fold differences are easily noticeable because they are based on a ratio or relationship judgment. As explained in this chapter and the next, ratio judgment is a natural function of the brain due to the way it calculates.

Multiplication, division, fraction, proportion, normalization, and gain have been recurring terms in the preceding chapters[5] because they are ubiquitous operations in brain circuits. All these operations utilize ratios, or *logos* in Latin. Hence the theme of this chapter: log-scaling in the brain.[6] My goal here is to demonstrate that brain connectivity is far from random and that the brain's plastic features do not conform to the assumed substrate of a *tabula rasa* brain. Instead, a highly structured connection matrix supports a skewed dynamic that constrains how neuronal groups respond and maintain self-organized activity. To perform effectively, physiological brain operations must occupy a

---

4. While this scenario is highly improbable in practice, it is not theoretically impossible. Domesticated dogs have the highest body size variation of any species, from the 5-inch tall Chihuahua to the 7-feet tall Great Dane, thanks to human breeders. Body size and brain size variations show a normal or bell-shaped (Gaussian) distribution in all natural species.

5. The relationship between addition and multiplication has been discussed in various cultures. In medieval Europe, Christian scholars claimed that the knowledge-based road to God was a product of learning the scriptures of the Bible and using logic. The argument in favor of multiplication was that no amount of logic will help without reading the scriptures. If scripture knowledge is zero, the total knowledge will be still zero. Conversely, if your logic value is zero, then your knowledge will be still zero even if you have memorized every page in the Bible (Harari, 2017). By a similar logic, yin and yang are inseparable and opposite forces according to Chinese philosophy. No amount of yin or yang alone means knowledge or balance. In contrast, addition would always yield some knowledge even if you do not read the Bible or you are ignorant of the yang.

6. *Logos* has multiple meanings, but in the mathematical context, it refers to *ratio* or *proportion*. "Logarithm" translates to numbers with ratios, as *arithmos* means number. The problem of logarithm evolved from astronomical observations that had to address ratios in spherical trigonometry, such as the sine, cosine, tangent, and cotangent of angles. In connection with logarithms, the Scot John Napier deserves to be remembered. Napier's main contribution was to introduce a practical device (the tables of logarithms). He recognized that multiplication or division of two positive variables can be calculated by adding or subtracting the logarithms and taking the antilog of that sum or difference, respectively. Unknown to Napier, the Swiss watchmaker Jost Bürgi had already prepared such log tables, which were used by Johannes Kepler (Clegg, 2016). [NeuroNote: Kepler's mother was accused of practicing witchcraft, and several of his family members had mental problems.]

wide dynamic range between silence and supersynchrony. This dynamic range is needed to allow many competing features to coexist in the brain, such as redundancy, resilience, degeneracy, homeostasis, plasticity, evolvability, robustness, and stability. These features constantly compete with each other.[7] In simpler terms, networks must be responsive to both weak and strong inputs. Weak stimuli should be amplified, whereas strong stimuli should be attenuated. At the same time, neuronal circuits should maintain their ability to evaluate differences in inputs. That is, networks must be both sensitive and stable. If these competing demands do not strike the right balance, the network becomes hyperexcitable and epileptic or remains dead silent forever. The delicate balance between these competing dynamic features is reached via components that are diverse, show a large quantitative variation, and combine in multiplicative ways. Systems with such properties are energy efficient and show high tolerance to component failures.[8] How to build and maintain such systems?

DIVERSITY YIELDS INTERESTING SYSTEMS PROPERTIES

Attending the annual Society for Neuroscience meeting always makes me feel anxious. That feeling arises from a specific uncertainty: What difference would it make if I were not part of this enterprise? In this stock exchange of ideas, 25,000 of my hard-working colleagues present their results. All the presentations contain something worth knowing, many of them are interesting, and there are always a handful that I cannot miss. The really interesting discoveries often come from the same top laboratories, year after year. I would guess that half of neuroscience knowledge comes from 10–20% of the labs. However, that also means

---

7. Some of these terms may not be clear to most readers. *Robustness* is the persistence of a function under perturbations. The somewhat related *resilience* describes the ability of a system to recover after challenges. *Redundancy* refers to components whose presence are not critical for a given function unless other components fail, in which case their role becomes critical (von Neumann, 1956). It is like having a chain on the door in addition to the lock. *Degeneracy* says that the same function can be achieved by a variety of different mechanisms. For example, a rhythm of the same frequency can be generated by a variety of structural-functional solutions. *Homeostasis* is a system's ability to maintain things (e.g., balance of inhibition and excitation through oscillations or other solutions). *Evolvability* is the capacity of a system to adapt to novel situations. One can call it *learning*. Learning, homeostasis, and adaptation to perturbation are achieved through plasticity.

8. Mitra et al. (2012); Ikegaya et al. (2013); Helbing (2013). An advantage of logarithmically linear computation in both electronic circuits and biochemical systems is its energy efficiency (Sarpeshkar, 2010; Daniel et al., 2013). This principle likely applies to brain circuits as well.

that the other half comes from the remaining 80–90% of investigators. The two halves of the distribution cannot function efficiently without each other.[9] Here is why.

Division of labor, made possible by diversity, is perhaps the most ubiquitous rule in living systems. The brain is not an exception. It contains many different neuron types and non-neuronal cells. In complex systems, this is known as *component diversity*. As an analogy, electronic circuits have a variety of different components, such as transistors, diodes, resistors, and capacitors. However, the proud transistor with its nonlinear amplification abilities cannot shine without the modest but critical contribution of the humble resistor. Even within components, there is a great variance. For example, the gain of a transistor can be small or large, and the resistance of resistors can vary by several orders of magnitude. Similarly, in the brain, the quantitative features within the same neuron types, synapses, and networks vary by ten- to a hundredfold and occasionally a thousandfold.

When many identical components interact with each other additively, we can get a good idea about their average behavior from macroscopic observations. For example, the properties of gases can be described by simply measuring their overall temperature without knowing the position and velocity of every molecule. As the number of similar components increases, the solutions typically become easier to compute.[10] In contrast, describing the precise behavior of one atom remains a huge challenge for physics. Yet the toughest realm to evaluate is when dozens of qualitatively different components interact. This is exactly the case in neuroscience. An analogy would be an orchestra, where the interactions of different instruments give rise to a virtually limitless variety of different compositions.

In an invertebrate neural circuit, such as the somatogastric ganglion of the crab, there are only a handful of neurons, yet each of them is different. Although the entire "wiring diagram" of this small circuit has been determined, we do not know the exact mechanisms through which these cell-to-cell interactions give rise to the sequential spatiotemporal patterns that produce a rhythmic

---

9. You may ask how such an allegedly skewed distribution of creativity relates to the normal distribution of the intelligence coefficient (originally *Intelligenzquotient* of William Stern; IQ), known to be a perfect bell curve (Herrnstein and Charles, 1994). The IQ curves are made symmetric by designing the tests to meet specific criteria. It does not measure creativity but aptitude to fit a particular niche.

10. This approach works because molecules of the same type are indistinguishable. We cannot label them and recognize them individually later. Neurons are different. They are individuals like we are. In fact, each neuron carries a distinguishable "barcode" that allows investigators to track its fate from birth to end of life (Kebschull et al., 2016; Mayer et al., 2016).

output.¹¹ Similarly, small brains have relatively few neurons compared to large brains, but their component diversity is high. However, component diversity grows only modestly across species.¹² Instead, a few neuron types, such as pyramidal cells, multiply in great numbers in larger brains, in a process akin to adding a million second violinists to an orchestra. The good news is that a system built from ten types of neurons, with one type multiplied a millionfold, is not vastly more challenging to understand than a tiny brain with just ten different individual neurons.¹³ The challenge is still huge (though not hopeless) because, even within the same type of component, quantitative variations are very large, as will be discussed later. Let us start the discussion on skewed distributions with an old problem in cognition and attempt to assign a neuronal mechanism to it.

THE LOG RULE OF PERCEPTION

Modern neuroscience, like early physics, is striving to describe and condense observations with mathematical equations. Compared to physics, neuroscience has only a modest number of laws. ¹⁴ Yet, one of these, the *Weber law* or also called the *Weber-Fechner law* has a breathtaking simplicity and generality. This law is named after two German scientists who laid the groundwork for human psychophysics, a quantitative investigation of the relationship between physical stimuli and the mental states they induce. Ernst Heinrich Weber was

---

11. Harris-Warrick et al. (1992); Marder et al. (2015). The degeneracy of this circuit is very rich, with hundreds of circuit solutions to the same rhythm in related species and within the members of the same species (Prinz et al., 2004).

12. For example, GABAergic interneuron types appear to be preserved in mammals (Klausberger and Somogyi, 2008). Principal cells of different cortical layers in different species also show strong similarities. Every now and then, novel neuron types, such as "spindle cells" in bonobos and whales are reported (Chapter 9), but they seem to be the exception rather than the rule.

13. A similar sentiment has been expressed by Gao and Ganguli (2015), who suggest that the number of neurons that need to be recorded in an experiment is proportional to the logarithm of the neuronal task complexity, although the role of component diversity was not explicitly addressed in this work.

14. Of course, there are no laws and rules in nature. In the original use of the term by Isaac Newton, a law refers to an event governed or caused by an external agent. According to Newton, all laws are enforced by God and never change. In reality, there are only reliable regularities that appear to obey some fictional laws, such as the law of gravity, Ohm's law, or the psychophysical Weber-Fechner law. These reliable regularities and recurrences of the world are exploited by the brain to predict the future.

a physician who was interested in how we can perceive differences in touch. After extensive experimentation, he concluded that "in observing the disparity between things that are compared, we perceive not the difference between the things, but the ratio of this difference to the magnitude of things compared." For example, if you are holding a 100-gram object, the second object needs to weigh at least 110 grams in order for you to notice a difference. If the object weighs 200 grams, you can detect a weight difference only if the other object weighs more than 220 grams or less than 180 grams. This threshold change, called the *Weber fraction*, is also called the *just noticeable difference*.[15]

Gustav Theodor Fechner did not conduct experiments. Instead, we can call him an early computational scientist. He trusted Weber's observations and calculated mathematically that sensation is a logarithmic function of physical intensity. Therefore, when stimulus strength multiplies, the strength of perception adds.[16] If the importance of a law depends on its generalizability, the Weber-Fechner rule is important. It applies to vision, hearing, and taste. Distance perception, time perception, and reaction time also vary logarithmically with the distance or time interval, respectively.[17] The distribution of intersaccadic intervals—that is, how long we inspect a visual scene—also shows a log-normal form. Similarly, it takes an increasingly longer time to discriminate between two numbers as the difference between them decreases. Decision-making and short-term memory error accumulation also obey the law.[18] This is an impressive list.[19] The Weber-Fechner law was conceived 150 years ago, but the reason that our subjective sense of many variables contains these particular systematic patterns has remained obscure.

15. The just noticeable difference can be considered as a threshold for subjective judgment.

16. "Simple differential sensitivity is inversely proportional to the size of the components of the difference; relative differential sensitivity remains the same regardless of size" (Fechner, 1860/1966). $S = k \ln I + C$, where S stands for sensation, I for stimulus intensity, C is the constant of integration, and $\ln$ is the natural logarithm. The constant k is sense-specific and varies across modalities. Fechner claimed that his insight into the relationship between material sensation and mental perception came to him while lying in bed half asleep (Heidelberger, 2004).

17. When we are asked to estimate the length of an interval, the variability in our responses scales with the magnitude of the interval to be timed (Gibbon, 1977; Gibbon et al., 1984; Wearden and Lejeune, 2008). Similarly, errors of distance estimation increase proportionally with the distance to be estimated.

18. Dehaene et al. (1998); Gold and Shadlen (2000); Deco and Rolls (2006); Buzsáki (2015). Stevens (1961) proposed that a power-law relation between sensation and stimulus intensity is more mathematically plausible than the log rule. However, Stevens's approach has been repeatedly criticized on methodological grounds (Mackay, 1963; Staddon, 1978).

19. Many nonbiological phenomena, such as sound level, earthquake intensity, pH scale, and entropy, also exhibit logarithmic behavior.

My hypothesis to explain these systematic patterns is based on the assumption that the subjective properties of our actions and perceptions are supported by the log-normal distribution of the meso- and microscopic "connectome" of the brain and its log rule–dominated dynamics.[20] To support this hypothesis, we must find a matrix of connections and a dynamic in the brain that match the statistical features found in psychophysics experiments. I agree that this explanation is not the only possibility, but it is worth pursuing. Before I can explain this idea further, we need to discuss the nature of some statistical distributions.

NORMAL AND SKEWED DISTRIBUTIONS

No person has a brain ten times larger than that of an average human. Brain volume within the same species shows a normal or Gaussian distribution.[21] A *normal distribution* is characterized by a bell-shaped curve that is symmetrical and can be quantified by two parameters—the mean and the standard deviation. In a true normally distributed population, the incidence of values several times larger at the left hand compared to the right hand is practically zero.

However, normal distributions are rare in biology. Driven by diversity, large variations are the norm rather than the exception in biological systems. Multiplication, rather than addition, of proportionate values gives rise to many "extreme" values, which make the distribution look asymmetric or "skewed." Such distributions can take many forms.[22] Perhaps the most common skewed

---

20. This idea was the major theme of my previous book (*Rhythms of the Brain*, 2006). Similarly, the Bayesian model of the brain is a quantitative attempt to describe the match between world events and constructions of patterns by the brain (Helmholtz, 1866/1962; Ashby, 1947, 1962; Dayan et al., 1995; Rao and Ballard, 1999; Friston, 2010; Friston and Buzsáki, 2016).

21. Physical quantities that are the sum of many independent variables often have normal distributions. According to the Central Limit Theorem from probability theory, the addition or subtraction of a large number of small independent random variables yields a frequency distribution that takes the form of a Gaussian curve. Not so long ago, scientists believed that all distributions in nature are normal, which is why it was called "normal." The idea was that there is a typical value, and the variability is due to some additive noise. Perhaps Henri Poincare was the first to question this belief: "Everyone believes in it: experimentalists believing that it is a mathematical theorem, mathematicians believing that it is an empirical fact" (as cited in Lyon, 2014).

22. Gamma distribution and log-normal distributions are similarly skewed, but gamma does not strictly rise from the multiplication and division of random variables. For log-normal, the skewness of a log plot is zero, whereas a gamma distribution has a heavier tail on the left. The fit of the different distributions to the data can provide clues to the generative mechanisms that give rise to the distribution.

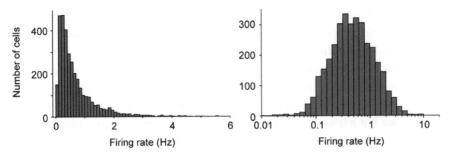

Figure 12.1. Relationship between linear and log display of a distribution. Firing rate of hippocampal pyramidal neurons during non–rapid eye movement (non-REM) sleep, plotted in the linear scale. Log-normal distribution is unimodal on the log scale.

distribution in biology is the logarithmic-normal or log-normal distribution.[23] This distribution is right-skewed on a linear scale but looks bell-shaped when the logarithms of the observed values are plotted. In other words, a log-normal distribution is a probability distribution of a random variable whose logarithm is normally distributed. Examples from biology include the number of species per family, survival times of species, sizes of fruits, pharmacological effects, times to first symptoms of infectious diseases, and blood pressure distribution within an age group, just to list a few. Basic physiologic processes, such as resting heart rate, visual acuity, and metabolic rate, also show a skewed distribution. In economics, incomes are often well fitted to a log-normal function.[24]

The log-normal distribution also describes the firing rates of cortical pyramidal neurons. The activity of these cells varies from nearly complete silence to several spikes per second (Figure 12.1). When the long-term average firing rates of many neurons are plotted on a linear scale, the distribution is strongly skewed, with many slowly firing neurons at the left end and a minority of highly

---

23. Aitchison and Brown (1957); Crow and Shimizu (1988); Ansell and Phillips (1994); Limpert et al. (2001).

24. Log-normal distributions describe data from diverse disciplines of science (Koch, 1966). Many of these distributions in biology have also been characterized by power laws. In a power-law relationship, a change in power of one quantity varies as a power of another, resulting in a line in a log-log plot. Power law and log-normal distributions connect naturally, and similar generative models can lead to one or the other distribution depending on minor variations (Mitzenmacher, 2003). The right tail of a log-normal distribution often follows a power law, so, in practice, the distinction between log-normal and power law distributions is often not trivial. If the variance in the left tail is large and "noisy" or thresholded at an arbitrary value, a log-normal distribution may appear as a line in a log–log plot, which is the telltale of the power law. However, a log-normal distribution has a finite mean and variance, in contrast to a power-law distribution in scale-free systems (Barabási, 2002; Barabási and Albert, 1999).

active neurons occupying the right tail of the plot. There is no "golden mean" or representative average neuron in this distribution because the mean is strongly biased by the fast firing minority. A median, which is the middle value in the distribution, is a bit more descriptive, but the shape of the distribution is more informative than any single value. When we replot the data on a logarithmic scale, we see the familiar bell-shaped curve reflecting a normal distribution of log values.

Scientists are interested in the frequency distribution of their observations for at least two reasons. The first reason is pragmatic. The distribution of values tells us how to verify the validity of the observations with appropriate mathematical methods because the choice of these methods depends on the nature of the distribution. There are numerous, so-called parametric methods (such as analysis of variance or regression) to quantify whether two or more normally distributed groups differ.[25] The second reason is important for a different goal: a distribution can offer clues about the mechanism that gives rise to those observations.[26] If the frequency distribution shows a symmetric bell-shaped curve, we can be confident that the data are produced by a mechanism in which many small independent factors are added to or subtracted from a representative mean value. Likewise, a bell-shaped distribution of log numbers is produced by multiplying or dividing random factors. I discuss the neuronal mechanisms that underlie skewed firing rate distributions later in this chapter. However, first I present a list of structural and dynamic variables in the brain to convince the reader that the log rule prevails in neuronal systems.

---

25. The frequency distribution of the data is a major factor in determining the type of statistical analysis that can be validly carried out on any data set. Yet frequency distribution is rarely tested rigorously, simply because there are not enough data in most cases. If the variance of a log-normal distribution is small compared to its mean, it can look very similar to a normal distribution. Some statisticians argue that the normal distribution is simply a special case of the log-normal distribution (Lyon, 2004).

26. In 2016, the news broke that the world's largest pumpkin was grown in Belgium. The super squash weighed more than a ton (2,624.6 pounds), easily a hundred times bigger than the ones my family carves for display at Halloween. Distribution of fruit and flower size has long been known to fit a log-normal rather than a Gaussian pattern, and the shape of the distribution has been used to argue that genes combine multiplicatively (Groth, 1914; Sinnot, 1937). Galton (1879) pointed out that variance is the "charms of statistics." He was the first to introduce the log-normal distribution, although it was not called log-normal until recently. Galton also showed that the response of the eye to light intensity changes follows a log-normal form (confirming the Weber-Fechner law; as described in Koch, 1966).

LOG ARCHITECTURE OF NEURONAL NETWORKS

Brains are among the most sophisticated scalable architectures in nature and come in a variety of sizes from super small to extra large in different species. *Scalability* refers to a property that allows the system to grow yet perform the same desired computation, often with increased efficacy. The largest brain, belonging to the sperm whale (7 kg), is several thousand times larger than that of the smallest mammal, the Etruscan shrew (65 mg). This huge variability raises the question of whether all brains share similar architectural rules. If they do, what are the brain's scaling laws and constraints?

Axon Diameter and Conduction Velocity

One important constraint discussed earlier is the preservation of brain rhythms in mammals (Chapter 6). Thus the computational and syntactic rules, set by these rhythms, should remain the same, irrespective of brain size.[27] In other words, neuronal networks need to preserve the brain's ability to temporally integrate large numbers of distributed local processes into globally ordered states so that the results of local computations can be broadcast to widespread brain areas simultaneously. In the reverse direction, local computation and the flow of signals to multiple downstream targets are under the control of global brain activity, often called "executive" or "topdown" control. A critical requirement for effective local–global communication is that the results of local computations in multiple areas are delivered within the integration time window of downstream reader mechanisms set by the various network oscillations. As brains grow larger, their parts inevitably become more distant from each other. To maintain temporal integration across brains of different sizes, hardware solutions are needed to compensate for the longer distances of transmission.

Long-distance communication occurs via the axon, which allows action potentials to travel from neuron to neuron via a synapse. Some axons can conduct spikes more quickly because myelin coats and insulates them, allowing the electrical impulses to move more efficiently. Because axon diameter and myelination determine the conduction velocity of neurons, evolutionary adjustment of these variables appears to be most important for brain size–invariant

---

27. In scalable systems, certain aspects of the system must be constrained if the same computational goals are to be achieved in the face of increasing organismal complexity (Buzsáki et al., 2013).

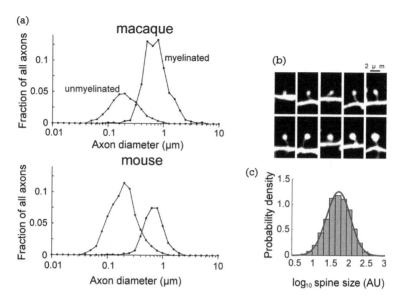

Figure 12.2. A: Distribution of diameters of unmyelinated and myelinated axons in the corpus callosum of the macaque monkey and mouse. B: Examples of spines imaged along the dendrites of one neuron. C: Lognormal distribution of spine sizes.
A: Reproduced from Wang et al. (2008); C: Reproduced from Loewenstein et al. (2011).

scaling. Increasing axon diameter enables signals to travel longer distances within a given time window and ensures that signals from various sources can be delivered to the target around the same time.

Axon diameters in the brain vary over several orders of magnitude, and their distribution is strongly skewed. In humans, the great majority of callosal axons have diameters smaller than 0.8 μm, but the thickest 0.1% of axons can have diameters as large as 10 μm (Figure 12.2). Large-diameter axons are typically found in the cross-hemisphere paths that carry sensory information, whereas in frontal cortical areas (where the speed of communication is slower), small diameter axons dominate. The thickness of axons emanating from the same neurons but targeting different brain regions can vary substantially, exemplifying a complex system in which lines of communication have a variety of geometrical and time-computing properties.

If all axons in larger brains increased proportionally in diameter, brain size and energy demands would increase enormously. Instead, a minority of axons with disproportionally increased diameter keeps timing relatively constant across brains of different species. Indeed, the thickest diameter tail of the distribution scales best with brain size. Adding a small fraction of axons with very large diameters can preserve cross-hemisphere conduction times across species

(Figure 12.2).[28] In summary, a disproportionate increase in larger diameter axons with fast conduction velocities keeps communication speed similar as brain size increases.

Microscopic Connectivity

The efficacy of communication between neurons, called *synaptic transmission strength*, depends on the anatomical features of the synapse. Excitatory inputs onto other excitatory neurons form synaptic connections with microscopic protrusions of the dendrites, known as *spines* because of their appearance under the microscope. Drumsticks or mushrooms would be perhaps more appropriate, as the head of the spine is connected to the dendrite by a narrow neck. This head is contacted by the presynaptic bouton at the end of the incoming axon. The volume of the spine is an important measure because it is correlated with the amount of postsynaptic density, which in turn is reliably correlated with the amplitude of the excitatory postsynaptic currents induced by the presynaptic release of excitatory transmitter.

Microscopic imaging studies have documented a large variety of spine sizes on single pyramidal neurons. Giant spines can be several hundred times larger than the smallest ones. The distribution of spine sizes in individual layer 5 neocortical neurons and hippocampal pyramidal cells is well described by a log-normal rule[29] (Figure 12.2). This large variation implies that some partner neurons exert a strong influence on the receiving neurons via large synapses, whereas most upstream partners establish weak synapses and therefore do not have much influence individually. This situation is thus quite similar to our social connections. A few individuals have a very strong impact on our behavior, while most of our acquaintances have little or no influence.

Mesoscopic Connectivity: A General Wiring Rule

If the brain were a *tabula rasa*, neurons randomly connected to each other would do a good job. Most network models use random connectivity

---

28. Swadlow (2000); Aboitiz et al. (2003); Wang et al. (2008); Innocenti et al. (2014). The host neurons of the giant caliber axons still need to be identified. At least a fraction of them are likely to be inhibitory neurons; the myelinated axon diameter of long-range inhibitory neurons in the rat can reach 3 μm (Jinno et al., 2007), several times thicker than that of pyramidal cells. In turn, theoretical and modeling studies suggest that long-range interneurons are critical for brain-wide synchronization of gamma, and potentially other, oscillations (Buzsáki et al., 2004).

29. Yasumatsu et al. (2008); Loewenstein et al. (2011).

statistics. Experience then could modify the connection strengths, as is often the case in artificial neural networks where every neuron has the same probability of connecting to every other neuron. In contrast to this expectation, careful microcircuit analyses in several laboratories have repeatedly demonstrated that neuronal motifs, such as that two neurons are connected reciprocally or that interconnected neurons often share inputs from the same third neuron, occur more frequently than expected from a random graph map.[30] However, general rules by which some such nonrandom connections are formed and how such rules scale across brains of different sizes have not been fully addressed yet.

Anatomical studies in macaque monkeys that trace the axonal projections back to their cell bodies show that the range of connection strengths between brain areas spans five orders of magnitude. A given cortical area is connected strongly to a few structures and weakly to many other structures. This connectivity profile is best described by a log-normal distribution.[31] Brain anatomy is similar in the mouse brain. An industrial-scale effort at the Allen Institute in Seattle showed that half of all axons sent out from each cortical area are directed toward only a handful of other areas. These few connections may represent half of all axons sent out from the cortical area in question. The remaining half of the axons weakly innervate a large number of areas. As in the monkey, the distribution of connections follows a log-normal form. In fact, the log rule applies not only to intracortical connections but to all the connections of any cortical patch, including their thalamic and other subcortical targets. Although this pattern has been established quantitatively in only two species so far, it is tempting to suggest that scaling of interareal connections is conserved across species and that the strength of interactions between regions obeys similar mesoscopic wiring rules.[32] With this information in hand, we can speculate how these structural foundations affect spike-mediated communication across neuronal assemblies.

---

30. Song et al. (2005); Yoshimura et al. (2005); Perin et al. (2011); Ko et al. (2013).

31. Markov et al. (2011, 2013); Wang et al. (2012); Ercsey-Ravasz et al. (2013). These findings considerably modify previous connectivity estimates in the neocortex. The average path length had been considered between 3 and 4, whereas the log-normal connectivity suggest that it is approximately 1.5. In other words, any cortical area is connected with any other through just one or two synaptic relays, although the long-distance excitatory connections are typically very weak.

32. Oh et al. (2014).

LOG DYNAMICS

As expected from the log-normal distribution of spine volumes in pyramidal cells, there is also a large variation in synaptic weights, which reflects the strengths of physiological connections among neurons. The most accurate way to assay connection strength is to record from pairs of neurons intracellularly and quantify the amplitude of the excitatory or inhibitory postsynaptic potentials in the postsynaptic neuron in response to evoked spikes in the presynaptic neuron. Because multiple intracellular recording is not practical in behaving animals, indirect methods are used to estimate synaptic strength. These involve calculating the probability of spike transmission between synaptically connected pairs of neurons (Figure 12.3). Typically, these two types of methods show good agreement.[33] Such experiments have established that spike transmission probability between neurons shows a log-normal distribution, and spike transmission, in general, is stronger in waking than in sleeping animals. Stronger synapses are not only stronger on average but also more reliable from event to event, reminiscent of the Weber-Fechner logarithmic law.

Log-normal Distribution of Firing Rates and Spike Bursts

The firing rate distribution of hippocampal neurons in Figure 12.1 is representative of all known neuron types in all cortical and subcortical regions of every species investigated to date, from turtles to humans.[34] The mean spontaneous firing rates of individual neurons of the same class span four orders of magnitude. In any given time window, 10–20% of neurons within each type contribute half of all spikes, whereas the rest are responsible for the remaining half of the spikes. These distributions can shift to the left or right across different classes of neurons and in different brain states, matched by similar shifts in several types of inhibitory interneurons. Importantly, the log firing rates of spontaneous spikes and spikes evoked by stimuli strongly correlate. If a stimulus evokes an extra spike in a neuron firing at a rate of one per second, the

---

33. Gerstein and Perkel (1969); Csicsvari et al. (1998); Fujisawa et al. (2008). To verify synaptic connections between neuron pairs, English, McKenzie et al. (2017) decoupled presynaptic neurons from ongoing network activity through single-cell juxtacellular current injection or optogenetic stimulation of small groups of cells in behaving mice. For direct paired recordings in vitro, see Song et al. (2005). Inhibitory neurons can also serve as hubs by either integrating larger numbers of inputs or contacting many targets in wide areas (Bonifazi et al., 2009).

34. Sometimes, rate distributions show a better fit with gamma than with log-normal distribution.

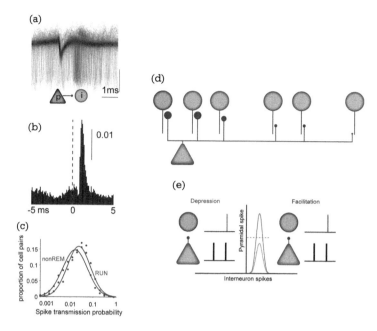

Figure 12.3. Dynamics of pyramidal cell–interneuron synaptic connections. A: Superimposed spikes of a pyramidal cell (*p*, small wide spikes) and an interneuron (*I*, large narrow spikes) triggered by spiking of the pyramidal cell. Note many spikes of the interneuron at 1 msec after the spike of the pyramidal cell. B: Temporal offset of spike transmission probability values between the CA1 pyramidal cell spikes (time 0) and spikes of the interneuron (peak of the histogram). C: Distribution of spike transmission probabilities (note log scale) during exploration (RUN) and non–rapid eye movement (non-REM) sleep. The right shift of the curve indicates stronger spike transmission during exploration. D: Connection strength decreases rapidly with distance between the neuron pairs. E: Spike transmission probability exhibits both activity-dependent short-term facilitation and depression, as indicated by the Gaussian curves above and below 'no change' (horizontal dashed line).
Modified from English, McKenzie et al. (2017).

same stimulus may induce ten excess spikes in another neuron whose baseline firing rate is ten per second. This is a huge difference in the number of induced spikes, yet proportionally they yield the same value. In other words, the response of a neuron to extrinsic inputs is proportional to its long-term firing rate, reminding us of the Weber-Fechner law.[35]

---

35. Remarkably, fast firing neurons also respond more robustly to optogenetic stimulation when the only or at least main source of excitation is channel rhodopsin-controlled depolarization. The firing rates of the artificially (light) induced spikes and spontaneous firing rates strongly correlate (Stark et al., 2015).

When the intervals between spikes are short, say 10 ms, that firing event is called a spike *burst*. The fraction of these short intervals relative to all interspike intervals is a quantitative measure of a neuron's burst propensity. Burst distribution also has a log-normal form. A small fraction of pyramidal neurons at the right tail of the distribution can be called *superbursters*, emitting a burst 40 to 50% of the time, whereas those at the left side hardly ever burst.

Skewed Distribution of Ensemble Size

The impact of a neuronal assembly on its targets depends on its degree of synchrony. There is a difference between a hundred neurons firing together in seconds or in a short gamma cycle (Chapter 4). Temporal synchrony can be measured by quantifying the fraction of spiking neurons in any given time window. Such measurements show that "ensemble size"—the number of neurons active in a particular time window—does not vary around a typical mean. Instead, the magnitude of synchrony follows a log-normal distribution: rare strongly synchronized (very large) events are interspersed irregularly among many medium and small events.[36] The skewed nature of the distribution implies that there is no characteristic neuronal ensemble size in cortical networks.[37] Rapidly firing neurons are more frequent participants in large population events as more frequently occurring spikes are expected to coincide more often with any other event than spikes of slow neurons.[38] However, what

---

36. For example, a small minority (1.5%) of CA1 pyramidal cells participates in half of hippocampal sharp wave ripple events, whereas half of all neurons fire in less than 10% of the sharp wave ripples (Mizuseki and Buzsáki, 2013).

37. The log-normal distribution of population synchrony is a challenge to Hebb's definition of cell assemblies. If the postulated assembly is a fixed set, it is not clear why the fraction of spiking neurons in successive time windows varies by orders of magnitude. However, if a good part of the neuronal messages can be carried by a minority of highly active neurons, then in each time window "good enough" information is present. Not only the fraction of active neurons but the correlation coefficients between neuron pairs also shows log-normal distribution, and coactivated neuron pairs largely correlate with the firing rates of the partners (Buzsáki and Mizuseki, 2014). Coactivation of calcium signals of CA3 neuron pairs in hippocampal tissue culture also shows a log-normal distribution (Takahashi et al., 2010). The factors that give rise to the strongly skewed distribution of population responses are not well understood. The log-normal variation of synaptic weights is a likely source because computational networks with dynamic synapses can respond robustly to external inputs yet return to baseline activity shortly after the perturbation (Sussillo and Abbott, 2009).

38. Okun et al. (2015) found that neighboring cortical neurons can differ in their coupling to the overall firing of the population. They called strongly coupled ones "choristers" and weakly coupled ones "soloists." It is possible that choristers and soloists represent neurons drawn from

is even more critical is their better connectedness to each other and everyone else in the population. This oligarchic "hub" nature of connectedness makes fast firing neurons more frequent partners and often leaders of large population events. Not only the magnitude but the duration of population events is also log-normally distributed, including that of hippocampal sharp wave ripples, neocortical slow oscillations, and thalamocortical sleep spindles.

Autocratic Organization of Neural Networks

The skewed distributions of so many structural and dynamic variables in neuronal networks constrain their operations in important ways. The fast firing minority has a special status. They not only work harder all the time, but they also have many more connections as well as being more strongly connected to their target partners, including other fast firing neurons.[39] This is the case for both excitatory and inhibitory neurons. They form *hubs*—an oligarchic organization of connections and information transmission. In network speak, such partnership is called a "rich club."[40] The highly active autocratic neurons have longer axons and send their axons to more structures than their slowly firing peers. Through these connections, the privileged club members have access to more information than nonmember neurons, and they share such information among themselves.

Again, it is important to reiterate that, despite the hard work of the hub neurons, they contribute only, say, half of the spikes and exert half of the influence. However, for normal brain operations, the other half is equally important. The remaining spiking activity is supplied by a large fraction of slowly discharging neurons, which are connected via weaker synapses in a more loosely formed large network. Although we like to anthropomorphize and talk about hubs and leaders, it is critical to emphasize that in the log-normal oligarchy, there is no definable demarcation line between the fast and slow and strong and weak ends. Therefore, the boundary between local and global connectivity

---

the right and left tails of the log-normal rate distribution, respectively. Population coupling is largely independent of sensory preferences, and it is a fixed cellular attribute, invariant to stimulus conditions. As expected from features of high-firing neurons, choristers respond to multiple features and are connected with larger numbers of other neurons compared to soloists (Mizuseki and Buzsáki, 2014).

39. Yassin et al. (2010); Ciocchi et al. (2015).

40. Olaf Sporns's book on brain networks (2010) has many links relevant to my discussion of the skewed organization of neuronal networks. See also van den Heuvel and Sporns (2011); Nigam et al. (2016).

is opaque. The advantage of networks with such local–global interactions is an optimal balance between computation speed and accuracy.

As it should be clear from the preceding discussion, a measurement as simple as the long-term firing rate goes a long way because rate is related to so many other features of the neuron. A pertinent analogy here is language. The frequency with which different words are used in everyday language affects their rate of meaning change across all languages and time scales. There is almost a hundredfold variation in rates of such lexical evolution among words in Indo-European languages, and the distribution shows a log-normal-like form. The most frequently used words, such as "one," "two," "who," and "night" are strongly conserved so it may take 10,000 years to diverge across languages, whereas rarely used words, such as "to turn," "to stab," and "guts" may change in just a few hundred years.[41] The frequently used words, similar to fast firing neurons, form rigid hubs and dominate our everyday conversations, while many words of our existing vocabulary are rarely used. It is tempting to conclude that the brain mechanisms that support such selection are related to the log rules that operate in brain structure and dynamics. How are such distributions generated and maintained, and what are their functional consequences?

CREATING LOG DYNAMICS

Perhaps the most exciting question is how the log patterns at so many levels of brain organization relate to each other. Skewed distributions can arise by a variety of nonlinear operations. Thus it is not obvious how the log distribution of anatomical connectivity gives rise to log dynamics or how skewing of synaptic weights relates to log-normal firing rates and skewed population cooperation. There are at least two options. Even if the inputs to a population of neurons show a bell-shaped distribution, the firing rate distribution of the responding neurons can become skewed due to nonlinearities in their membrane properties. Alternatively, the neurons may respond linearly to a log-normally distributed synaptic input, in which case the source of the skewness is the input.[42]

---

41. I borrowed these examples from Pagel et al. (2007), which describes several further consequences of rate of occurrence. The rate of word usage may be analogous to the externalization of brain processes. The more frequently we encounter similar consequences of an action, the more explicit its brain representation becomes.

42. Several papers have dealt with this topic: Hromádka et al. (2008); Koulakov et al. (2009); Roxin et al. (2011); Ikegaya et al. (2013).

As discussed earlier, the strengths of the hundreds to thousands of synapses on a single pyramidal cell have a log-normal distribution. Yet, given that a presynaptic neuron alone rarely causes a spike in a postsynaptic cell, we also need to know the rules of presynaptic cooperation and postsynaptic integration of inputs. We also learned that the magnitude of population synchrony follows a log-normal pattern. Yet even this combined knowledge is not sufficient to predict how neurons will respond without knowing their integration rules. Indirect evidence and modeling results support the first hypothesis; namely that the log-normal distribution of firing rates across neurons emerges from the skewed distribution of their intrinsic, nonlinear properties. While it is clear that variation in firing rates is coupled to variation in synaptic efficacies, stronger evidence is needed to establish which one is of primary importance and drives the second.[43]

We can only speculate how skewed distributions emerge in the first place. During neurogenesis, early-born neurons have a higher chance of innervating each other, and the synapses they make on the budding dendrites are placed nearest to the soma. Thus later-born neurons can only innervate the more distant segments of the dendrites. Based on this observation, we can conjecture that early-born neurons are more effective in discharging their peers, so the firing rate distribution may have a simple chronological origin.

Maintaining Log-Dynamics

How are these firing patterns maintained? Consider the following two options and guess. First, after the fast-firing neurons get tired, they take a long nap while previously dormant ones are invigorated, keeping the overall activity distribution the same but with revolving active participants. Second, the rate distributions remain correlated across brain states and across various testing situations, implying that some neurons must work harder than others forever. If you cannot decide, no worries. When I asked many of my colleagues, the majority favored the first option, but some argued in favor of the second, which is correct. Individual neurons maintain their firing rate ranks over days, weeks, and months, as if they sense their own firing outputs and adjust them to a set point customized to each cell. The distribution of intrinsic firing rates reflects a fundamental biophysical heterogeneity in neuronal populations, of which

---

43. In a neural network model, it is the spiking rate that seems to be of primary importance (Kleberg and Triesch, 2018). When spike timing–dependent plasticity rules and multiplicative normalization (Chapter 11) are incorporated into the network, the initially similar synaptic weights will soon form a skewed distribution and hub neurons with strong outgoing synapses.

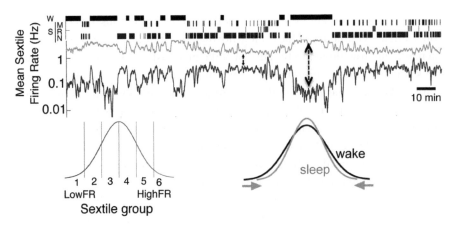

Figure 12.4. Sleep and wake states oppositely affect fast and slow firing neurons. *Bars*: Wake–sleep state scoring. W, wake; S, sleep; R, rapid eye movement (REM) sleep; N, non-REM sleep; M: microarousal. Firing rate of fastest and slowest sextile groups over wake and sleep change in opposite directions (*large and small double-headed arrows*; sextile distribution is shown in the left inset). *Bottom right*: Log-gaussian curves illustrate the log firing rate distributions at the end of waking period (*black*) and at the end of sleep (*gray*). Sleep narrows the rate distribution of firing rates.

the neuron's firing rate acts as a readily identifiable marker.[44] Of course, the firing rates of individual neurons vary transiently (e.g., in response to relevant stimuli), but the magnitude of the response is correlated with the neuron's long-term default rate. Cortical pyramidal cells can increase their spiking activity up to about 80/s. However, when the rate of a neuron is sampled over a long period of time, it will be dominated by the activity in its "idling" state.

The neurons' firing rate rank in the population is maintained through sleep. Neuronal excitability, as expressed by firing rates, and local network-related sources of neuronal heterogeneity, as expressed by the skewed distribution of synaptic weights, are related. This relationship arises from, and is maintained by, neuronal dynamics during alternating sleep–wake cycles (Figure 12.4).

---

44. Even after a sustained change in the firing rate due to altered afferent drive, the neuron returns to its previous rate (Koulakov et al., 2009; Mitra et al., 2012; Mizuseki and Buzsáki, 2013; Fu et al., 2016; Hengen et al., 2016). The log-normal distribution of spike burst rates favors the idea that membrane channel properties are important because the likelihood of burst occurrence depends largely on the intrinsic properties of neurons. Yet, in addition to neuron-autonomous rate homeostasis regulating intrinsic biophysical properties and channel density variations, the networks in which neurons are embedded can also exert a sustained influence on neuronal spiking.

Sleep has long been viewed as a brain state necessary to recuperate exhausted neurons from their hard work during waking.[45] The details of this hypothesized "sleep homeostatic plasticity" are still debated, but inroads have been made. The regulatory forces of sleep serve to restore population-wide dynamics to maintain a stable neuronal system. During sleep, release of subcortical neuromodulators is diminished, the neocortex is isolated from the influences of the outside world, and the thalamocortical system sustains its activity in a self-organized manner.[46] Sustained firing of a neuronal population (called an "up state") is interrupted by brief (30–200 ms) population-wide silent periods ("down states"). During the down state, nearly the entire cortex, thalamus, and many subcortical sites are silent. The brain has to reboot itself from this silence thousands of times each night. The durations of both the up and down states follow a log-normal distribution, and their relative durations shift over the course of a night's sleep.[47]

Experimental observations show that, during sleep, neuronal population activity transitions from an externally driven to an internally driven regime, complemented by a global increase in drive from the local network. During waking, the log-normal distribution of firing rates widens, meaning that many neurons at the left tail become silent or reduce their discharge activity, while the fast-firing neurons increase their spiking. During non–rapid eye movement (non-REM) sleep (i.e., the greater part of sleep when saccadic or rapid eye movements do not occur), the opposite happens. Many wake-silent neurons become active, while the very active neurons gradually decrease their firing rates over the course of sleep. As a result, the brain wakes up with a much narrower and sharper log-normal curve than before sleep.

What mechanisms could widen and narrow the firing rate distribution? As discussed in Chapter 11, subcortical neurotransmitters modulate gain and control the responsiveness of cortical neurons to sensory inputs. Thus, sensory

---

45. In contrast to memory consolidation (Chapter 8), in which specific synapses are thought to be "selectively" modified to strengthen particular memory traces, homeostatic function is assumed to involve a general, population-wide modification of synapses to maintain a stable neuronal system. An influential model by Tononi and Cierelli (2003, 2014; Vyazovskiy and Harris, 2013) suggests that most active cells and synapses during waking continue to be active during sleep, whereas weak synapses and slow-firing neurons are "down-selected" (i.e., eliminated from network participation). The model would predict, therefore, that log-normal distribution would be asymmetrically affected during sleep by reducing, rather than increasing, the rate of slowly firing neurons (i.e., opposite to the experimentally observed changes).

46. Pioneers of sleep research suggested that the power delta activity during non-REM sleep is a reliable measure of sleep pressure and serves as a homeostatic mechanism (Feinberg, 1974; Borbely, 1982).

47. Watson et al. (2016); Levenstein et al. (2017).

drive could explain the elevation of firing rates among fast neurons during waking. The faster responding neurons more effectively recruit inhibitory neurons, which in turn would further reduce the firing rates of the already slow neurons. As a result, both ends of the distribution would widen.

Another way to widen the rate distribution is via synaptic plasticity. At least one mechanism, called *spike timing–dependent plasticity*, depends on the temporal order of spikes. If neurons A and B are mutually connected, and A consistently fires before neuron B within the duration of a gamma cycle, the synapse A to B will be strengthened but B to A will be weakened (Chapter 4). Because the probability of temporal coincidence increases with the firing rates of the coactive neurons, this synaptic plasticity rule favors preferential strengthening of synapses involving neurons with higher firing rates. In the waking brain, the asynchronously active heterogeneous population leads to asymmetric strengthening of synapses on neurons with higher spontaneous firing rates. The consequence of this process is easy to imagine: excitable neurons become progressively more excitable, which can eventually destabilize the network. Unless some mechanism combats this "plasticity pressure" in the awake brain, the resulting positive feedback loop could increase synaptic weights to maximum saturation. One mechanism that counteracts this pressure is the up-state transition of non-REM sleep.

During the transitions from down state to up state in non-REM sleep, neurons fire in a sequence. A neuron's place in this order is correlated with its baseline firing rate, such that neurons with higher firing rates tend to spike before those with lower firing rates (Figure 12.5).[48] A simple but important consequence is that neurons with high and low firing rates get temporally segregated. Because high-firing neurons tend to fire earlier than low-firing neurons after the down-to-up transition, the plasticity rule tends to increase the weights of synapses from high-firing onto low-firing neurons while decreasing the weights of synapses from low-firing onto high-firing neurons. This redistribution of synaptic weights during non-REM sleep pulls both ends of the log-normal firing rate distribution closer to the mean and acts as a homeostatic counter to changes in the asynchronous wake state.[49] This normalization mechanism is an important function of sleep.

---

48. Luczak et al. (2007); Peyrache et al. (2011); Watson et al. (2016); Levenstein et al. (2017).

49. The skewed organization of neuronal circuits has important implications for disease. Homeostatic maintenance of the status quo between the two ends of the distribution is critical for the normal brain. A shift to either direction will inevitably impair performance. For example, stronger than needed dominance of the hub neurons can form an epileptic focus.

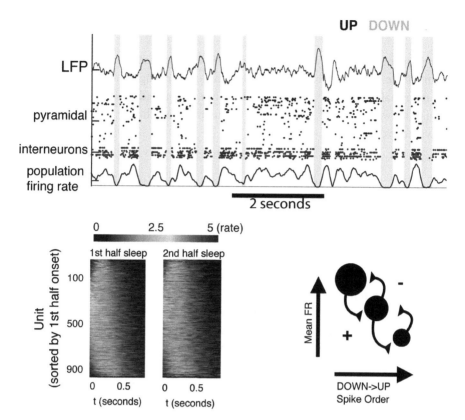

Figure 12.5. Homeostatic regulation of firing rates by "down" states of non–rapid eye movement (non-REM) sleep. *Top*: Illustration of up and down states. Note that during down states both pyramidal cells and interneurons are silent (dots spikes of neurons; each row is a single neuron). *Bottom left*: Sequential activity from high to low firing rate cells at the down–up transitions of non-REM sleep (> thousand neurons). Each gray line is the firing event of a single cortical pyramidal neuron, sorted by the occurrence of its first spike after the up state onset, shown separately for first and second halves of non-REM sleep. The density of gray reflects the firing rate of the neuron. Note that faster firing neurons tend to fire earlier at the up state onset (neurons on the top). *Right*: Spontaneous plastic pressure between low (*smaller circles*) and high (*larger circles*) firing rate cells during non-REM sleep due to high-to-low firing rate sequential activity at the down–up transition. Synaptic plasticity rules increase and decrease the firing rates of slow and fast firing neurons, respectively.
From Watson et al. (2016); Levenstein et al. (2017).

CONSEQUENCES OF LOG RULES

I began this chapter by explaining that skewed connections and activity dynamics are necessary to maintain the status quo among competing properties of neuronal networks.[50] Next, I will illustrate that a wide range of firing rates, synaptic weights, and degrees of population synchrony provides several advantages.

Familiar–Novel Continuum

The terms "old and new" or "familiar and novel" are designated as discrete categories, yet they may also be viewed as the two extremes of a continuous quantity. Log-normal distributions may explain how these categories are represented in neuronal networks because the two ends of the distributions are associated with multiple distinguishable properties.

In novel and familiar environments, most hippocampal pyramidal neurons have no or only one place field, while a small minority of neurons have many. Thus, the distribution of maze coverage is strongly skewed. When a hippocampal neuron does not fire in a given maze, we expect that it may fire in another one. That is, every hippocampal neuron is viewed as a potential place cell. This idea was tested by training rats to run novel maze tracks with lengths of 3, 10, 22, and 48 meters. Some place cells that fired on the short tracks formed additional fields on the larger tracks, but most new place cells were recruited from the pool of initially silent cells. The number of fields formed was strongly skewed: a few neurons had many fields, whereas many neurons had one or none, similar to the behavior of hippocampal neurons in a single maze. Extrapolation from the observed log distributions suggested that nearly all hippocampal pyramidal cells would be active in an environment with a diameter of approximately 1 kilometer, which is believed to correspond to the ecological niche of rats. In another experiment, rats were tested in multiple rooms. Most pyramidal neurons fired only in a single room, but a small minority fired in multiple rooms or all rooms, producing

---

50. The *spike timing–dependent plasticity* (STDP) rule can segregate synaptic weights into a bimodal distribution of weak and strong synapses, but the network may destabilize without additional constraints. However, when STDP works in a synaptic weight-dependent manner (called log-STDP), it can produce a skewed weight distribution and sustain stable network dynamics. In such a regime, long-term depression exhibits linear weight dependence for weak synapses but much less so for strong synapses. The long-term depression of strong synapses enables synaptic weights to grow, generating a long tail in the synaptic weight distribution (Omura et al., 2015).

a log-normal distribution of the overlap of neuronal activity in the different rooms.[51] Overall, these experiments demonstrate that the skewed distribution of place fields is a general rule, irrespective of the nature or size of the testing environment.

The generalizer–specialist continuum is not an exclusive feature of higher order cortical areas but is also prominently present in primary sensory cortices. The majority of neurons in the superficial layers of the primary visual cortex respond relatively selectively to orientation of the stimulus (e.g., vertical or horizontal bars) and/or to the direction of the moving stimulus (e.g., left to right or right to left). Some neurons are true specialists and respond selectively to a single feature. However, across the population, the magnitude of selectivity varies considerably. A small group at the right end of the distribution responds to several visual features. Similarly, a minority of neurons in the somatosensory cortex shows selective preference to specific directions of whisker deflection. Yet another minority responds to multiple directions, with most neurons responding with intermediate specificity. In the auditory cortex, superficial neurons can have highly specific and narrow frequency tuning properties (such as one neuron selectively responding to a 10 kHz beep, whereas another responds to a 20 kHz beep), where others can respond to multiple overlapping bands. The overall wisdom that we can draw from the many physiological experiments performed in numerous laboratories and in different parts of the cortex is that some neurons may appear to treat multiple stimuli and situations as the same and similar; that, is they generalize input features. Other neurons, on the other hand, appear to be super-specialists and respond only to a single feature of the many options. However, the most important message of this discussion is that the generalizers and specialist form a continuum.[52] Whether an object or situation is judged as the same, similar, or different depends on how the downstream observer neurons and circuits sort out their responses to the spike output of a skewedly distributed upstream population.

Nonegalitarian Votes of Neurons

What could be the advantage of such skewed distributions, for example, for 'encoding' spatial relationships? From the perspective of independent (also called "orthogonal" in geometry) coding, the minority of neurons with multiple

---

51. Alme et al. (2014); Rich et al. (2014). For a review, see Buzsáki and Mizuseki (2014).
52. Ohki et al. (2005); Rothschild et al. (2010); Kremer et al. (2011).

place fields may be regarded as "noise" or imperfection in the system.[53] However, when other physiological features of this minority are considered, a different picture emerges. Not only are they more active in multiple environments, but their firing rates also are higher, they emit more spike bursts, and their individual place fields are larger compared with those of the majority neurons. The higher mean firing rates of the active minority in the maze strongly correlate with their peak firing rates within their place fields and even with their firing rates during sleep in the animal's home cage. Furthermore, the diligent minority fires synchronously with other neurons more frequently across all brain states than the slower firing majority and produces a stronger and more effective excitation in their targets. As a result, in physiological time frames, such as theta oscillations and sharp wave ripples (Chapters 6 and 8), about half of the spikes in the hippocampus are contributed by this active minority, with the other half from the majority of neurons with no or a single place field.

Thus, the neurons in cell assemblies make strongly skewed contributions, suggesting that the cell assembly is a nonegalitarian organization. For example, the consensus among the low-firing majority of hippocampal place cells could be that "the owner of the brain is now in a different room." If votes are represented by spikes, each of them may have only a single vote, although there are many voters. In contrast, the high-firing minority may represent the view that "this is the same room." Despite their small number, they have many spikes, possibly stronger synapses, and thus many votes. Downstream reader neurons have to interpret this mixed output to make a judgment and generate action potentials (Figure 12.6). The high-firing neurons may exert a stronger influence by emitting more spikes within the same time window. But we should also consider the option that the reader neuron responds to the ratio of change; that is, it normalizes the spike messages from both fast and slow neurons, in which case the proportional impact of the two upstream neuronal populations is equal.

Rigid-to-Plastic Continuum

If neurons with different activity rates exert a nonegalitarian influence on their downstream partners, new experience might also have a differential effect on the two tails of the log-normal distributions. In addition to homeostasis, non-REM sleep is also critical for memory consolidation. As discussed

---

53. The idea of orthogonalization was introduced to neuroscience by Marr (1969) to support his hypotheses that independent constellations of neurons can represent the numerous stimuli we encounter.

Chapter 12. Everything Is a Relationship

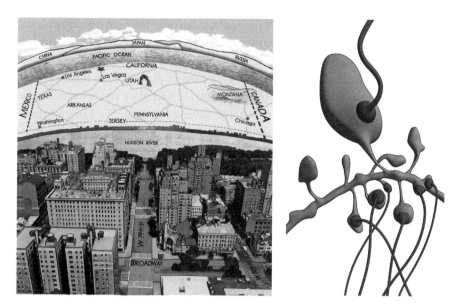

Figure 12.6. *Left*: New Yorkers' skewed view of the world, inspired by Saul Steinberg's 1976 illustration in the *New Yorker* magazine. *Right*: Neurons receive mixed messages from upstream partner neurons. The majority of weakly active neurons convey detailed information about the distinctiveness of a situation, whereas the strongly firing minority with strong and rigid synapses generalize across situations.

in Chapter 8, memory replay is often conceptualized as occurring via higher order interactions within populations of neurons that have similar properties. However, as demonstrated earlier, cortical networks contain neurons with highly diverse biophysical properties, characterized by skewed distributions of synaptic weights, long-term firing rates, and spike bursts. Furthermore, their temporal correlations are largely preserved across brain states and environmental situations, suggesting that learning-induced changes are constrained within a dynamically stable network.

In the widely used population analysis methods, the individual differences among neurons are obscured. This is because a small group of highly active neurons can strongly bias the conclusions reached by such methods, as is typical of populations with skewed distribution. [54] To counter this problem, every neuron in the population should be characterized individually. Just as in sport, where one would like to know how a team would perform if one of the players is pulled out, in the team performance of neuronal actions we need

---

54. This is similar to human interactions: "Remove one highly assertive member from a group of eight candidates and everyone else's personalities will appear to change." (Kahneman 2011).

to measure the impact of individual neurons because they are not equal. Such individual neuron analyses demonstrate that learning-induced plasticity is not equally distributed among hippocampal pyramidal neurons after an encounter with a novel spatial task. Instead, the sequences of place cells in the novel environment are formed from neurons that already had place cell features before the experience and do not change much during learning. We can call them "rigid" cells. These rigid neurons belong to the high firing end of the population, and they are more strongly connected to each other. Many of their afferent connections may already be too strong, also called "saturated," and therefore cannot get much stronger.[55] At the other end of the spectrum are the "plastic" neurons. They also may have a place field from early on in a new environment, but they can modify their firing patterns and can become incorporated into the backbone of the rigid group after experience.[56] Thus, the hippocampal network contains a continuum of neurons spanning a range of rigid to plastic features. The plastic cells show lower mean firing rates, more specific firing fields, and larger changes in firing rates and field specificity during the first minutes of exploration than the rigid cells. Fast-firing rigid cells provide a backbone into which slow-firing plastic cells can be rapidly incorporated (Chapter 8).[57]

55. Perin et al. (2011), investigating interactions among neocortical neurons in vitro, have found that synaptic strength correlates with connection probability among neurons. Furthermore, the synaptic strength can saturate after only a small fraction of the maximal clustering is reached. Thus, they comprise a rigid cluster.

56. Grosmark and Buzsáki (2016). The rigid versus plastic ends of the distribution relate to the long debate about whether place fields preexist or require experience (Chapter 13). Early studies suggested that hippocampal pyramidal cells display place fields when the rat first enters a novel environment (Hill, 1978; Samsonovich and McNaughton, 1997). Before our work, Dragoi and Tonegawa (2011, 2013a, 2013b) reported that place cell sequences in a novel part of a maze were correlated with sharp wave ripple–related sequences during sleep before the novel experience. They called such sequences "preplay" (Dragoi and Tonegawa, 2011). The existence of preplay was challenged on statistical grounds (Silva et al., 2015) and by suggesting that the rat's visual survey of the novel parts of the maze before the animal entered those corridors induced place cell sequences (Ólafsdóttir et al., 2015). This line of reasoning is also supported by other studies showing that stable firing fields require at least a few passes through novel corridors (Wilson and McNaughton, 1993; Frank et al., 2004) and perhaps postexploration sharp wave ripples. In our studies, the rats were tested in an entirely new room. Yet it is possible that visual inspection of the environment before the first run contributes to the selection of preexisting rigid generalizer cells to form a scaffold of sequential activity that unfolds during exploration.

57. Experiments by George Dragoi in my laboratory (Dragoi et al., 2003) demonstrated that place cells with high within-field firing rates are robust against perturbation of the synaptic strengths of hippocampal networks. In contrast, slow-firing place cells had a high propensity to express new place fields after we artificially affected their synaptic inputs (see also Lee et al., 2012; Bittner et al., 2015).

Plastic neurons also have higher place-specific indices compared with rigid cells and typically only one place field. The firing patterns of the plastic and rigid neurons change differently during learning. While both fast- and slow-firing neurons can have place fields from the beginning of maze exploration, slow- but not fast-firing neurons increase their spatial specificity steadily during learning, as measured by the ratio of the number of spikes emitted inside versus outside the place field. This change in behavior is what makes these cells "plastic." Learning thus does not necessarily create new place fields but increases the "signal-to-noise" ratio, so that the hippocampal spatial map becomes more reliable. Plastic hippocampal neurons also increase their burst emission during sleep after experience, especially when they fire together with other neurons during sharp wave ripples (Chapter 8).[58] Overall, these experimental findings reveal that replay sequence-forming neurons during both waking and sleep are drawn from a skewed distribution with varying coding and plasticity properties and that replay events send a synthesis of preexisting and recently learned information to downstream observer neurons.

SKEWED DISTRIBUTION OF TERRITORY

So far, the picture I have painted is that skewed distributions of both structural components and dynamics of neurons and neuronal networks may contribute to many important functions of the brain, including perception and memory.[59] The brain's skewed organization could also influence our behavior and social interactions. Building a mechanistic link between structural, functional levels of brain organization and behavior, especially social behavior, is acknowledged to be difficult problem. Yet we can look at some examples. Territorial behavior is well studied by ethologists, but quantitative experiments involving large populations are rare. Figure 12.7 illustrates an early attempt to do just that by examining a closed mouse village in which food and water were supplied by the researchers.

Of the fourteen available shelter boxes, one male owned three, another male owned two, and five males owned one each. These males were heavier and stronger than the remaining 21 males. Five of the strong males had one female companion each, whereas two had two companions each. Another eight males lived together in one small box without female partners, and the remaining

---

58. Preventing place cells from firing during sharp wave ripples interferes with place cell stability (Roux et al., 2017).

59. Fusi et al. (2007).

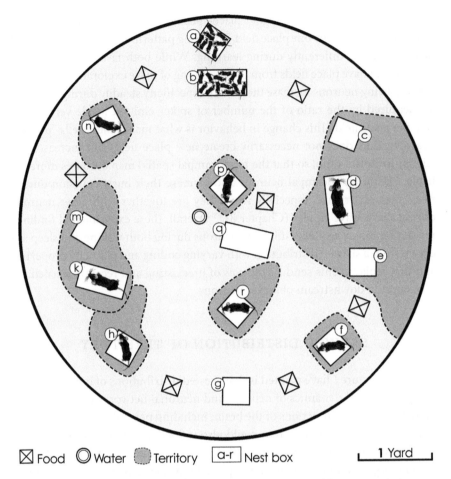

Figure 12.7. Skewed distribution of territories in a mouse town. The great majority of available space were owned by a few mice (one male each in box d, f, h, k, n, p, r), whereas the majority of mice were crowded into two boxes (a: eight males and b: thirteen males and nineteen females—only males are shown). Boxes c, e, g, m, and q were unoccupied but patrolled by territorial males. The size of the mouse indicates its magnitude of dominance in the social hierarchy.

thirteen males and nineteen females were crowded into a single house. There were no nests in the unoccupied boxes, and the territorial males did not allow other mice to settle in them. The territorial males defended their specific boxes against all comers. This social structure persisted for months.

The males' territories were not confined to the boxes alone. Large floor areas also belonged to certain males. The invisible boundaries between their respective domains were nevertheless very real and recognizable to all mouse citizens. The most aggressively dominant mouse lived in a large nest-box (d), and he spent most of his waking hours patrolling his territory or guarding over it from a vantage point at one corner of the top of his home box. "Here he would crouch, sniffing in the air, and peering about below, and if a strange mouse entered its territory, he would hurl himself through the air, like a tiger onto the intruder and drive it away."[60] His attacks were invariably successful, and afterward he always made the rounds of his territory.

The eight males that lived together in box (a) without any nest material were the weakest and most downtrodden of the colony. Box (b) housed two-thirds of the colony's population. This box was dominated by a single male, which constantly tyrannized most other members of his house, although he rarely challenged the status of the seven males who guarded private houses. These "slum" mice rarely mated, and most offspring derived from the territorial males.

Extrapolation to Human Interactions

Compare these observations on mice with the statistics from another species, *Homo sapiens*. The wealthiest 10% of households own half of the total wealth in the United States and many industrialized countries. Almost all of rest is distributed among half of the population, while the poorest 40% of people own as little as 3% of the wealth. Income distribution is well described by a log-normal distribution, which allows for an unambiguous quantification of inequality.[61] The socioeconomic difference tends to grow over time, transmitting advantages to the minority and disadvantages to the majority from one generation to the next. In the 1980s, the average disposable income of the richest 10% in Europe was seven times higher than that of the poorest 10%; today, it is almost ten times higher. This "Great Divide" not only segregates the rich and the poor in terms of incomes and accumulated wealth but also affects education, access to jobs, well-being, and life expectancy.[62]

---

60. Crowcroft's book (1966) is perhaps the best and most entertaining I ever read on mouse behavior. It contains careful observations of the social and territorial behavior of the house mouse (*Mus musculus musculus*), with lots of detailed and remarkable observations. In repeated experiments, it remained typical that two-thirds or more of the mice were crowded into one shelter, independent of the size of the pen or whether food was scarce or abundant.

61. Vilfredo Pareto (see Koch, 1998) observed that 20% of the population owned about 80% of the land in Europe. This came to be known as the 80–20 law. However, today, the richest 10% owns more than 85% of world wealth. Pareto's power law is only applicable to the tail of income distribution. Incomes follow a log-normal distribution, not power laws (Glomm and Ravikumar, 1992; Deininger and Squire, 1996; Benabou, 2000).

62. Rising economic inequality can also lower social trust in institutions, increase gaps in education, generate a perception of injustice, lead to intolerance and discrimination, and fuel

Like the "rich club" of highly active neurons with better connections, more effective access to information, and the collective ability to make quick decisions, human group formation may be guided by similar rules. Well-disciplined armies and police controlled by an organized elite have always effectively dominated large but disorganized masses of people. At the turn of the twentieth century, 3 million noblemen and their officials lorded over 180 million peasants and workers in Russia. Then along came the super well-organized Communist Party, with just 23,000 members in 1917, which gained control over the vast Russian Empire overnight and eventually dominated one-sixth of the world's population. As one of the "beneficiaries" of this system, I had a hard time understanding how such a small fraction of people can maintain such powerful authority. Now I understand it better: organization is the key.[63] Paradoxically, all their social injustice was done under the banner of equality.

I began this chapter by explaining that skewed distributions offer stability when many competing factors are in play, which is the case in biology in general and in the brain specifically. This principle may apply to human social groups as well. Egypt with its pharaohs existed for more than three millennia. India's caste system is among the world's oldest surviving social hierarchies.[64] The long-lived Chinese dynasties, the Holy Roman Empire, the Khmer Empire, and the Mayan civilization were all highly stratified systems that bestowed many privileges on the upper classes while sanctioning repression of the lower ones by the privileged minority. In contrast, Greek democracy and communist "democracies" (or the "dictatorship of the majority") were ephemeral (even if it did not feel that way for many). The moment the Soviet grip on these countries was lifted, the artificially maintained wealth distribution shifted back to the skewed form in less than a decade.

Can we learn anything from the similar distribution of owned stuff in species as different as mice and humans? Some economists argue that artificially increasing the income of the poor alone will not produce the desired impact. I agree. It is not only the magnitude of wealth differences that matters, but also how people feel about it. Inequality is relational at both economic and emotional levels. In addition to compressing the left end of the log-normal distribution,

---

political instability. The social contract varies considerably across nations. For supporting statistics, see Centre for Opportunity and Equality. http://oe.cd/cope-divide-europe-2017.

63. The emphasis on organization was the key point of Yuval Harari's recent book *Homo Deus* (2017).

64. The division of labor creates specialized behavioral groups and occurs frequently in communal animal societies, including some mammals. Group membership determines lifestyle and heredity of privileges. In contrast, the mainly hunter-gatherer societies of Aboriginal Australians have existed for 60,000 years, perhaps because they live in small segregated communities.

we must compress the right end as well. We need a societal version of the homeostatic compensatory mechanism of non-REM sleep to combat the intrinsic drive of technological development and global economy toward widening the gap between the two tails (Chapter 9). Unfortunately, history shows that such compensatory mechanisms are exceptionally rare. Inequality is usually reduced suddenly only by natural disasters, plagues, wars, and revolutions.[65]

Skewed distributions in human societies may be regarded as an "is–ought problem," the problem of identification of natural properties with moral properties.[66] Just because things are as they are, it does not mean that they ought to stay that way. Our social contract, which has matured over hundreds of thousand years by externalizing brain function, should be different from that of the rodents. We have the ability to introduce rules and enforce them wisely so that they counter the forceful tendencies of skewed distributions. Promulgating the idea that we are all born with equal abilities, as the Marxist doctrine naïvely and consistently did, will never lead to social justice or equal opportunities. On the other hand, acknowledging that things are the way they are but do not have to stay that way in human societies is what we should strive for. Unless this idea is understood and acted upon, market economies in their present form will only continue to reinforce social injustice described by log-normal distribution.

Neuronal organization driven by skewed distribution is remarkably different from the expectation of the outside-in framework and is more compatible with a preconfigured yet flexible brain, which is the topic of the last chapter.

SUMMARY

Unlike physics, neuroscience has few generalizable rules that cut across multiple domains of brain organization. The Weber-Fechner law, established in human psychophysics, is an exception. The law states that for any sensory modality, perceptual intensity is a logarithmic function of physical intensity. Therefore, as stimulus strength multiplies, the strength of perception only adds. Distance perception, time perception, and reaction time also vary logarithmically with the distance and time interval, respectively. Error accumulation in short-term memory, and other phenomena also obey this law. In short,

---

65. Several excellent recent books are available on this important topic. Perhaps not unexpectedly, they differ extensively in their analyses and recommendations (Atkinson, 2015; Payne, 2017; Scheidel, 2017). A concise summary is by Reeves (2017).

66. The "is–ought problem" was most prominently articulated by David Hume in his *A Treatise of Human Nature* (Stanford Encyclopedia of Philosophy; https://plato.stanford.edu/entries/hume-moral/).

a variety of sensory and cognitive domains may be governed by similar neuronal mechanisms.

I discussed the hypothesis that the strongly skewed nature of our perceptions and memory result from log-normal distributions of anatomical connectivity at both micro- and mesoscales, synaptic weight distributions, firing rates, and neuronal population activity. Nearly all anatomical and physiological features of the brain are part of a continuous but wide distribution, typically obeying a log-normal rule. This rule implies that the interactions that give rise to this distribution involve multiplication or division of random factors, resulting in values that can span several orders of magnitude. Neuronal networks with such broad distributions are needed to maintain stability against competing needs, including wide dynamic range, redundancy, resilience, homeostasis, and plasticity.

Neurons on the two ends of the log-normal distribution of activity organize themselves differently. Fast-firing neurons are better connected with each other and burst more than slow-firing neurons. The more strongly connected faster firing neurons form a "rich club" with better access to the entire neuronal population, share such information among themselves, and, therefore, generalize across situations. In contrast, slow firing neurons keep their independent solitude and elevate their activity only in unique situations. The two tails of the distribution are maintained by a homeostatic process during non-REM sleep. The emerging picture is that a simple measure, such as the baseline firing frequency, can reveal a lot about a neuron's role in computation and its wiring properties.

The log-normal distribution rule acting at so many different levels of neuronal organization allows the brain to make relational judgments and, in humans with language, to discretize the extreme ends of such distributions with words. The left and right ends of the log distribution appear to convey different or opposing effects, as if they were discrete groups with distinct features.[67] In reality, though, they are part of a continuum with long tails. An ever-active minority of neurons at the right end may be responsible for generalizing across environments, which enables the brain to regard no situation as completely unknown. On the other hand, the majority of less-active neurons can be mobilized to precisely distinguish one situation from another and label each as distinct.

---

67. This may explain the philosopher Georg W. F. Hegel's mysterious law of the transformation of quantity into quality and vice versa. Hegel argued that qualitative differences in nature arise from either differences of composition or differences in quantities of energy. From the point of view of log-normal distribution, this point appears almost obvious.

Neuronal systems organized according to these principles may form brain networks with opposing but complementary features, producing good-enough and fast decisions in most situations using only a subset of the brain's resources. In contrast, extensive brain networks are needed to learn important details.

13

The Brain's Best Guess

Better a diamond with a flaw than a pebble without.
—Confucius[1]

Le mieux est l'ennemi du bien. [The perfect is the enemy of the good.]
—Attributed to Voltaire[2]

Philosophy is written is this grand book, the universe, which stands continually open to our gaze. But the book cannot be understood unless one first learns to comprehend the language and read the characters in which it is written.
—Galileo Galilei[3]

We have two brains in our skull or at least two virtual divisions. First, there is the "good-enough" brain. This is largely prewired and acts quickly via a minority of highly active and bursting neurons connected by fast-conducting axons and strong synapses into a network. The good-enough brain judges the events in the world in a fast and efficient way but is not particularly precise. The privileged minority of this virtual division is responsible for perhaps half of the spikes in the brain at any given time and share information among themselves and have faster access to the rest of the brain than the remaining majority of neurons. The strongly interactive circuits that form the good-enough brain can generalize across situations but with less

---

1. https://www.goodreads.com/quotes/63434-better-a-diamond-with-a-flaw-than-a-pebble-without.
2. https://fr.wiktionary.org/wiki/le_mieux_est_l%E2%80%99ennemi_du_bien.
3. https://www.princeton.edu/~hos/h291/assayer.htm.

than perfect fidelity. In short, the strongest 10% or so of synapses and fast-firing neurons in the brain do the heavy lifting at all times. They do a good-enough job under most conditions, just as a few key players from a large orchestra can produce an enjoyable recital.[4]

The fast execution of plans by the good-enough brain is amply illustrated by Wesley Autrey, the New York City "Subway Hero." He saved the life of a young person who had suffered a seizure and stumbled from the platform, falling onto the tracks in 2007. Without hesitation, Autrey dove onto the tracks and protected the man by throwing himself over his body while the subway cars passed over them. In an interview, Autrey summarized perfectly what I am trying to say about the good-enough brain: "I just saw someone who needed help. I did what I felt was right."[5] All of us do such acts all the time, although in a less heroic form.

However, good enough is far from perfect. We would not want to drive a car with 60–80% accuracy or submit a scientific paper with such precision. To perform better, we also need to deploy the second virtual brain: a large fraction of slow-firing neurons with plastic properties that occupy a large brain volume connected by weaker synapses into a more loosely formed giant network. Their work is absolutely critical for increasing the accuracy of brain performance.

Of course, I am not thinking about two discrete brains in one skull, but instead a continuum of a broad distribution of mixed networks that performs apparently different qualitative computations at the left and right tails. This distribution allows the brain to implement anything from preexisting rigid patterns to highly flexible solutions. Thus, there is no definable boundary between the "fast decision, low precision" and "slow decision, high precision" networks.[6] Brain performance is a tradeoff between speed and accuracy.

After one of my talks on this topic, somebody commented: "The dichotomy you were talking about reminded me of Daniel Kahneman's book *Thinking, Fast and Slow*. This was an embarrassing moment for me. I was familiar with

---

4. A common observation in physiological experiments is that the activity of just a dozen or so strongly firing neurons is often more informative about an animal's behavior than members of the majority of slow-firing neurons. For example, in brain–machine interface applications, just ten strongly task-related neurons of a hundred recorded neurons can predict 60% accuracy for limb position or gripping force. Adding spikes from the remaining ninety neurons improves the prediction by only a few percent. The "backbone" of task-relevant neurons provides a "best guess" for appropriate action. To increase task performance by another 15% would require a thousand simultaneously recorded neurons, and to achieve 30% improvement would perhaps need ten thousand (Nicolelis and Lebedev, 2009). A key question is what would happen in the network if the committed task-related minority, the backbone, was silenced.

5. https://www.nytimes.com/2007/01/03/nyregion/03life.html.

6. These paragraphs summarize the review by Buzsáki and Mizuseki (2014).

Kahneman's interesting work on cognitive biases with Amos Tversky,[7] but I had never heard of the book. The first thing I did after returning to my hotel room that night was to order the paperback. The Table of Contents did not reveal any obvious relationship to my thesis. On the other hand, the blurb on the back of the book was explicit: the mind has two systems. "System 1 is fast, intuitive, and emotional; System 2 is slower, more deliberative, and more logical." These short sentences induced panic. Is it possible that I had read about this research somewhere, and my poor source memory tricked me into believing that Kahneman's ideas were mine? I left the book on my desk and glanced at it every night but did not have the courage to delve into it. Then, after two weeks or so, I surrendered and read it over a weekend from the beginning to the end. Chapter after chapter, my heart rate went up, as I dreaded finding an outline of how brain mechanisms support the alleged two systems. But there was no mention of the brain.

Kahneman discussed numerous psychological experiments that effectively highlight the differences between his two phenomenologically defined systems and explained eloquently how they could produce different behavioral results even given the same input. After finishing the last page, I made peace with the book and with myself. Kahneman and I arrived at similar conclusions even though we approached the same problem from very different directions. His interest is the mind and cognition, and he took an economics point of view. I started my research program inside the brain and made my way slowly outward to conclude something similar; namely, that the wide distributions of brain dynamics allow it to perform complementary operations that may often appear fundamentally different. These dedicated networks can quickly make good-enough decisions, but precision requires a more protracted process across a large area of the brain. There are no two systems, just one system with two tails. Thus, our views are complementary, and we are looking at the two sides of the same coin.[8] In this last chapter, I will recapitulate the headlines of the previous chapters and attempts to fill the inside-out framework with content.

---

7. Tversky and Kahneman (1974, 1981). Daniel Kahneman received the Nobel prize in Economic Sciences in 2002. *Thinking, Fast and Slow* (2011) was published three years before our "log-dynamic brain" paper (Buzsáki and Mizuseki, 2014). In my defense, it took us two years of struggle to publish the experimental basis of the log distributions (Mizuseki and Buzsáki, 2013). Kahneman's book won the National Academies Communication Award. It illustrates how our cognitive biases, judgments, and decisions shape everything from planning an experiment to playing the stock market.

8. The two approaches are not just distinct at the level of investigation (psychology vs. neuroscience). Kahneman's approach falls under the umbrella of discrete and nonhierarchical general contrast models (Tversky, 1977), in contrast to the log scale-based continuous distributions I propose here.

PREFORMED BRAIN DYNAMICS

Perhaps no self-respecting neuroscientist subscribes to the *tabula rasa* view, which explicitly states that the mind is a whiteboard on which experiences are written. Yet experiments in most neuroscience laboratories are still being performed according to this outside-in tradition or its modern-day version, *connectionism*.[9] The concept of *tabula rasa* is alive but well hidden.

Perhaps the best illustration of the outside-in perspective is the field of memory research. Nearly all computational memory models tacitly start out as a blank slate onto which new experiences or representations are engraved.[10] Most artificial neural networks suffer from an inconvenient bug known as "catastrophic interference." The failure is called "catastrophic" because new learning can effectively erase all stored memories in connectionist models, which never happens to people (short of severe head trauma). The problem of overwriting remains a significant challenge for the development of computational models.[11] This would be an issue in a hypothetical *tabula rasa* brain as well, which is essentially a connectionist model where the dynamic is built up during the course of learning because adding new information would inevitably destabilize its dynamic state. Furthermore, brain wiring and physiology of members of the same species with different degrees of experience would be dramatically different in *tabula rasa* brains because, in such models, the properties of the network are scaled by the amount of experience. Yet it is not clear how neurons in such a brain would know where to direct attention or what events in the world to process. Equally importantly, there is no easy way to understand how a blank slate, passive observer brain, embedded in an outside-in framework, can become a doer and creator of goals.

---

9. It has been repeatedly suggested that preconfigured connectivity and an internally regulated dynamic come at a cost of redundancy and constrain the combinatorial possibilities of neurons, mainly based on the observation that improvement of sensory performance is often accompanied by reduced spike correlation among cortical neurons (Cohen and Maunsell, 2009; Mitchell et al., 2009; Gutnisky and Dragoi, 2008). This redundancy is usually referred to as "noise correlation" (Zohary et al., 1994; Averbeck et al., 2006; Cohen and Maunsell, 2009; Renart et al., 2010; Ecker et al., 2010). The term "noise" in the brain, of course, only reflects our ignorance of refined mechanisms.

10. See e.g., Marcus et al. (2014). "Representationism" cannot explain the relationship between "the thing to be represented" and "the thing that represents." Is it one of similarity? Is it metaphoric (symbolic)? Or is it a numerical map, like the representation of pictures in a digital camera? In normal perception, the *tabula rasa* brain is impacted by external objects and perceived by the mind. Illusions are then distorted representations, and hallucinations are unfounded representations (Berrios, 2018).

11. McCloskey and Cohen (1989); McClelland et al. (1995).

Although views about neuronal mechanisms of perception underwent several important transformations over the past two decades, these new developments remained largely within the realm of the outside-in framework. Problems in today's perceptual research have been strongly connected to decision-making and are couched in the language of Bayesian estimation, named after the eighteenth-century thinker Thomas Bayes.[12]

One should distinguish between the "Bayesian method" as a body of mathematics, which can effectively deal with observations and contrast them with outcomes, and the "Bayesian brain model," which is based on a metaphoric interpretation between observations and experimenter-assumed goals of the brain, such as an ideal observer and efficient evaluator of situations. Formally, the Bayesian method is a disciplined statistical method to collect and evaluate data. For example, given the sequential activity of hippocampal neurons during sharp wave ripples during sleep, how well can this observation explain the animal's position in the maze, given the sequence (Chapter 8)? The Bayesian brain model is a representational framework with a strong emphasis on evaluation of perceptual inputs and decision-making with no or very little role attributed to action or internal motor reafferentation to sensory processing.[13] Its predictions rest on an internal or a generative model of how sensory inputs unfold. It posits that the brain—more precisely, our perceptual system—makes assumptions about the objective world. More precisely, it suggests that representation of all possible values of the parameters is computed along with associated probabilities in successive stages, and this process leads to the most efficient prediction. When a decision-maker of the Bayes theorem needs to make a prediction, it factors in the prior outcomes, such as previous successes and failures in similar situations. The Bayes model allows the brain to integrate different sensory cues and modalities gradually and carefully without committing too early to a particular judgment.

The Bayesian model presumes that more complex brains, such as ours, can accurately estimate the properties of the objective world on the basis of sensory information alone. It assumes that sensory information can provide an observer-independent truth, yielding veridical perception.[14] Decisions are

---

12. Bayes's theorem was generalized and expanded by the French mathematician Pierre-Simon Laplace (in his Analytic theory of probability, 1812) to address problems such as celestial mechanics and medical statistics. The Bayesian inference methods gained enormous popularity in the field of machine learning (e.g., Bishop, 2007; Stone, 2013) and more recently in virtually all fields of neuroscience. Bayesian statistical inference is particularly effective in the dynamic analysis of neuronal sequences (Pfeiffer and Foster, 2013).

13. Knill and Pouget (2004); Doya (2007); Friston (2010); Friston and Kiebel (2009).

14. The "truth" has been a long-standing theme in Western culture and it is believed to be absolute. The truth is either assumed to be objective (out there) or subjective (created by the

made on accumulated evidence that also includes previous experience, and the model tacitly assumes that evolution gradually perfected our perceptual systems. But flies, frogs, mice, and humans may see and evaluate the same environment differently. Thus, postulating that the human's brain view is more veridical than that of other species is hard to justify.

The idea of actively examining the sources and meaning of sensory signals in the Bayesian model resonates with the hypothesis-testing and future-predicting brain view of the inside-out framework. Furthermore, the important role of previous experience and phylogenetic hardwiring ("priors") link aspects of the Bayesian model to the inside-out framework as well. However, in contrast to the Bayesian model, grounding is a key feature in the inside-out framework and is provided by a second opinion of action-based experience. There is no assumption or need that the perceptual system describes the true state of affairs in the outside world. As servants of the action system, the sensors are tuned to effectively guide the animal in its ecological niche by supplying sufficient clues. Sufficiency is not a feature of the objective world but reflects a relationship between the organism's needs and its environment.

Empty Versus Prewritten Book

A view opposite to the *tabula rasa* has long been advanced, perhaps starting with Plato, although he was agnostic about the brain. According to Platonic philosophy, "ideal forms" exist independent of the observer, perhaps not too far from the idea of preexisting brain dynamics. We mortals are chained in a cave, and all we ever perceive of the world and the beings in it are the shadows they cast on the wall of our cave as they pass its entrance. While Platonic and empiricist philosophies are diametrically opposite in many ways, they are virtually identical when it comes to the tacitly assumed source of knowledge: passive observation, either from directly sensing the signals or indirectly observing their shadows. Thus, both views fall under the outside-in framework. To make the preexisting neuronal patterns useful, Plato's cave dwellers must break loose of their chains, leave the cave, explore the world, and get a second opinion. We need to venture out to bring back knowledge. The only way we can learn that sticks that "look bent" in water are not really broken is to touch and move them.

---

perceiving mind through understanding). However, understanding a situation always depends on our interaction with the situation. Different versions of truth give rise to different accounts of meaning because what one considers as truth always depends on one's conceptual system.

If brain networks and dynamics are preformed, what advantages do they offer over the blank slate model? First and foremost, its preexisting "ideal forms" provide the necessary balance to keep the brain's dynamical landscape stable and robust against other competing needs, such as wide dynamic range, sensitivity, and plasticity (Chapter 12). There is no threat of catastrophic interference because preformed brain networks are not significantly perturbed by new experiences. Indeed, computational models attempting to avoid catastrophic interference include two different synaptic populations. One set can change rapidly but decays to zero rapidly ("fast" weights); the remaining set is hard to change but decays only slowly back to zero ("slow" weights). The weighting used in the learning algorithm is a combination of slow and fast weights, reminiscent of the two ends of the log distribution of synaptic weights in real brains.[15] Second, newly acquired experience is not created in the sense of adding new words to a vocabulary list. Instead, the preformed brain is an already existing dictionary, although its numerous words and sentences are initially meaningless. As a sharp piece of lifeless stone has the potential to become an essential tool for scraping meat or a weapon, neuronal words have the potential to become meaningful to the organism after experience attaches utility to them. To paraphrase Galileo, the book (i.e., the brain) cannot be understood unless one first learns to comprehend its language and read the characters in which it is written. The meaning of the neuronal words in this brain-book is acquired through exploration.

In the preformed brain model, neural words and sentences are neither built from scratch nor fully controlled by perceptual inputs—they preexist. They are combined from preexisting components with a skewed distribution rule. These components, such as neuronal assemblies (letters) contained in gamma or ripple cycles (Chapters 4 and 8) are fully active even when the brain disengages from the environment or falls asleep. Evolving cell assemblies, therefore, reflect

---

15. Hinton and Plaut (1987). In this model, the strong and weak synapses are not distributed, but each connection has two weights: a slowly changing, plastic weight that stores long-term knowledge and a fast-changing, elastic weight that stores temporary knowledge and spontaneously decays toward zero. When old memories are "blurred" by subsequent training on new associations, the memories can be "deblurred" by rehearsing just a few of them. As soon as the fast weights have decayed back to zero, the newly learned associations are restored. Fusi and Abbott (2007) followed a similar rationale, using a model with logarithmic dependence of memory capacity on the number of synapses used to store the memories. A system-level solution to interference is to have a rapid learning structure with a mechanism for rapid acquisition of new information that does not interfere with previously stored information (e.g., the hippocampus). Subsequently, the hippocampus serves as a teacher to the slow-learner neocortex, where details of all existing memories are stored (McClelland et al., 1995; Chapter 8).

the default functional mode of the brain. In fact, it would be hard to imagine nondynamic, stationary circuits in which neurons would be silent or idling.

To summarize this section, I suggest that brains come with a preexisting dynamic even without prior experience, providing a scaffold that allows it to make guesses about the consequences of the actions of the body it controls and to filter which aspects of the world are worth attending. The brain is not a blank tablet to be filled gradually by the truths of the world but an active explorer with a preformed dynamics ready to incorporate events from its points of view. The brain's only job is to assist the survival and prosperity of the body it interacts with, independent of whether, in the process, it learns "objective reality" of the outside world or not.

EXPERIENCE AS A MATCHING PROCESS

For the *tabula rasa* brain, knowledge is synthesized from scratch. For the inside-out model, it is experience that adds meaning to preformed neuronal trajectories and their combinations. We discussed abundant experimental evidence, particularly the skewed distribution of many anatomical and physiological parameters, in previous chapters that favor the latter view. Rich brain dynamics can emerge from the genetically programmed wiring circuits and biophysical properties of neurons during development even without any sensory input or experience. One such fundamental, experience-independent source of dynamics is brain rhythms. The brain's hierarchically related oscillations serve a dual purpose: they maintain stability and robustness on the one hand and offer a needed substrate for syntactical organization of neural words and sentences on the other (Chapter 6). This is the organization I call the *preformed or preconfigured brain*: a preexisting dictionary of nonsense words combined with internally generated syntactical rules. The neuronal syntax with its hierarchically organized rhythms determines the lengths of neuronal messages and shapes their combinations. Thus, brain syntax preexists prior to meaningful content.

Neuronal circuits with recurrent connections, the prototypical organization of the mammalian cortex, cannot be silent for long. Instead, their default dynamic is the maintenance of neuronal sequences through self-organization (Chapter 7). All brain networks are characterized by a rich dynamic of constantly evolving cell assembly sequences that can be combined by the rhythm syntax in a virtually infinite number of ways.[16] This process produces a rich

---

16. Similar views have been expressed by others. Ken Harris, a previous postdoctoral fellow from my lab, refers to the "realm of possibilities" offered by cell assembly sequences in the

realm of potential neuronal words and sentences, although the words acquire meaning only by action-based experience. Meanings are action-calibrated neuronal trajectories.

Only through actions can neurons relate their responses to sensory inputs to something else supported by the corollary discharge mechanism (Chapters 1, 3, and 5). Mainly through such mechanism can sensory signals acquire meaning, defined as their significance to the organism. Because the brain's action signals are always copied to sensory circuits (Chapter 3), meaningless neuronal words can become meaningful by comparing actions with sensations. The copy of the output provides sensory circuits with a second opinion, a sort of a reality check against what comes into the brain through the sensors. Every self-generated action, even a glance of the eye, can be regarded as the brain's hypothesis testing.

The null hypothesis of brain circuits is that nothing strange happens outside there. However, when the comparator mechanisms detect a difference between action-produced expectation and sensory inputs, the null hypothesis is rejected. This discrepancy triggers further investigation, and those preexisting neuronal patterns that coincide with the attended unexpected event are marked as important. As more knowledge is accumulated, a larger fraction of neuronal trajectories become meaningful to the organism (Figure 13.1). With increasing experience, the comparator mechanisms can boost their ability to notice even more subtle differences. Thus, in contrast to the general wisdom that "everything is novel to a newborn baby," I think the opposite might be the case. Instead of moving from the specific to the general, for a newborn, every human face is just a phylogenetically biased vague face cue. It takes several months of brain training to differentiate among the many possible faces and eventually identify the mother and father as explicit identities and different from everybody else and to distinguish novel from the familiar.

Let me try to explain this idea differently. Whereas it is intuitive for an adult to think that concepts are formed by extracting joint features from individual elements, in reality, we learn categories and gestalt concepts faster than we do specifics. When a child sees a dog and names it as a "dog," it is neither a

---

cortex (Luczak et al., 2009). George Dragoi, a previous student of mine, calls it "preplay" of future, never before-experienced, sequences (Dragoi and Tonegawa. 2011, 2013a, 2013b; Liu et al., 2018). The similarity between spontaneous and sensory-evoked spatial patterns in the visual cortex of anesthetized monkeys and ferrets has also been interpreted in the framework of preconfigured networks (Kenet et al., 2003; Tsodyks et al., 1999; see also Fiser et al., 2004 and MacLean et al., 2005). My first assignment as a student was to examine the variability of click-evoked responses in the auditory cortex of the cat. The conclusion of the experiment was that the main source of variability of the evoked events is the background spontaneous activity (Karmos et al., 1971).

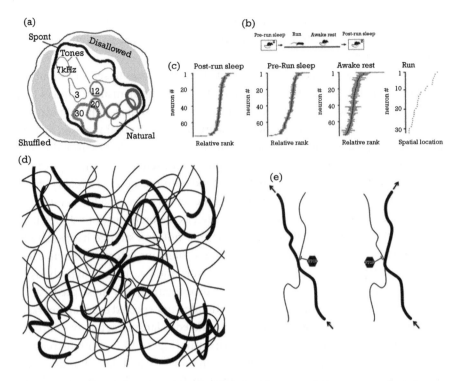

Figure 13.1. Neuronal responses to stimuli and environmental changes are selected from a large reservoir of internally generated neuronal patterns. A: Cartoon illustrating the geometrical interpretation of the relationship between spontaneous and evoked events in the rodent auditory cortex. Of all theoretical scenarios (outer boundary), neuronal connectivity constrains the realm of possible self-organized patterns (area outlined by thick black line). The gray area outside this boundary (disallowed) does not occur spontaneously. Evoked responses to 3–30 kHz tones (outlined by various shapes) and natural sounds (lower, right ovals) are drawn from the preexisting spontaneous patterns. B: Comparison of behavioral states in which similarity of neuronal sequences in the hippocampus were examined: pre-experience and post-experience sleep in the home cage (pre-run and post-run), running on a maze track, and resting. C: Neurons with place fields on the track were ordered according to their place field peaks (*rightmost panel*). Neuronal sequences with the same sequential order as during run were found in abundance not only during resting and post-experience sleep but also during sleep *before* the rat entered the novel track, indicating that the pattern of activity during the very first experience was drawn from a reservoir of self-organized spontaneous patterns. D: Illustration of a small fraction of the large number of possible neuronal trajectories (thin lines). Segments of the trajectories that have already gained meaning to the animal are marked by thick lines (experience-matched trajectories). E: Inhibition (STOP sign) can affect the movement of neuronal traffic and prevent/allow cross-over from one trajectory segment to another.
A: Reproduced with permission from Luczak et al. (2009); B, C: Reproduced with permission from Liu et al. (2018).

specific nor a general category but a mixture. Confronted with a sheep, he may say "dog." A superordinate category, for example "mammal," can be readily matched to a preexisting neuronal trajectory and generalized to many similar appearances. Specifying a subordinate, such as your specific pet dog, needs several extra features which are learned slowly by concatenating preexisting segments of neuronal trajectories (Figure 13.1). The first utterances of a baby are generalized concepts rather than specific references. "Bana" may refer to seeing a banana, requesting a banana, or daddy eating a banana. Linguists call these early generalized utterances "holophrases." The general to specific or good-enough to precise transition is also more economical. It is not realistic to assume that we learn all things in our surrounding world and give them names. Even when we name something (e.g., a house), it is most often vague (e.g., it may or may not include a hut or an igloo). The brain's economic solution seems to be a compromise between precision and the brain's storage and processing capacity.

Even an inexperienced brain has a huge reservoir of unique neuronal trajectories with the potential to acquire real-life significance. But only exploratory, active experience can attach meaning to the largely preconfigured firing patterns of neuronal sequences. The reservoir of neuronal sequences contains a wide spectrum of high-firing, rigid and low-firing, plastic members that are interconnected via preformed rules (Chapter 12). The strongly interconnected preconfigured backbone, with its highly active member neurons, enables the brain to regard no situation as completely unknown. As a result, the brain always takes its "best guess" in any situation and tests its most plausible hypothesis. Each situation—novel or familiar—can be matched with a highest probability neuronal state, a reflection of the brain's best guess. The brain just cannot help it; it always instantly compares relationships rather than identifying explicit features. There is no such a thing as unknown for the brain. Every new mountain, river, or situation has elements of familiarity, reflecting previous experiences in similar situations, which can activate one of the preexisting neuronal trajectories. Thus, familiarity and novelty are not complete strangers. They are related to each other and need each other in order to exist. In case of a discrepancy—when there is a mismatch between an input and the highest probable existing trajectory—a novel combination is constructed from existing trajectories and added to the brain's "knowledge base" (thick lines in Figure 13.1D). For example, after multiple corridors of a maze have already been matched to neuronal trajectories by the process of exploration, a novel combination of a never-taken path between two corridors can be achieved by concatenating the neuronal trajectories of the two corridors (Chapter 8; Figure 13.1E). This process is how I imagine that the initially nonsensical neural words of the brain's vocabulary become meaningful and useful. Meaning is thus neither objective not absolute but relative to the grounding action.

I take the risk of moving my speculation one step further. Once a neuronal pattern acquires meaning through action-mediated grounding experience, its spontaneous or imagination-assisted recurrence can be used as a simulated stimulus. In turn, manipulating such internalized patterns and combining them in various constellations can generate new ideas and abstract constructs, possibly assisted by the mixing-embedding ability of hippocampal sharp wave ripples (Chapter 8) or related mechanisms in the neocortex. This is possible because each pattern in question has already been anchored by active exploration, and each has acquired some utility for the organism. But the resulting abstract constructs also need to be grounded by action to verify the validity of the inference (Chapter 9). Otherwise, the generated ideas may remain in the domain of entertaining fictional thoughts, such as Santa Claus, or pathological ones, such as hallucinations.[17]

CONSTRAINTS HAVE ADVANTAGES

From our discussion of the log rules of connectivity and dynamics (Chapter 12), the constraints on circuit dynamics should not be surprising. Neurons rarely signal anything alone, as any individual neuron functions mainly as part of an ensemble (Chapter 4). Such collective cohesiveness decreases the freedom of assembly members. Self-organized networks create their own constraints and dynamics that guide the evolution of neuronal trajectories rather than being imposed on them by external stimuli (Chapter 5). Neuronal networks with such internally controlled and continually changing dynamics can respond differently to the same sensory stimulus at different times. Emergence of the statistical features of such evolving firing dynamics does not require external control. Sharp wave bursts are present at birth, before any exploratory experience.[18] Their incidence increases modestly postnatally, perhaps due to circuit maturation. Intrinsic processes can also explain why the synaptic strengths and firing rates of neurons are so strongly preserved across behavioral states (Chapter 12). These circuit constraints also influence the order of firing in sequences of member neurons during different brain states and at different temporal scales. For example, during the sharp wave ripples of sleep (Chapter 8),

17. Feinberg (1978).

18. Leinekugel et al. (2002).

the neurons during exploration of a novel environment may become active in a similar order as during sharp wave ripples of the preceding sleep. This arrangement offers the somewhat counterintuitive prediction that sequential dynamics should exist before experience. [19] Indeed, this idea fits comfortably with observations that firing rates, the strength of synaptic connections between neurons, and their order of firing, in general, are impervious to brain state (Chapters 8 and 12). The preconfigured bias of neuronal circuits for certain activity patterns is also demonstrated by the observation that their artificial activation can induce firing sequences that are related to their native sequences.

In further support of preformed brain dynamics, when an adult animal visits a novel environment for the first time, a particular constellation of neurons is selected from the existing repertoire of sequences and maps. The hippocampus and entorhinal cortex always produce a map in any situation, known or unknown. Exploratory locomotion is randomly interrupted by immobility periods associated with patterns of fast neuronal sequences, replaying the past trajectory of the animal or predicting a new one.[20]

Matching preconfigured patterns with experience is not unique to the hippocampus. In the motor cortex, neuronal firing sequences observed after learning a particular movement or intention of movement are remarkably similar to preexisting neuronal patterns produced before learning, thus demonstrating that neuronal populations are constrained to give rise to neuronal sequences from a large reservoir of internally induced patterns. Instead of random and unlimited combinations, a particular preexisting pattern from the large reservoir is assigned to a new movement and perceived pattern, and thereby the neuronal sequence acquires a behavioral meaning.[21] This framework can be contrasted with the

---

19. In this experiment, CA1 pyramidal neurons were locally activated by optogenetic drive. The induced firing rates and sequences were correlated with those of native rates and sequences, illustrating strong circuit property constraints in any given situation (Stark et al., 2015). This circuit constraint may explain the remarkable observation that place cell sequences in a novel environment can be predicted from the sharp wave ripple-associated "preplay" sequences during sleep *prior* to the experience (Dragoi and Tonegawa, 2011, 2013a, 2013b; Liu et al., 2018).

20. Replay sequences were even observed after the first lap on a novel track (Foster and Wilson, 2006). The center of the current debate is whether the model of the world that informs replay is developed very rapidly, after only one or two experiences (Silva et al., 2015) or whether at least part of the map existed before exploration (Dragoi and Tonegawa, 2011).

21. In these experiments, monkeys were trained to move a cursor on the computer screen by a "brain-machine interface" in which firing patterns in the motor cortex was the input. Thus, the monkey commanded the direction and speed of the cursor simply by intention. After learning, the monkeys produced roughly the same set of activity patterns across all movements as before learning (Golub et al., 2018). These experiments are in support of the model of neuronal group

outside-in *tabula rasa* model, in which neuronal assemblies and sequences are constructed largely from sensory instructions.

Human language metaphorically illustrates the possibilities and constraints of the log-dynamic brain. The more than 6,000 mutually unintelligible languages are spoken by humans exemplify the rich generative abilities of neuronal circuits. A large part of human language is semantic knowledge acquired through externalization and societal interactions (Chapter 9). Any proposition, intention, or thought can be conveyed by any language and readily translated to another one. Yet, irrespective of the meaning of utterances and written words in different languages, all languages have the same basic generative rules (i.e., a preexisting syntax),[22] and guided by the constraints of the hierarchical syntactic organization of brain rhythms with their log rule.

Good-Enough Guesses

The matching process works fastest if it does not have to be exact. Instead of identifying each item separately in all its details, weighting the evidence in a precise manner, and deliberating on the smartest solution, the brain produces a good-enough action by rapidly assessing the relationships among the input variables. The brain senses relationships in the world effectively due to its skewed dynamical organization.

Gestalt psychologists recognized that the mind has an innate ability to perceive patterns based on similarity, proximity, continuity, closure, and connectedness. All these terms describe a relationship.[23] Detailed knowledge of the parts of an image is not necessary to recognize their sum (i.e., the whole). No amount of analysis of the parts in a Jackson Pollock painting would give a clue to why the whole is perceived as it is, as an aesthetic experience derived exclusively from relationships of form and color likely matching the perceiving brain's statistical dynamics.

The illusions discussed in this book also inevitably emerge from the dominance of the good-enough brain, which evaluates the relationship among the constituent parts and ignores often meaningless elements. Only when previously irrelevant details acquire some behavioral significance to the perceiver does the precision brain step in and identify them. Another way to illustrate how

---

selection or the "neural Darwinism" of Gerry Edelman (1987), a framework closely related to the inside-out approach.

22. Noam Chomsky proposed that a Universal Grammar binds all known languages (1980). The system of brain rhythms may be the foundation of such grammar (Buzsaki 2010; Chapter 6).

23. Eichenbaum and Cohen (2014).

the preconfigured log-dynamic brain organization affects the good enough–precision spectrum is through our sense of quantities.

RELATIONSHIPS AND SENSE OF QUANTITIES

Math enthusiasts often assume that math is as natural as language and therefore should be taught as early as possible in school. Part of the argument is based on behavioral research, which claims that counting does not need a complicated brain because even simple animals have a sense of numbers. Starting with the infamous counting horse Clever Hans,[24] the list of animal species with an alleged ability to do basic math, such as counting, has been increasing, now including chimpanzees, orangutans, macaques, dogs, parrots, New Zealand robins, coots, salamanders, newborn chicks, and even bees. Some behaviorists go so far as stating that counting is innate in many species.[25] This is strange. If the number sense is as natural to us as the sense of color, why did it take a few million years of hominid evolution to conceive the idea of numbers and use them to count anything (Chapter 9)?[26] My math-expert friends would counter that objection by explaining that mathematics is not about memorizing tables but spotting patterns. A person might be terrible at arithmetic, yet her pattern-spotting ability would allow her to judge quantities efficiently. I agree.

The debate about the innate versus acquired form of math may be a reflection of the good-enough versus precision extremes of a log-dynamic brain because two apparently different worlds are lumped into one discipline. As discussed in the previous chapter, ratio judgment is a natural function of neuronal circuits. Every animal, including young children, can effortlessly estimate ratios. Such estimation is not the exact arithmetic of division or multiplication, but, for practical purposes, it is good enough. Estimating the size difference between two antelopes can save a lion quite a bit of extra running and fighting, even if

---

24. Hans, of course, turned out to be "cheating" by taking cues from his owner or other observers of his "counting" act (Pfungst, 2000).

25. Matsuzawa (1985); Pepperberg (1994); Jordan et al. (2008); Tennesen (2009). See http://www.scientificamerican.com/article/how-animals-have-the-ability-to-count/

26. Stanislas Dehaene, the French mathematician-neuroscientist, believes that mathematics is language-dependent (1997). In this case, there would be no such thing as preverbal or nonverbal counting (Gallistel and Gelman, 1992), and animals other than humans could not have mathematical skills. Others approach the same problem more radically. Animals "never conceive absolute quantities because they lack the faculty of abstraction" (Shepard et al., 1975; Ifrah, 1985).

she does not calculate the exact ratio of 1.6666. The good-enough brain can make fast and efficient judgments about quantities under most conditions, even without any clues about numbers.

Quantity judgment is different from number-based counting. This view explains why monkeys can "count" only up to five, robins look first to the holes with the most mealworms, and newly hatched chicks go looking for the larger number of objects. They do not count but *estimate* ratios. However, the ability to precisely distinguish between 20 and 21 objects is a different thing. The difference between numbers cannot be innately understood; instead, we have to go to school and learn the invented rules and formalized logic.[27]

Linear Versus Log Scales

Giving each number equal importance by spacing numbers equally along a line is a convention that people agreed upon. This aspect of math is not natural to the brain. Numbers, such as "threeness" and "fourness," are explicit abstract symbols, a human invention that must be learned through hard education. This convention is a contrast to the logarithmic number line in the brain, in which bigger quantities gets progressively more compressed.[28]

In support of the log versus linear division, brain imaging studies show that the estimation aspect of number sense is present from an early age, as reflected by activation of the parietal cortical area. Recall from Chapter 10 that this region overlaps or is remarkably close to areas that represent spatial and geometrical dimensions such as magnitude, location, gaze direction as well as duration. Neurological patients with brain damage in this area show not only spatial neglect but also an idiosyncratic mental number line. For example, they might choose five as the midpoint of a numerical interval between two and six, perhaps biased by the brain's natural tendency to estimate logarithmic values.[29] In contrast to good-enough estimation, problems

---

27. Geary (1994); Gardner (1999); Devlin (2000).

28. Humans without formal math education, such the Mundurucu, an Amazonian indigene group, map symbolic and nonsymbolic numbers onto a logarithmic, instead of linear, scale (Dehaene et a., 2008). Their performance is somewhat similar to that of Western kindergarteners, who also devote more space to small numbers, imposing a compressed logarithmic mapping. For instance, they might place the number 10 near the middle of the 0 to100 segment (Siegler and Opfer, 2003). My guess is that children's difficulties with education-based arithmetic may result from their persistent reliance on preexisting logarithmic representations of numerical magnitudes.

29. Brannon and Terrace (1998); Zorzi et al. (2002); Sawamura et al. (2002).

requiring exact calculation activate the frontal cortical regions, including language areas.[30] Thus, "natural numbers" are not so natural to the brain. They and their combinatorial rules have to be acquired through social interaction, similar to language.

In line with the imaging studies, electrophysiological studies in rhesus monkeys also demonstrate that firing of single neurons in the parietal and prefrontal cortex obeys a log rule of quantity representation. The experimenters 'asked' the monkeys to judge whether consecutive images containing one to five dots showed the same or different numbers of dots. There were two surprising findings. First, many neurons responded most strongly to pictures containing the same number of dots (e.g., 2 or 4), irrespective of their sizes and positions. Second, when neuronal firing responses were plotted as the function of the log number of dots, they conformed to a Gaussian curve with a fixed variance, obeying the Weber-Fechner law. Thus, this experiment finds, again, that neuronal firing corresponds to the perceived quantities on a logarithmic rather than by a linear scale.

MEANING IS ACQUIRED THROUGH EXPLORATION

One can argue that acquisition of new knowledge, such as a map, is a result of generalization from previous experience with similar situations. What about the brain of a newborn animal?
Rats, even more than humans, have limited sensory abilities at birth. Rudimentary sniffing and whisking can be observed at postnatal day 6, and by day 10 they reach adult levels. During the first week of life, pups can only crawl. Ambulation on four limbs emerges only by the end of the second week of life, after which the pups begin to leave the nest. The length of such exploratory trips increases abruptly at the beginning of the third week, and the pups become independent of their mother by the end of the third week.

The head direction system is already stable early in life. The first exploratory trip beyond the confines of the nest takes place between postnatal day 15 to 17. At this stage of development, hippocampal pyramidal neurons already display adult-like place fields on the very first trip and ordered sequences, while emergence of stable entorhinal grid cells is delayed until weaning. Thus, the hippocampus possesses many possible neuronal trajectories before exploration has begun. Because of these preformed trajectories, new maps can be retrieved

---

30. Dehaene et al. (1999).

instantly when pups enter into novel environments.[31] Hippocampal neurons do not fire because the cues at particular parts of the environment make them fire but because the internally evolving neuronal trajectory reaches the point that it is their turn to become active. Think of a dictionary of artificial words. We can map English words to this artificial language by randomly opening a page in the dictionary and blindly pointing to a word.[32] The knowledge of the English words can then ground the meaning of the artificial words, similar to the grounding ability of action-based experience. Of course, the brain is not a simple dictionary that can be opened randomly at any page. It has built-in constraints, as we discussed in Chapter 12. Thus, there seems to be a preorganized priority to which experience can be linked.

The first spatial map may be only a crude, yet good-enough sketch (a "protomap")—the brain's best guess about the new situation. With further exploration, plastic neurons from the large reservoir of the slow firing majority can be recruited to the protomap (Chapter 8). Under this admittedly speculative framework, the physiological correlates of learning from experience involve rapid matching to the familiar aspects of the environment based on pre-existing dynamics of the fast-firing minority of neurons, followed by gradual refinement of firing patterns in a subgroup of the slow-firing reserve population to reflect the novel aspects of the situation. For example, the cycles of a sharp wave ripple event contain a sequential series of cell assemblies, each of which can be conceptualized as a letter of the hypothetical neuronal vocabulary so that multiple cycles form a neuronal word. In clusters of sharp wave ripples, the words can be reassembled into novel sentences to match the extended sequences of discrete segments of travel paths, situations, or events (Chapter 8). This "chunking" or parsing role of chained sharp wave ripples

---

31. Like head-direction cells, border cells are also in place before spontaneous ambulation. Grid cells have irregular and variable fields until the fourth week, after the emergence of place cells. Together, they form a functional map at the time when rat pups leave the nest for the first time in their life, and the maps get fine-tuned thereafter (Langston et al., 2010; Wills et al., 2014; Bjerknes et al., 2014). Muessig et al. (2016) state that: "What drives remapping in young pups remains an open question." Perhaps nothing really "drives" remapping, it is a matching process between already existing maps and ambulatory exploration.

32. For example, cat = xx.u. By building an initial set this way, when next time we have a word in mind (dog), we should not only find another random word (n-zk) but somehow indicate a relationship between them. This analogy has a major weakness as well. The initially meaningless neuronal words in the brain may already have preexisting relationships to each other. You can call it prejudice. These preconfigured constraints bias how the pages open up in our hypothetical dictionary.

allows the flexible generation of a large number of combinations from a finite number of preformed neuronal words.[33]

Overall, learning with neurons involves a synthesis of preexisting neuronal events, which only modestly and transiently affects network dynamics. In each learning episode, a small number of the many possible preexisting neuronal trajectories acquire behavioral meaning and thus start to represent a particular constellation of relationships. I have used spatial learning as an example because the most quantitative experimental data are available in this field. The same process likely applies when neuronal trajectories are matched to events and situations. This view echoes the idea that perception is hypothesis-testing in the brain.[34] Matching is the brain's attempt to pull out the best-fitting neuronal trajectory in any situation from its preformed vast reservoir and refine it only when mismatch occurs.

SUMMARY

The Weber-Fechner law describes not only weight, loudness, brightness, and odorant intensity, but also more abstract dimensions, such as representation of sizes, distances, owned territory, intervals, durations, numbers, and other symbolic representation of quantities. In this chapter and the previous one, I proposed that these mental metrics are based on a skewedly organized brain structure and dynamics. Behavior-based calibration of perceptions and even abstract representations stem directly from the constraints of a preconfigured brain. This view is different from the philosophical idea of isomorphism,[35] which assumes that things in the brain correspond to shapes and properties in the world. Instead, the nervous system may have evolved to mimic the statistical probabilities of the physical world and thus become an efficient predictor

---

33. For more details on concatenation of neuronal "chunks" by sharp wave ripples, see Buzsáki (2015) and Pfeiffer (2018). Sharp wave ripples during post-experience sleep are also critical for reinstating the default network dynamics after experience-induced perturbation (Grosmark et al., 2012). After learning and even after plasticity-inducing electrical stimulation, such as long-term potentiation, synapses may get stronger, and some neurons increase their firing rates but only at the expense of others. Within the same brain state, the total synaptic weights and the mean excitability of the system remains constant over time (Dragoi et al., 2003).

34. Helmholtz (1866/1962); Gregory (1980). Bayesian models of the brain express similar sentiments Friston (2010, 2012). However, the Bayesian brain falls short in explaining how the brain creates new knowledge (Friston and Buzsáki, 2016).

35. Ullman (1980); Biederman (1987).

of events. As a result, neurophysiological and perceptual brain dynamics, both spanning several orders of magnitude, share a common mathematical foundation: the log rule.

Skewed distributions at multiple levels of brain organization come with both advantages and disadvantages. The tails of these quantitative distributions have apparently distinct qualitative features, which we describe by discrete words, such as familiar and novel, rigid and plastic, good-enough and precise.[36] Yet every novel situation contains elements of familiarity. Similarly, the plastic versus rigid division is really a continuum. Brain correlates of newly acquired experience are not created in the sense of inventing new words from nothing to an ever-expanding vocabulary. Instead, the preformed brain is a dictionary in which initially there is no meaning assigned to its preexisting words and sentences supported by the numerous neuronal trajectories that brain networks perpetually generate. The behavioral significance or meaning of neuronal words in the brain dictionary is acquired through exploration. Thus, experience is primarily a process of matching preexisting neuronal dynamics to events in the world.

---

36. The Babel of terms describing the physiological attributes of neurons in the hippocampal-entorhinal system (such as place cells, time cells, grid cells, head direction cells, boundary vector cells, border cells, object cells, object vector cells, distance cells, reward cells, concept cells, view cells, speed cells, memory cells, goal cells, goal direction cells, splitter cells, prospective cells, retrospective cells, island cells, ocean cells, band cells, object cells, pitch cells, trial order cells, etc.) might be explained by the apparent distinctiveness of a few sets of continuous distributions. Relations are often quantitative rather than qualitative. Naming elements of a distribution creates the illusion that they are distinct (i.e., qualitatively different).

Epilogue

The outside is always an inside.

—Le Corbusier[1]

It's what's inside that counts.

—CubeSmart (subway ad)

All enquiry and all learning is but recollection.

—Socrates in Plato's *Meno*

I did not aim to write a perfect book—just a story good enough that the reader can understand my views and challenge them. My goal was not so much to convince but to expose the problems and highlight my offered solutions. Perfection and precise solutions will have to wait for numerous experiments to be performed and reported in detail in scientific journals. I analyzed how an undefined and unagreed-upon terminology, which we inherited from our pre-neuroscience ancestors and never questioned, has become a roadblock to progress. The neuronal mechanisms of invented terms with ill-defined content are hard to discover. Such conceptual confusion is perhaps the primary reason why "my scientist" could not explain to me my pig friend's cognitive abilities (see the Preface). This message is especially important today, when newly invented terms are again popping up like mushrooms after a rain. I do not insist that my inside-out framework is right or the only way to go, but I hope I presented enough evidence in this book to convince the attentive reader that the outside-in strategy has reached its limits in neuroscience research.

1. Le Corbusier (1923).

I discussed two major themes. First, I addressed why understanding brain mechanisms requires a reader/observer-based, inside-out approach. My hunch is that an action-centered approach will eliminate the problem of the "middle," the decision-maker homunculus, which is an inevitable ingredient of the outside-in, perception-decision-action framework with its tacitly implied *tabula rasa* brain. Just a century ago, *élan vital*, or "vital impetus," an expression from the philosophy of the nineteenth-century vitalist movement, was considered the key to understanding how life arises from inanimate matter. Coined by the French philosopher Henri Bergson,[2] this term provided a hypothetical explanation for evolution and how an adult organism can arise from an egg and a sperm. Bergson also thought consciousness was the *élan vital* of the brain, needed to drive human thought. When molecular genetics entered the scene, an essentially inside-out approach, *élan vital* was eliminated from our vocabulary. After the discovery of DNA, this hypothetical expression was no longer needed to explain how a seed becomes a tree. Different frameworks require different vocabularies.

Second, I examined the implications of a brain with preformed connectivity and dynamic operations. We know the unknown not because we remember it, but because our brains are programmed to make some kind of a guess under all conditions. Even in the most unexpected situation, our brain generalizes by relating the novel situation to something else. Only after some discrepancies with importance to the organism are detected does the brain try to identify circumstances that can differentiate the current event from previously experienced similar ones.

I drew several conclusions in various chapters and suggested new directions, the way I see them. However, this is not the end of the story of the inside-out brain but closer to its beginning. A new book should start right here, perhaps discussing the implications of outside-in versus inside-out frameworks for neurological and especially mental diseases and for brain-inspired artificial intelligence. Let me offer here a few sentences for potential future directions.

Mainstream psychiatry, like cognitive neuroscience, is also under the historical influence of the empiricist representational framework. The *Diagnostic and Statistical Manual of Mental Disorders* (DSM-5) in the United States is another prominent example of how human-conceived terms are used in attempts to draw boundaries among mental disorders. It is the "James's list" of psychiatry. Each new edition is an attempt to fix problems with the previous ones. According to the American Psychiatric Association, the "goal in developing DSM-5 is an evidence-based manual that is useful to clinicians in helping them accurately

---

2. Bergson (1911).

diagnose mental disorders." This aim reflects a collective subconscious desire to reduce a high-dimensionality problem, such as a complex psychiatric condition, to a single word or phrase: the diagnosis. Dimensionality reduction in statistics and information theory refers to a mathematical process of reducing the large number of variables under consideration to extract a set of principal variables or features using methods such as principal component analysis or nonlinear dimensionality reduction.[3] Of course, these methods can only be as good as the data we supply to them. A further problem is that "humans are incorrigibly inconsistent in making summary judgment of complex information," as expressed succinctly by the economist Daniel Kahnemann.

Here are a few examples of DSM-5 category reductions: adjustment disorder, reactive attachment disorder, disruptive mood dysregulation disorder, intermittent explosive disorder, and somatic symptom disorder. I doubt that any practicing psychiatrist believes that these phrases correspond to distinct mechanisms in the brain. It is more likely that disorders with widely different etiologies are lumped together into the same diagnosis (e.g., autism spectrum disorder), and different diagnoses correspond to the same brain mechanisms, with some variations (e.g., schizophrenia and certain forms of undetected network seizures). Psychiatrists are aware of these problems, but, without grounding mechanisms, there is no way to tell how to draw the boundaries.

Similar to our garden-variety cognitive terms, these conceived diagnostic categories with their arbitrary boundaries are useful in casual conversations and for insurance companies, but they should not be assumed to relate directly to particular brain structures or mechanisms. Just ask any practicing psychiatrist, such as my wife, how well the boundaries set up by the DSM translate into the diseases they treat. Psychiatrists treat symptoms and care about behavioral outcomes, not about conceived boundaries. Yet neuroscientists and geneticists search for animal models of depression, schizophrenia, autism, and so forth. Funding agencies support efforts that fit the DSM criteria. As we do not yet understand the biological foundations and circuit mechanisms of these disorders, we cannot really judge which of these dreamed-up terms will approximate neuronal mechanisms. It is more likely that variations in the disorders reflect a broad distribution of brain mechanisms. This could explain why seemingly unrelated disorders respond to similar drugs. For example, while gabapentin[4] is approved by the US Food and Drug Administration for the treatment of certain forms of epileptic seizures and neuralgia, it is also widely used to treat a variety of neurological and psychiatric conditions.

3. Jolliffe (1986); Lee and Seung (1999); Roweis and Saul (2000); Tenenbaum et al. (2000).

4. Also known by the trade names of Neurontin, Gralise, Pregabalin.

The inside-out framework may also offer an alternative approach to brain-inspired artificial intelligence (AI). Instead of *tabula rasa* connectionist models, researchers could build models that mimic the firing and population statistics of neuronal networks, with their rich repertoire of syntactic rules, and learn by action rather by than a multitude of reinforcement-supervised repetitions.[5] Such models, a fusion of AI and robotics, might be used to investigate how their vast repertoire of stable and preformed dynamic trajectories can be matched with real-world events and situations. A fraction of these trajectories can be grounded by action-based "experience," as machines are trained to communicate with the environment in which they are embedded. There is a chance that machines built in this way (like brains) could disengage their operations from their inputs and create novel solutions. Such hybrid co-evolution of neuroscience and AI would be extraordinarily exciting because the potentially creative ideas they might produce could be useful for us, not just the machines.

Of course, the most important implications of the inside-out framework are those that have not occurred to me. This challenge I pass on to the readers who made it to this page. Some of you may be able to show us the way how to get "there." The first task, of course, is to redefine where "there" is. I have in mind especially those creative young scientists whose brains possess an extraordinary repertoire of preexisting neuronal trajectories and the ability to match them with astonishing ideas and combine them to form even more prodigious ones. As expressed eloquently by William Shakespeare "What's past is prologue."

---

5. LeCun et al. (2015).

REFERENCES

Abbott LF, Varela JA, Sen K, Nelson SB (1997) Synaptic depression and cortical gain control. Science 275:220–224.
Abeles M (1991) Corticonics: Neural Circuits of the Cerebral Cortex. New York: Cambridge University Press.
Aboitiz F, Lopez J, Montiel J (2003) Long distance communication in the human brain: timing constraints for inter-hemispheric synchrony and the origin of brain lateralization. Biol Res 36:89–99.
Acharya L, Aghajan ZM, Vuong C, Moore JJ, Mehta MR (2016) Causal influence of visual cues on hippocampal directional selectivity. Cell 164:197–207.
Adolphs R, Tranel D, Damasio AR (1998) The human amygdala in social judgment. Nature 393:470–474.
Adolphs R, Tranel D, Damasio H, Damasio A (1994) Impaired recognition of emotion in facial expressions following bilateral damage to the human amygdala. Nature 372:669–672.
Adouane L (2016) Autonomous Vehicle Navigation: From Behavioral to Hybrid Multi-Controller Architectures. Boca Raton, FL: CRC Press.
Afraimovich VS, Tristan I, Varona P, Rabinovich M (2013) Transient dynamics in complex systems: heteroclinic sequences with multidimensional unstable manifolds. Int J Discontig, Nonlin Complex 2:21–41.
Ahissar E, Arieli A (2001) Figuring space by time. Neuron 32:185–201.
Ahissar E, Assa E (2016) Perception as a closed-loop convergence process. eLife 5:e12830.
Ahissar E, Nagarajan S, Ahissar M, Protopapas A, Mahncke H, Merzenich MM (2001) Speech comprehension is correlated with temporal response patterns recorded from auditory cortex. Proc Natl Acad Sci U S A 98:13367–13372.
Aitchison J, Brown JAC (1957) The Log-Normal Distribution. Cambridge: Cambridge University Press.
Akhlaghpour H, Wiskerke J, Choi JY, Taliaferro JP, Au J, Witten I (2016) Dissociated sequential activity and stimulus encoding in the dorsomedial striatum during spatial working memory. eLife 5:e19507.
Alger BE, Nicoll RA (1979) GABA-mediated biphasic inhibitory responses in hippocampus. Nature 281:315–317.

Allen DC (1960) The predecessors of Champollion. Proc Amer Phil Soc 104:527–547.

Allman J, Hakeem A, Watson K (2002) Two phylogenetic specializations in the human brain. Neuroscientist 8:335–346.

Allman JM, Hakeem A, Erwin JM, Nimchinsky E, Hof P (2001) The anterior cingulate cortex. The evolution of an interface between emotion and cognition. Ann N Y Acad Sci 935:107–117.

Alme CB, Miao C, Jezek K, Treves A, Moser EI, Moser MB (2014) Place cells in the hippocampus: eleven maps for eleven rooms. Proc Natl Acad Sci U S A 111:18428–18435.

Alonso A, Llinás RR (1989) Subthreshold Na+-dependent theta-like rhythmicity in stellate cells of entorhinal cortex layer II. Nature 342:175–177.

Alyan S, McNaughton BL (1999) Hippocampectomized rats are capable of homing by path integration. Behav Neurosci 113:19–31.

Amaral DG, Lavenex P (2007) Hippocampal neuroanatomy. In: The Hippocampus Book (Andersen P, Morris R, Amaral D, Bliss T, O'Keefe J, eds), pp. 37–114. New York: Oxford University Press.

Amit DJ (1988) Modeling Brain Function. Cambridge: Cambridge University Press.

Anastassiou CA, Montgomery SM, Barahona M, Buzsáki G, Koch C (2010) The effect of spatially inhomogeneous extracellular electric fields on neurons. J Neurosci 30:1925–1936.

Andersen RA (1997) Multimodal integration for the representation of space in the posterior parietal cortex. Philos Trans R Soc Lond B Biol Sci 352:1421–1428.

Andersen RA, Essick GK, Siegel RM (1985) Encoding of spatial location by posterior parietal neurons. Science 230:450–458.

Andreasen NC (1987) Creativity and mental illness: prevalence rates in writers and their first-degree relatives. Am J Psychiatry 144:1288–1292.

Angelaki DE, Cullen KE (2008) Vestibular system: the many facets of a multimodal sense. Annu Rev Neurosci 31:125–150.

Ansell JI, Phillips MJ (1994) Practical Methods for Reliability Data Analysis. Oxford: Clarendon Press.

Araki M, Bandi MM, Yazaki-Sugiyama Y (2016) Mind the gap: neural coding of species identity in birdsong prosody. Science 354:1282–1287.

Arbib MA (2005) From monkey-like action recognition to human language: an evolutionary framework for neurolinguistics. Behav Brain Sci 28:105–167.

Aristotle (1908) Metaphysica. (Ross WD, eds). Oxford: Clarendon Press.

Aronov D, Nevers R, Tank DW (2017) Mapping of a non-spatial dimension by the hippocampal-entorhinal circuit. Nature 543:719–722.

Aru J, Aru J, Priesemann V, Wibral M, Lana L, et al. (2015) Untangling cross-frequency coupling in neuroscience. Curr Opin Neurobiol 31:51–61.

Ascher M, Ascher R (1981) Code of the Quipu: A Study in Media, Mathematics, and Culture. Ann Arbor: University of Michigan Press.

Ashby WR (1947) Principles of the self-organizing dynamic system. J Gen Psychology 37:125–128.

Ashby WR (1962) Principles of the self-organizing system. In: Principles of Self-Organization: Transactions of the University of Illinois Symposium (Von Foerster H, Zopf GW, eds), pp. 255–278. London: Pergamon Press.

Ashton BJ, Ridley AR, Edwards EK, Thornton A (2018) Cognitive performance is linked to group size and affects fitness in Australian magpies. Nature 554:364–367.

Aston-Jones G, Cohen JD (2005) An integrative theory of locus coeruleus-norepinephrine function: adaptive gain and optimal performance. Annu Rev Neurosci 28:403–450.

Atallah BV, Scanziani M (2009) Instantaneous modulation of gamma oscillation frequency by balancing excitation with inhibition. Neuron 62:566–577.

Atance CM, O'Neill DK (2001) Episodic future thinking. Trends Cogn Sci 5:533–539.

Atkinson A (2015) Inequality. Cambridge: Harvard University Press.

Attneave F (1954) Some informational aspects of visual perception. Psychol Rev 61:183–193.

Averbeck BB, Latham PE, Pouget A (2006) Neural correlations, population coding and computation. Nat Rev Neurosci 7:358–366.

Axmacher N, Henseler MM, Jensen O, Weinreich I, Elger CE, Fell J (2010) Cross-frequency coupling supports multi-item working memory in the human hippocampus. Proc Natl Acad Sci U S A 107:3228–3233.

Baars BJ (1988) A Cognitive Theory of Consciousness. New York: Cambridge University Press.

Bach-y-Rita P, Collins CC, Saunders FA, White B, Scadden L (1969) Vision substitution by tactile image projection. Nature 221:963–964.

Bailey DH, Geary DC (2009) Hominid brain evolution: testing climatic, ecological and social competition models. Human Nature 20:67–79.

Ball P (2017) A world without cause and effect. Nature 546:590–592.

Ballard DH (2015). Brain Computation as Hierarchical Abstraction. Cambridge, MA: MIT Press.

Banquet JP (1973) Spectral analysis of the EEG in meditation. Electroencephalogr Clin Neurophysiol 35:143–251.

Barabási AL (2002) Linked: How Everything Is Connected to Everything Else. New York: Perseus Publishing.

Barabási AL, Albert R (1999) Emergence of scaling in random networks. Science 286:509–512.

Barbour J (1999) The End of Time: The Next Revolution in Physics. New York: Oxford University Press.

Barlow HB (1972) Single units and sensation: a neuron doctrine for perceptual psychology? Perception 1:371–394.

Barnett L, Barrett AB, Seth AK (2009) Granger causality and transfer entropy are equivalent for Gaussian variables. Phys Rev Lett 103:238701.

Barrett-Feldman L (2017) How Emotions Are Made: The Secret Life of the Brain. New York: Houghton Mifflin Harcourt.

Barrow-Green J, Siegmund-Schultze R (2016) "The first man on the street"— tracing a famous Hilbert quote (1900) back to Gergonne (1825). Historia Mathematica 43:415–426.

Barthó P, Hirase H, Monconduit L, Zugaro M, Harris KD, Buzsáki G (2004) Characterization of neocortical principal cells and interneurons by network interactions and extracellular features. J Neurophysiol 92:600–608.

Bartos M, Vida I, Jonas P (2007) Synaptic mechanisms of synchronized gamma oscillations in inhibitory interneuron networks. Nat Rev Neurosci 8:45–56.

Basso G, Nichelli P, Frassinetti F, di Pellegrino G (1996) Time perception in a neglected space. Neuroreport 7:2111–2114.

Bastos AM, Vezoli J, Bosman CA, Schoffelen JM, Oostenveld R, et al. (2015) Visual areas exert feedforward and feedback influences through distinct frequency channels. Neuron 85:390–401.

Batey M (2009). Dilly: The Man Who Broke Enigmas. Dialogue. London: Biteback Publishing, Ltd.

Bayley PJ, Squire LR (2002) Medial temporal lobe amnesia: gradual acquisition of factual information by nondeclarative memory. J Neurosci 22:5741–5748.

Beckmann P (1971) A History of Pi. London: MacMillan.

Beer RD (1990). Intelligence as Adaptive Behavior: An Experiment in Computational Neuroethology. New York: Academic Press.

Bell C (1811) An Idea of a New Anatomy of the Brain; submitted for the observations of his friends. privately printed pamphlet: London. Strahan & Preston.

Benabou R (2000) Unequal societies: income distribution and social contract. Am Econ Rev 90:96–129.

Berger H (1929) Ueber das Elektroenkephalogramm des Menschen. Arch Psychiatr Nervenkrankh 87:527–570.

Bergson H (1911) Creative Evolution. New York: Henry Holt and Company.

Bergson H (1922/1999) Duration and Simultaneity. Manchester: Clinamen Press.

Berkeley G (1710/1982) A Treatise Concerning the Principles of Human Knowledge. Kenneth Winkler edition. Indianapolis, IN: Hackett Publishing Company, Inc.

Bernstein N (1947/1967) The Coordination and Regulation of Movements. Oxford: Pergamon Press.

Berridge KC, Whishaw IQ (1992) Cortex, striatum and cerebellum: control of serial order in a grooming sequence. Exp Brain Res 90:275–290.

Berrios GE (2018) Historical epistemology of the body-mind interaction in psychiatry. Dialogues Clin Neurosci 20:5–12.

Berthoz A (1997) Le sens du mouvement. Paris: Odile Jacob.

Berthoz A, Israel I, Georges-Francois P, Grasso R, Tsuzuku T (1995) Spatial memory of body linear displacement: what is being stored. Science 269:95–98.

Berwick RC, Friederici AD, Chomsky N, Bolhuis JJ (2013) Evolution, brain, and the nature of language. Trends Cogn Sci 17:89–98.

Berwick RC, Okanoya K, Beckers GJ, Bolhuis JJ (2011) Songs to syntax: the linguistics of birdsong. Trends Cogn Sci 15:113–121.

Betz W (1874) Anatomischer Nachweis zweier Gehirncentra. Centralblatt für die medizinischen Wissenschaften 12:578–580, 595–599.

Bi GQ, Poo MM (1998) Synaptic modifications in cultured hippocampal neurons: dependence on spike timing, synaptic strength, and postsynaptic cell type. J Neurosci 18:10464–10472.

Bickerton D, Szathmáry E (2009) Biological Foundations and Origin of Syntax. Cambridge, MA: MIT Press.

Biederman I (1987) Recognition by components: a theory of human image understanding. Psychol Rev 94:115–147.

Bienenstock E (1994) A model of neocortex. Network 6:179–224.
Binfield K (2004) The Luddites: Machine Breaking in Regency England. New York: Shocken Publishers.
Bird BL, Newton FA, Sheer DE, Ford M (1978) Behavioral and electroencephalographic correlates of 40-Hz EEG biofeedback training in humans. Biofeedback Self Regul 3:13–28.
Bishop G (1933) Cyclic changes in excitability of the optic pathway of the rabbit. Am J Physiol 103:213–224.
Bishop CM (2007) Pattern Recognition and Machine Learning. New York: Springer.
Bisiach E, Luzzatti C (1978) Unilateral neglect of representational space. Cortex 14:129–133.
Bittner KC, Grienberger C, Vaidya SP, Milstein AD, Macklin JJ, et al. (2015) Conjunctive input processing drives feature selectivity in hippocampal CA1 neurons. Nat Neurosci 18:1133–1142.
Bjerknes TL, Moser EI, Moser M-B (2014) Representation of geometric borders in the developing rat. Neuron 82:71–78.
Blasi DE, Wichmann S, Hammarström H, Stadler PF, Christiansen MH (2016) Sound-meaning association biases evidenced across thousands of languages. Proc Natl Acad Sci U S A 113:10818–10823.
Bocca E, Antonelli AR, Mosciaro O (1965) Mechanical co-factors in olfactory stimulation. Acta Otolaryngol 59:243–247.
Bolhuis JJ, Okanoya K, Scharff C (2010) Twitter evolution: converging mechanisms in birdsong and human speech. Nat Rev Neurosci 11:747–759.
Bonasia K, Blommesteyn J, Moscovitch M (2016) Memory and navigation: compression of space varies with route length and turns. Hippocampus 26:9–12.
Bonifazi P, Goldin M, Picardo MA, Jorquera I, Cattani A, et al. (2009) GABAergic hub neurons orchestrate synchrony in developing hippocampal networks. Science 326:1419–1424.
Borbély AA (1982) A two process model of sleep regulation. Hum Neurobiol 1:195–204.
Borges JL (1994) Ficciones. (Kerrigan A, ed). New York: Grove Press.
Botvinick M (2004) Probing the neural basis of body ownership. Science 305:782–783.
Boyd R, Richerson PJ, Henrich J (2011) The cultural niche: why social learning is essential for human adaptation. Proc Natl Acad Sci USA 108:Suppl 2:10918–25.
Boyden ES, Zhang F, Bamberg E, Nagel G, Deisseroth K (2005) Millisecond-timescale, genetically targeted optical control of neural activity. Nat Neurosci 8:1263–1268.
Bragin A, Jandó G, Nádasdy Z, Hetke J, Wise K, Buzsáki G (1995). Gamma (40–100 Hz) oscillation in the hippocampus of the behaving rat. J Neurosci 15:47–60.
Braitenberg V (1971) Cell assemblies in the cerebral cortex. In: Theoretical Approaches to Complex Systems (R. Heim, G. Palm (eds.), pp 171–188. Berlin: Springer.
Branco T, Clark BA, Häusser M (2010) Dendritic discrimination of temporal input sequences in cortical neurons. Science 329:1671–1675.
Branco T, Staras K (2009) The probability of neurotransmitter release: variability and feedback control at single synapses. Nat Rev Neurosci 10:373–383.
Brannon EM, Terrace HS (1998) Ordering of the numerosities 1 to 9 by monkeys. Science 282:746–749.
Brecht M (2017) The body model theory of somatosensory cortex. Neuron 94:985–992.

Brecht M, Schneider M, Sakmann B, Margrie TW (2004) Whisker movements evoked by stimulation of single pyramidal cells in rat motor cortex. Nature 427:704–710.

Breland K, Breland M (1961) The misbehavior of organisms. American Psychologist 16:681–684.

Bressler SL, Freeman WJ (1980) Frequency analysis of olfactory system EEG in cat, rabbit and rat. Electroencephalogr Clin Neurophysiol 50:19–24.

Bressler SL, Kelso JAS (2001) Cortical coordination dynamics and cognition. Trends Cogn Neurosci 5:26–36.

Brette R (2015) Philosophy of the spike: rate-based vs. spike-based theories of the brain. Front Syst Neurosci 9:151.

Brette R (2017) Is coding a relevant metaphor for the brain? BioRxiv. https://doi.org/10.1101/168237.

Brodmann K (1909) Vergleichende Lokalisationslehre der Gro -hirnrinde. Leipzig: Barth.

Brooks RA (1991) Intelligence without representation. Artificial Intelligence 47:139–159.

Brooks, AS, Yellen JE, Potts, R, Behrensmeyer AK, Deino, AL, et al. (2018) Long-distance stone transport and pigment use in the earliest Middle Stone Age. Science 360:90–94.

Broome BM, Jayaraman V, Laurent G (2006) Encoding and decoding of overlapping odor sequences. Neuron 51:467–482.

Brotchie PR, Andersen RA, Snyder LH, Goodman SJ (1995) Head position signals used by parietal neurons to encode locations of visual stimuli. Nature 375:232–235.

Brown PL, Jenkins HM (1968) Auto-shaping of the pigeon's key-peck. J Exp Anal Behav 11:1–8.

Brukner C (2014) Quantum causality. Nature Physics 10:259–263.

Bucher D, Marder E (2013) SnapShot: neuromodulation. Cell 155:482.

Buckner RL (2010) The role of the hippocampus in prediction and imagination. Annu Rev Psychol 61:27–48.

Buckner RL, Carroll DC (2007) Self-projection and the brain. Trends Cogn Sci 11:49–57.

Buhl DL, Buzsáki G (2005) Developmental emergence of hippocampal fast-field "ripple" oscillations in the behaving rat pups. Neurosci 134:1423–1430.

Buhusi CV, Meck WH (2005) What makes us tick? Functional and neural mechanisms of interval timing. Nat Rev Neurosci 6:755–765.

Bullmore E, Sporns O (2009) Complex brain networks: graph theoretical analysis of structural and functional systems. Nature Rev Neurosci 10:186–198.

Bullock TH (1970) Operations analysis of nervous functions. In: The Neurosciences; Second Study Program (Schmitt FO, ed), pp. 375–383. New York: Rockefeller University Press.

Buonomano D (2017) Your Brain Is a Time Machine: The Neuroscience and Physics of Time. New York: W. W. Norton and Company.

Buonomano DV, Karmarkar UR (2002) How do we tell time? Neuroscientist 8:42–51.

Buonomano DV, Maass W (2009) State-dependent computations: spatiotemporal processing in cortical networks. Nature Rev Neurosci 10:113–125.

Burak Y, Fiete IR (2012) Fundamental limits on persistent activity in networks of noisy neurons. Proc Natl Acad Sci U S A 109:17645–17650.

Burgess N, O'Keefe J (2011) Models of place and grid cell firing and theta rhythmicity. Curr Opin Neurobiol 21:734–744.
Burr DC, Morrone MC (2010) Vision: keeping the world still when the eyes move. Curr Biol 20:R442–444.
Buzsáki G (1982) The "where is it?" reflex: autoshaping the orienting response. J Exp Anal Behav 37:461–484.
Buzsáki G (1983) Situational conditional reflexes. Physiologic studies of the higher nervous activity of freely moving animals: P. S. Kupalov. Pavlovian J Biol Sci 18:13–21.
Buzsáki G (1989) Two-stage model of memory trace formation: a role for "noisy" brain states. Neuroscience 31:551–570.
Buzsáki G (1996) The hippocampo-neocortical dialogue. Cereb Cortex 6:81–92.
Buzsáki G (1998) Memory consolidation during sleep: a neurophysiological perspective. J Sleep Res 7:17–23.
Buzsáki G (2002) Theta oscillations in the hippocampus. Neuron 33:325–340.
Buzsáki G (2004) Large-scale recording of neuronal ensembles. Nat Neurosci 7:446–451.
Buzsáki G (2005) Theta rhythm of navigation: link between path integration and landmark navigation, episodic and semantic memory. Hippocampus 15:827–840.
Buzsáki G (2006) Rhythms of the Brain. New York: Oxford University Press.
Buzsáki G (2010) Neural syntax: cell assemblies, synapsembles, and readers. Neuron 68:362–385.
Buzsáki G (2015) Neuroscience. Our skewed sense of space. Science 347:612–613.
Buzsáki G (2015) Hippocampal sharp wave-ripple: a cognitive biomarker for episodic memory and planning. Hippocampus 25:1073–1188.
Buzsáki G, Anastassiou CA, Koch C (2012) The origin of extracellular fields and currents: EEG, ECoG, LFP and spikes. Nat Rev Neurosci 13:407–420.
Buzsáki G, Bragin A, Chrobak JJ, Nadasdy Z, Sik A, Hsu M, Ylinen A (1994) Oscillatory and intermittent synchrony in the hippocampus: relevance to memory trace formation. In: Temporal Coding in the Brain (Buzsáki G, Llinás R, Singer W, Berthoz A, Christen Y, eds). Berlin: Springer, pp. 83–96.
Buzsáki G, Buhl DL, Harris KD, Csicsvari J, Czeh B, Morozov A (2003) Hippocampal network patterns of activity in the mouse. Neuroscience 116:201–211.
Buzsáki G, Chrobak JJ (1995) Temporal structure in spatially organized neuronal ensembles: a role for interneuronal networks. Curr Opin Neurobiol 5:504–510.
Buzsáki G, Czopf J, Kondakor I, Bjorklund A, Gage FH (1987) Cellular activity of intracerebrally transplanted fetal hippocampus during behavior. Neuroscience 22:871–883.
Buzsáki G, Draguhn A (2004) Neuronal oscillations in cortical networks. Science 304:1926–1929.
Buzsáki G, Geisler C, Henze DA, Wang XJ (2004) Interneuron diversity series: circuit complexity and axon wiring economy of cortical interneurons. Trends Neurosci 27:186–193.
Buzsáki G, Horvath Z, Urioste R, Hetke J, Wise K (1992) High-frequency network oscillation in the hippocampus. Science 256:1025–1027.
Buzsáki G, Kaila K, Raichle M (2007) Inhibition and brain work. Neuron 56:771–783.

Buzsáki G, Leung LW, Vanderwolf CH (1983) Cellular bases of hippocampal EEG in the behaving rat. Brain Res 287:139–171.

Buzsáki G, Llinás R (2017) Space and time in the brain. Science 358:482–485.

Buzsáki G, Logothetis N, Singer W (2013) Scaling brain size, keeping timing: evolutionary preservation of brain rhythms. Neuron 80:751–764.

Buzsáki G, Mizuseki K (2014) The log-dynamic brain: how skewed distributions affect network operations. Nat Rev Neurosci 15:264–278.

Buzsáki G, Moser EI (2013) Memory, navigation and theta rhythm in the hippocampal-entorhinal system. Nat Neurosci 16:130–138.

Buzsáki G, Stark E, Berényi A, Khodagholy D, Kipke DR, Yoon E, Wise KD (2015) Tools for probing local circuits: high-density silicon probes combined with optogenetics. Neuron 86:92–105.

Buzsáki G, Tingley D (2018) Space and time: the hippocampus as a sequence generator. Trends Cogn Sci 22:853–869.

Buzsáki G, Vanderwolf CH (eds) (1985) Electrical Activity of the Archicortex. Budapest: Akadémiai Kiadó.

Buzsáki G, Wang XJ (2012) Mechanisms of gamma oscillations. Annu Rev Neurosci 35:203–225.

Calcott B (2017) Causal specificity and the instructive–permissive distinction. Biol Philos 32:481–505.

Calder AJ, Young AW, Rowland D, Perrett DI, Hodges JR, Etcoff NL (1966) Facial emotion recognition after bilateral amygdala dmage: differentially severe impairment of fear. Cogn Neuropsychol 13:699–745.

Canales TJ (2015) The Physicist and the Philosopher: Einstein, Bergson, and the Debate That Changed Our Understanding of Time. Princeton,NJ: Princeton University Press.

Cannon W (1927) The James-Lange theory of emotions: a critical examination and an alternative theory. Am J Psychol 39:106–124.

Canolty RT, Edwards E, Dalal SS, Soltani M, Nagarajan SS, et al. (2006) High gamma power is phase-locked to theta oscillations in human neocortex. Science 313:1626–1628.

Canolty RT, Knight RT (2010) The functional role of cross-frequency coupling. Trends Cogn Sci 14:506–515.

Carandini M, Heeger DJ (1994) Summation and division by neurons in primate visual cortex. Science 264:1333–1336.

Carandini M, Heeger DJ (2011) Normalization as a canonical neural computation. Nat Rev Neurosci 13:51–62.

Carpenter RHS (1980) Movements of the Eyes. London: Pion.

Carreiras M, Seghier ML, Baquero S, Estévez A, Lozano A, et al. (2009) An anatomical signature for literacy. Nature 461:983–986.

Carroll JM, Solity J, Shapiro LR (2016) Predicting dyslexia using prereading skills: the role of sensorimotor and cognitive abilities. J. Child Psychol Psychiatry 57:750–758.

Carroll SM (2000) From Eternity to Here: The Quest for the Ultimate Theory of Time. New York: Dutton, Penguin Group.

Catania AC, Cutts D (1963) Experimental control of superstitious responding in humans. J Exp Anal Behav 6:203–208.

Chacron MJ (2007) Electrolocation. Scholarpedia 2:1411.

Chance FS, Abbott LF, Reyes AD (2002) Gain modulation from background synaptic input. Neuron 35:773-782.

Chemero A (2009) Radical Embodied Cognitive Science. Cambridge, MA: MIT Press.

Chen G, Manson D, Cacucci F, Wills TJ (2016) Absence of visual input results in the disruption of grid cell firing in the mouse. Curr Biol 26:2335-2342.

Chen, X., Kebschull, J. M., Zhan, H., Sun, Y.-C., Zador, A. M. (2018). Spatial organization of projection neurons in the mouse auditory cortex identified by in situ barcode sequencing. bioRxiv. https://doi.org/10.1101/294637.

Chiel HJ, Beer RD (1997) The brain has a body: adaptive behavior emerges from interactions of nervous system, body and environment. Trends Neurosci 20:553-557.

Childe VG (1956) Piecing Together the Past: The Interpretation of Archeological Data. London: Routledge and Kegan Paul.

Cho A (2016) Gravitational waves, Einstein's ripples in spacetime, spotted for first time. Science—online. February 11, 2016. https://www.sciencemag.org/news/2016/02/gravitational-waves-einstein-s-ripples-spacetime-spotted-first-time.

Choe Y, Yang H-F, Eng DC-Y (2007) Autonomous learning of the semantics of internal sensory states based on motor exploration. Int J Humanoid Robotics 4:211-243.

Chomsky N (1980) Rules and Representations. New York: Columbia University Press.

Chang SW, Papadimitriou C, Snyder LH (2009) Using a compound gain field to compute a reach plan. Neuron 64:744-755.

Chrobak JJ, Buzsáki G (1998) Gamma oscillations in the entorhinal cortex of the freely behaving rat. J Neurosci 18:388-398.

Church RM (1984) Properties of the internal clock. Ann N Y Acad Sci 423:566-582.

Churchland P (2002) Brain-Wise: Studies in Neurophilosophy. Cambridge, MA: MIT Press.

Churchland PS, Sejnowski TJ (1992) The Computational Brain. Cambridge, MA: MIT Press.

Churchland MM, Cunningham JP, Kaufman MT, Foster JD, Nuyujukian P, et al. (2012) Neural population dynamics during reaching. Nature 487:51-56.

Ciocchi S, Passecker J, Malagon-Vina H, Mikus N, Klausberger T (2015) Brain computation. Selective information routing by ventral hippocampal CA1 projection neurons. Science 348:560-563.

Clark A, Chalmers DJ (1998) The extended mind. Analysis 58:7-19.

Clarke AC (1987) July 20, 2019: Life in the 21st Century. New York: Omni Book.

Clegg B (2016) Are Numbers Real? The Uncanny Relationship of Mathematics and the Physical World. New York: St. Martin's Press.

Cohen N, Eichenbaum H (1993) Memory, Amnesia, andThe Hippocampal System. Cambridg, MA: MIT Press.

Cohen Kadosh R, Gertner L, Terhune DB (2012) Exceptional abilities in the spatial representation of numbers and time: insights from synesthesia. Neuroscientist 18:208-215.

Cohen MR, Maunsell JH (2009) Attention improves performance primarily by reducing interneuronal correlations. Nat Neurosci 12:1594-1600.

Colgin LL, Denninger T, Fyhn M, Hafting T, Bonnevie T, et al. (2009) Frequency of gamma oscillations routes flow of information in the hippocampus. Nature 462:353–357.

Collinger JL, Wodlinger B, Downey JE, Wang W, Tyler-Kabara EC, et al. (2013) High-performance neuroprosthetic control by an individual with tetraplegia. Lancet 381:557–564.

Collinwood RG (1946) The Idea of History. Oxford: Clarendon Press.

Connor CE, Knierim JJ (2017) Integration of objects and space in perception and memory. Nat Neurosci 20:1493–1503.

Constantinescu AO, O'Reilly JX, Behrens TEJ (2016) Organizing conceptual knowledge in humans with a gridlike code. Science 352:1464–1468.

Constantinidis C, Goldman-Rakic PS (2002) Correlated discharges among putative pyramidal neurons and interneurons in the primate prefrontal cortex. J Neurophysiol 88:3487–3497.

Constantino JN, Kennon-McGill S, Weichselbaum C, Marrus N, Haider A, et al. (2017) Infant viewing of social scenes is under genetic control and is atypical in autism. Nature 547:340–344.

Conway JH, Guy RK (1996) The Book of Numbers. Gottenburg, GER: Copernicus.

Cooke M, Hershey JR, Rennie SJ (2010) Monaural speech separation and recognition challenge. Comput Speech Lang 24:1–15.

Cooper BG, Mizumori SJY (1999) Retrosplenial cortex inactivation selectively impairs navigation in darkness. Neuroreport 10:625–630.

Corkin S (2013) Permanent Present Tense: The Unforgettable Life of the Amnesic Patient, H.M. New York: Basic Books.

Cowey A (2010) Visual system: how does blindsight arise? Curr Biol 20:1–3.

Cowey A, Stoerig P (1995) Blindsight in monkeys. Nature 373:247–249.

Craik FIM, Tulving E (1975) Depth of processing and the retention of words in episodic memory. J Exp Psychol: General 104:268–294.

Crapse TB, Sommer MA (2008) Corollary discharge across the animal kingdom. Nat Rev Neurosci 9:587–600.

Crapse TB, Sommer MA (2012) Frontal eye field neurons assess visual stability across saccades. J Neurosci 32:2835–2845.

Cromwell HC, Berridge KC (1996) Implementation of action sequences by a neostriatal site: a lesion mapping study of grooming syntax. J Neurosci 16:3444–3458.

Crow EL, Shimizu K (1988) Log-normal Distributions: Theory and Applications. New York: Dekker.

Crowcroft P (1966) Mice All Over. London: G. T. Foulis and Co.

Csicsvari J, Hirase H, Czurkó A, Buzsáki G (1998) Reliability and state dependence of pyramidal cell-interneuron synapses in the hippocampus: an ensemble approach in the behaving rat. Neuron 21:179–189.

Csicsvari J, Hirase H, Czurkó A, Mamiya A, Buzsáki G (1999) Oscillatory coupling of hippocampal pyramidal cells and interneurons in the behaving rat. J Neurosci 19:274–287.

Csicsvari J, Hirase H, Mamiya A, Buzsáki G (2000) Ensemble patterns of hippocampal CA3-CA1 neurons during sharp wave-associated population events. Neuron 28:585–594.

Csicsvari J, Jamieson B, Wise KD, Buzsáki G (2003) Mechanisms of gamma oscillations in the hippocampus of the behaving rat. Neuron 37:311–322.

Czurkó A, Hirase H, Csicsvari J, Buzsáki G (1999) Sustained activation of hippocampal pyramidal cells by 'space clamping' in a running wheel. Eur J Neurosci 11:344–352.

Dalai Lama, Chodron T (2017) Approaching the Buddhist Path. Somerville, MA: Wisdom Publications.

Damasio AR (1989) Time-locked multiregional retroactivation: a systemslevel proposal for the neural substrates of recall and recognition. Cognition 33:25–62.

Damasio AR (1994) Descartes' Error: Emotion, Reason and the Human Brain. New York: Grosset/Putnam.

Damasio AR (1995) Toward a neurobiology of emotion and feeling: operational concepts and hypotheses. The Neuroscientist 1:19–25.

Damasio H, Grabowski TJ, Tranel D, Hichwa RD, Damasio AR (1996) A neural basis for lexical retrieval. Nature 380:499–505.

Daniel R, Rubens JR, Sarpeshkar R, Lu TK (2013) Synthetic analog computation in living cells. Nature 497:619–623.

Das NN, Gastaut H (1955) Variations de l'activite electrique du cerveau, du coeur et des muscles squellettiques au cours de la meditation et de l'extase yogique. Electroencephal Clin Neurophysiol 6:211–219.

Davidson TJ, Kloosterman F, Wilson MA (2009) Hippocampal replay of extended experience. Neuron 63:497–507.

Dayan P, Hinton GE, Neal RM, Zemel RS (1995) The Helmholtz machine. Neural Computation 7:889–904.

Deco G, Rolls ET (2006) Decision-making and Weber's law: a neurophysiological model. Eur J Neurosci 24:901–916.

Dede AJ, Frascino JC, Wixted JT, Squire LR (2016) Learning and remembering real-world events after medial temporal lobe damage. Proc Natl Acad Sci U S A 113:13480–13485.

DeFelipe J, Lopez-Cruz PL, Benavides-Piccione R, Bielza C, Larranaga P, et al. (2013) New insights into the classification and nomenclature of cortical GABAergic interneurons. Nat Rev Neurosci 14:202–216.

Dehaene S (1997) The Number Sense: How the Mind Creates Mathematics. New York: Penguin Group.

Dehaene S, Changeux JP (2011) Experimental and theoretical approaches to conscious processing. Neuron 70:200–227.

Dehaene S, Dehaene-Lambertz G, Cohen L (1998) Abstract representations of numbers in the animal and human brain. Trends Neurosci 21:355–361.

Dehaene S, Izard V, Spelke E and Pica P (2008) Log or linear? Distinct intuitions of the number scale in Western and Amazonian indigene cultures. Science 320:1217–1220.

Dehaene S, Pegado F, Braga LW, Ventura P, Filho GN, et al. (2010) How learning to read changes the cortical networks for vision and language. Science 330:1359–1364.

Dehaene S, Spelke E, Pinel P, Stanescu R, Tsivkin S (1999) Sources of mathematical thinking: behavioral and brain-imaging evidence. Science 284:970–974.

Deininger K, Squire L (1996) A new data set measuring income inequality. World Bank Econ Rev 10:565–591.

Delage Y (1919) Le Réve. Etude psychologique, philosophique et litteraire. Paris: Presses Universitaires de France.

de Lavilléon G, Lacroix MM, Rondi-Reig L, Benchenane K (2015) Explicit memory creation during sleep demonstrates a causal role of place cells in navigation. Nat Neurosci 18:493–495.

DeLong, AJ (1981) Phenomenological spacetime: toward an experiential relativity. Science 213:681–683.

Demarse TB, Wagenaar DA, Blau AW, Potter SM (2001) The neurally controlled animat: biological brains acting with simulated bodies. Autonomous Robots 11:305–310.

Dennett DC (1991) Consciousness Explained. Boston, MA: Little, Brown & Co.

Dennett DC, Kinsbourne M (1992) Time and the observer. Behavi Brain Sci 15:183–247.

Descartes R (1984) The Philosophical Writings of Descartes. J. Cottingham, D. Murdoch and R. Stoothoff (eds) 2 volumes. Cambridge: Cambridge University Press.

Desimone R (1991) Face-selective cells in the temporal cortex of monkeys. J Cogn Neurosci 3:1–8.

Destexhe A, Rudolph M, Paré D (2003) The high-conductance state of neocortical neurons in vivo. Nat Rev Neurosci 4:739–751.

Dethier VG (1987) Sniff, flick, and pulse: an appreciation of interruption. Proc Amer Philos Soc 131:159–176.

Devlin K (2000) The Math Gene: How Mathematical Thinking Evolved and Why Numbers Are Like Gossip. New York: Basic Books.

De Volder AG, Catalan-Ahumada M, Robert A, Bol A, Labar D, et al. (1999) Changes in occipital cortex activity in early blind humans using a sensory substitution device. Brain Res 826:128–34.

Diekelmann S, Born J (2010) The memory function of sleep. Nat Rev Neurosci 11:114–126.

Di Pellegrino G, Fadiga L, Fogassi L, Gallese V, Rizzolatti G (1992) Understanding motor events: a neurophysiological study. Exp Brain Res 91:176–180.

Diaconis P, Mostelle F (1989) Methods for studying coincidences. J Am Statist Assoc 84:853–861.

Diba K, Buzsáki G (2007) Forward and reverse hippocampal place-cell sequences during ripples. Nat Neurosci 10:1241–1242.

Diba K, Buzsáki G (2008) Hippocampal network dynamics constrain the time lag between pyramidal cells across modified environments. J Neurosci 28:13448–13456.

Ding N, Simon JZ (2012) Neural coding of continuous speech in auditory cortex during monaural and dichotic listening. J Neurophysiol 107:78–89.

Ditchburn RW, Ginsborg BL (1952) Vision with a stabilized retinal image. Nature 170:36–37.

Dohrn-van Rossum G (1996). History of the Hour. Chicago, IL: University of Chicago Press.

Dorris MC, Paré M, Munoz DP (1997) Neuronal activity in monkey superior colliculus related to the initiation of saccadic eye movements. J Neurosci 17:8566–8579.

Downes JJ, Mayes AR, MacDonald C, Hunkin NM (2002) Temporal order memory in patients with Korsakoff's syndrome and medial temporal amnesia. Neuropsychologia 40:853–861.

Doya K (2007) Bayesian Brain: Probabilistic Approaches to Neural Coding. Cambridge, MA: MIT Press.

Dragoi G, Buzsáki G (2006) Temporal encoding of place sequences by hippocampal cell assemblies. Neuron 50:145–157.

Dragoi G, Harris KD, Buzsáki G (2003) Place representation within hippocampal networks is modified by long-term potentiation. Neuron 39:843–853.

Dragoi G, Tonegawa S (2011) Preplay of future place cell sequences by hippocampal cellular assemblies. Nature 469:397–401.

Dragoi G, Tonegawa S (2013a) Selection of preconfigured cell assemblies for representation of novel spatial experiences. Philos Trans R Soc Lond B Biol Sci 369:20120522.

Dragoi G, Tonegawa S (2013b) Distinct preplay of multiple novel spatial experiences in the rat. Proc Natl Acad Sci U S A 110:9100–9105.

DuBrow S, Davachi L (2013) The influence of context boundaries on memory for the sequential order of events. J Exp Psychol: General 142:1277–1286.

DuBrow S, Davachi L (2016) Temporal binding within and across events. Neurobiol Learning Memory 134:107–114.

Duhamel JR, Bremmer F, BenHamed S, Graf W (1997) Spatial invariance of visual receptive fields in parietal cortex neurons. Nature 389:845–848.

Duhamel JR, Colby CL, Goldberg ME (1992) The updating of the representation of visual space in parietal cortex by intended eye movements. Science 255:90–92.

Dupret D, O'Neill J, Csicsvari J (2013) Dynamic reconfiguration of hippocampal interneuron circuits during spatial learning. Neuron 7:166–180.

Dupret D, O'Neill J, Pleydell-Bouverie B, Csicsvari J (2010) The reorganization and reactivation of hippocampal maps predict spatial memory performance. Nat Neurosci 13:995–1002.

Durrant-Whyte H, Bailey T (2006) Simultaneous localization and mapping: part I. IEEE Robotics Automation Mag 13:99–110.

Dusek JA, Eichenbaum H (1997) The hippocampus and memory for orderly stimulus relations. Proc Natl Acad Sci U S A 94:7109–7114.

Eagleman DM, Tse PU, Buonomano D, Janssen P, Nobre AC, Holcombe AO (2004) Time and the brain: how subjective time relates to neural time. J Neurosci 25:10369–10371.

Ecker AS, Berens P, Keliris GA, Bethge M, Logothetis NK, Tolias AS (2010) Decorrelated neuronal firing in cortical microcircuits. Science 327:584–587.

Economo MN, Clack NG, Lavis LD, Gerfen CR, Svoboda K, et al. (2016). A platform for brainwide imaging and reconstruction of individual neurons. eLife, 5:e10566.

Eddington AS (1928) The Nature of the Physical World. New York: Cambridge University Press.

Edelman GM (1987) Neural Darwinism: The Theory of Neuronal Group Selection. New York: Basic Books.

Eggermont, JJ (2007) Correlated neural activity as the driving force for functional changes in auditory cortex. Hear Res 229:69–80.

Ego-Stengel V, Wilson MA (2010) Disruption of ripple-associated hippocampal activity during rest impairs spatial learning in the rat. Hippocampus 20:1–10.

Ehrsson HH, Holmes NP, Passingham RE (2005) Touching a rubber hand: feeling of body ownership is associated with activity in multisensory brain areas. J Neurosci 25:10564–10573.

Eichenbaum H (2000) A cortical-hippocampal system for declarative memory. Nat Rev Neurosci 1:41–50.

Eichenbaum H (2014) Time cells in the hippocampus: a new dimension for mapping memories. Nat Rev Neurosci 15:732–744.

Eichenbaum H, Cohen NJ (2014) Can we reconcile the declarative memory and spatial navigation views on hippocampal function? Neuron 83:764–770.

Eichenbaum H, Dudchenko P, Wood E, Shapiro M, Tanila H (1999) The hippocampus, memory, and place cells: is it spatial memory or a memory space? Neuron 23:209–226.

Einevoll GT, Kayser C, Logothetis NK, Panzeri S (2013) Modelling and analysis of local field potentials for studying the function of cortical circuits. Nature Rev Neurosci 14:770–785.

Einstein A (1989). The Collected Papers of Albert Einstein, Volume 2: The Swiss Years: Writings, 1900–1909 (English translation supplement; translated by Anna Beck, with Peter Havas, consultant ed.). Princeton, NJ: Princeton University Press.

Einstein A (1997). The Collected Papers of Albert Einstein, Volume 6: The Berlin Years: Writings, 1914–1917 (English translation supplement; translated by Alfred Engel, with Engelbert Schucking, consultant ed.). Princeton, NJ: Princeton University Press.

Einstein A, Infeld L (1938/1966) The Evolution of Physics: From Early Concepts to Relativity and Quanta. New York: Simon and Schuster.

El-Bizri N (2000) The Phenomenological Quest Between Avicenna and Heidegger. Bristol, UK: Global Academic Publishing.

Eldar E, Cohen JD, Niv Y (2013) The effects of neuronal gain on attention and learning. Nat Neurosci 16:1146–1153.

Eliades SJ, Wang X-J (2008) Neural substrates of vocalization feedback monitoring in primate auditory cortex. Nature 453:1102–1106.

Eliasmith C, Anderson CH (2003) Neural Engineering Computation, Representation, and Dynamics in Neurobiological Systems. Cambridge, MA: MIT Press.

Ellender TJ, Nissen W, Colgin LL, Mann EO, Paulsen O (2010) Priming of hippocampal population bursts by individual perisomatic-targeting interneurons. J Neurosci 30:5979–5991.

Elsinger CL, Rao SM, Zimbelman JL, Reynolds NC, Blindauer KA, Hoffmann RG (2003) Neural basis for impaired time reproduction in Parkinson's disease: an fMRI study. J Int Neuropsychol Soc 9:1088–1098.

Emerson RW (1899) The Conduct of Life: Emmerson's complete works. New York: T. Y. Crowell & Company.

Engel AK, Fries P, Singer W (2001) Dynamic predictions: oscillations and synchrony in top-down processing. Nat Rev Neurosci 2:704–716.

English DF, McKenzie S, Evans T, Kim K, Yoon E, Buzsáki G (2017) Pyramidal cell-interneuron circuit architecture and dynamics in hippocampal networks. Neuron 96:505–520.

English DF, Peyrache A, Stark E, Roux L, Vallentin D, Long MA, Buzsáki G (2014) Excitation and inhibition compete to control spiking during hippocampal ripples: intracellular study in behaving mice. J Neurosci 34:16509–16517.

Ercsey-Ravasz M, Markov NT, Lamy C, Van Essen DC, Knoblauch K, et al. (2013) A predictive network model of cerebral cortical connectivity based on a distance rule. Neuron 80:184–197.

Ermentrout GB, Kleinfeld D (2001) Traveling electrical waves in cortex: insights from phase dynamics and speculation on a computational role. Neuron 29:33–44.

Etchamendy N, Desmedt A, Cortes-Torrea C, Marighetto A, Jaffard R (2003) Hippocampal lesions and discrimination performance of mice in the radial maze: sparing or impairment depending on the representational demands of the task. Hippocampus 13:197–211.

Etienne AS, Jeffery KJ (2004) Path integration in mammals. Hippocampus 14:180–192.

Evarts EV (1964) Temporal patterns of discharge of pyramidal tract neurons during sleep and waking in the monkey. J Neurophysiol 27:152–171.

Evarts EV (1973) Brain mechanisms in movement. Sci Am 229:96–103.

Fadiga L, Fogassi L, Pavesi G, Rizzolatti G (1995) Motor facilitation during action observation: a magnetic stimulation study. J Neurophysiol 73:2608–2611.

Falk JL, Bindra D (1954) Judgment of time as a function of serial position and stress. J Exp Psychol 47:279–282.

Fechner GT (1860/1966) Howes DH, Boring EG, eds. Elements of Psychophysics [Elemente der Psychophysik]. volume 1. Translated by Adler HE. New York: Rinehart and Winston.

Fee MS, Kozhevnikov AA, Hahnloser RH (2004) Neural mechanisms of vocal sequence generation in the songbird. Ann N Y Acad Sci 1016:153–170.

Feinberg I (1974) Changes in sleep cycle patterns with age. J Psychiatr Res 10:283–306.

Feinberg I (1978) Efference copy and corollary discharge: implications for thinking and its disorders. Schizophr Bull 4:636–640.

Ferbinteanu J, Shapiro ML (2003) Prospective and retrospective memory coding in the hippocampus. Neuron 40:1227–1239.

Ferbinteanu J, Kennedy PJ, Shapiro ML (2006) Episodic memory: from brain to mind. Hippocampus 16:691–703.

Ferezou I, Haiss F, Gentet LJ, Aronoff R, Weber B, Petersen CC (2007). Spatiotemporal dynamics of cortical sensorimotor integration in behaving mice. Neuron 56:907–923.

Fernández-Ruiz A, Oliva A, Nagy GA, Maurer AP, Berényi A, Buzsáki G (2017) Entorhinal-CA3 dual-Input control of spike timing in the hippocampus by theta-gamma coupling. Neuron 93:1213–1226.

Feynman R (1965) The Character of Physical Law. Cambridge, MA: MIT Press.

Feynman RP, Leighton RB, Sand Matts (1963) The Feynman Lectures on Physics. Reading, MA: Addison-Wesley Co.

Fields C (2014) Equivalence of the symbol grounding and quantum system identification problems. Information 5:172–189.

Finnerty GT, Shadlen MN, Jazayeri M, Nobre AC, Buonomano DV (2015) Time in cortical circuits. J Neurosci 35:13912–13916.

Fischer B, Ramsperger E (1984) Human express saccades: extremely short reaction times of goal directed eye movements. Exp Brain Res 57:191–195.

Fiser J, Chiu CY, Weliky M (2004) Small modulation of ongoing cortical dynamics by sensory input during natural vision. Nature 431:573–578.

Fisher SE, Scharff C (2009) FOXP2 as a molecular window into speech and language. Trends Genet 25:166–177.

Fitch WT (2016) Sound and meaning in the world's languages. Nature 539:39–40.

Flash T, Hogan H (1985) The coordination of arm movements: an experimentally confirmed mathematical model. J Neurosci 5:1688–1703.

Fogassi L, Gallese V, di Pellegrino G, Fadiga L, Gentilucci M, et al. (1992) Space coding by premotor cortex. Exp Brain Res 89:686–690.

Ford JM, Mathalon DH (2004) Electrophysiological evidence of corollary discharge dysfunction in schizophrenia during talking and thinking. J Psychiatr Res 38:37–46.

Fortin NJ, Agster KL, Eichenbaum HB (2002) Critical role of the hippocampus in memory for sequences of events. Nat Neurosci 5:458–462.

Foster DJ (2017) Replay comes of age. Annu Rev Neuorsci 40:581–602.

Foster DJ, Wilson MA (2006) Reverse replay of behavioural sequences in hippocampal place cells during the awake state. Nature 440:680–683.

Frank LM, Brown EN, Wilson M (2000) Trajectory encoding in the hippocampus and entorhinal cortex. Neuron 27:169–178.

Frank LM, Stanley GB, Brown EN (2004) Hippocampal plasticity across multiple days of exposure to novel environments. J Neurosci 24:7681–7689.

Frankland PW, Bontempi B (2005) The organization of recent and remote memories. Nat Rev Neurosci 6:119–130.

Freeman D (1997) Imagined Worlds. Cambridge, MA: Harvard University Press.

Freeman WJ (1999) How Brains Make Up Their Minds. New York: Columbia University Press.

Frégnac Y, Carelli PV, Pananceau M and Monier C (2010) Stimulus-driven coordination of subcortical cell assemblies an propagation of Gestalt belief in V1. In: Dynamic Coordination in the Brain: from Neurons to Mind (von der Malsburg C, Phillips WA, Singer W, eds). Cambridge, MA: MIT Press.

Freund TF, Buzsáki G (1996) Interneurons of the hippocampus. Hippocampus 6:347–470.

Friederici AD, Singer W (2015) Grounding language processing on basic neurophysiological principles. Trends Cogn Sci 19:329–338.

Fries P (2005) A mechanism for cognitive dynamics: neuronal communication through neuronal coherence. Trends Cogn Sci 9:474–480.

Fries P, Reynolds JH, Rorie AE, Desimone R (2001) Modulation of oscillatory neuronal synchronization by selective visual attention. Science 291:1560–1563.

Friston K (2010) The free-energy principle: a unified brain theory? Nat Rev Neurosci 11:127–138.

Friston K (2012) Prediction, perception and agency. Int Psychophysiol 83:248–252.

Friston K, Buzsáki G (2016) The functional anatomy of time: what and when in the brain. Trends Cogn Sci 20:500–511.

Friston K, Kiebel S (2009) Predictive coding under the free-energy principle. Philos Trans R Soc Lond B Biol Sci 364:1211–1221.

Friston K, Moran R, Seth AK (2012) Analysing connectivity with Granger causality and dynamic causal modelling. Curr Opin Neurobiol 23:172–178.

Fu TM, Hong G, Zhou T, Schuhmann TG, Viveros RD, Lieber CM (2016) Stable long-term chronic brain mapping at the single-neuron level. Nat Methods 13:875–882.

Fu Y, Tucciarone JM, Espinosa JS, Sheng N, Darcy DP, et al. (2014) A cortical circuit for gain control by behavioral state. Cell 156:1139–1152.

Fujisawa S, Amarasingham A, Harrison MT, Buzsáki G (2008) Behavior-dependent short-term assembly dynamics in the medial prefrontal cortex. Nat Neurosci 11:823–833.

Funahashi S, Bruce CJ, Goldman-Rakic PS (1989) Mnemonic coding of visual space in the monkey's dorsolateral prefrontal cortex. J Neurophysiol 61:331–349.

Fusi S, Abbott LF (2007) Limits on the memory storage capacity of bounded synapses. Nat Neurosci 2007 Apr;10(4):485–493.

Fusi S, Asaad WF, Miller EK, Wang XJ (2007) A neural circuit model of flexible sensorimotor mapping: learning and forgetting on multiple timescales. Neuron 54:319–333.

Fuster JM (1995) Temporal processing. Ann NY Acad Sci769:173–181.

Fuster JM (2004) Upper processing stages of the perception-action cycle. Trends Cogn Sci 8:143–145.

Fuster JM, Alexander GE (1971) Neuron activity related to short-term memory. Science 173:652–654.

Gabbott PL, Warner TA, Jays PR, Salway P, Busby SJ (2005) Prefrontal cortex in the rat: projections to subcortical autonomic, motor, and limbic centers. J Comp Neurol 492:145–177.

Gadagkar V, Puzerey PA, Chen R, Baird-Daniel E, Farhang AR, Goldberg JH (2016) Dopamine neurons encode performance error in singing birds. Science 354:1278–1282.

Galilei G (1623/1954) Il Saggitore (The Assayer). English translation, Danto AC. Introduction to Contemporary Civilization in the West (2nd ed). New York: Columbia University Press, 1954, vol. I, p. 721.

Gallese V, Fadiga L, Fogassi L, Rizzolatti G (1996) Action recognition in the premotor cortex. Brain 119:593–609.

Gallistel CR (1990) The Organization of Learning. Cambridge, MA: MIT Press.

Gallistel CR, Gelman R (1992) Preverbal and verbal counting and computation. Cognition 44:43–74.

Gallistel CR, Gibbon J (2000) Time, rate, and conditioning. Psychol Rev 107:289–344.

Galton F (1879) The geometric mean, in vital and social statistics. Proc Roy Soc 29:365–367.

Gao P, Ganguli S (2015) On simplicity and complexity in the brave new world of large-scale neuroscience. Curr Op Neurobiology 32:148–155.

Gardner H (1999) Intelligence Reframed: Multiple Intelligences for the 21th Century. New York: Basic Books.

Geary DC (1994) Children's Mathematical Development. Washington DC: American Psychological Association.

Geisler C, Robbe D, Zugaro M, Sirota A, Buzsáki G (2007) Hippocampal place cell assemblies are speed-controlled oscillators. Proc Natl Acad Sci USA 104:8149–8154.

Geisler C, Diba K, Pastalkova E, Mizuseki K, Royer S, Buzsáki G (2010) Temporal delays among place cells determine the frequency of population theta oscillations in the hippocampus. Proc Natl Acad Sci U S A 107:7957–7962.

Gelbard-Sagiv H, Mukamel R, Harel M, Malach R, Fried I. (2008) Internally generated reactivation of single neurons in human hippocampus during free recall. Science 322:96–101.

Gelinas JN, Khodagholy D, Thesen T, Devinsky O, Buzsáki G (2016) Interictal epileptiform discharges induce hippocampal-cortical coupling in temporal lobe epilepsy. Nat Med 22:641–648.

Geller U (1996) Uri Geller's Mindpower Kit. New York: Penguin Books.

Georgopoulos AP, Lurito JT, Petrides M, Schwartz AB, Massey JT (1989) Mental rotation of the neuronal population vector. Science 243:234–236.

Georgopoulos AP, Schwartz AB, Kettner RE (1986) Neuronal population coding of movement direction. Science 233:1416–1419.

Gershman SJ (2017) Predicting the past, remembering the future. Curr Opin Behav Sci 17:7–13.

Giannitrapani D (1966) Electroencephalographic differences between resting and mental multiplication. Percept Motor Skills 22:399–405.

Gibbon J (1977) Scalar expectancy-theory and weber's law in animal timing. Psychol Rev 84:279–325.

Gibbon J, Church RM, Meck WH (1984) scalar timing in memory. Ann N Y Acad Sci 423:52–77.

Gibbon J, Malapani C, Dale CL, Gallistel C (1997) Toward a neurobiology of temporal cognition: advances and challenges. Curr Opin Neurobiol 7:170–184.

Gibson J (1977) The theory of affordances. In: Perceiving, Acting, and Knowing: Toward and Ecological Psychology (Shaw R, Brandsford J, eds), pp. 62–82. Hillsdale, NJ: Lawrence Erlbaum Associates.

Gibson JJ (1979) The Ecological Approach to Visual Perception. Boston, MA: Houghton Mifflin.

Gielow MR, Zaborszky L (2017) The input-output relationship of the cholinergic basal forebrain. Cell Rep 181817–1830.

Gilboa A, Winocur G, Rosenbaum RS, Poreh A, Gao F, et al. (2006) Hippocampal contributions to recollection in retrograde and anterograde amnesia. Hippocampus 16:966–980.

Gilchrist ID, Brown V, Findlay JM (1997) Saccades without eye movements. Nature 390:130–131.

Gilzenrat MS, Nieuwenhuis S, Jepma M, Cohen JD (2010) Pupil diameter tracks changes in control state predicted by the adaptive gain theory of locus coeruleus function. Cogn Affect Behav Neurosci 10:252–269.

Girardeau G, Benchenane K, Wiener SI, Buzsáki G, Zugaro MB (2009) Selective suppression of hippocampal ripples impairs spatial memory. Nat Neurosci 12:1222–1223.

Giraud AL, Poeppel D (2012) Cortical oscillations and speech processing: emerging computational principles and operations. Nat Neurosci 15:511–517.

Giurgea C (1974) The creative world of P.S. Kupalov. Pavlov J Biol Sci 9:192–207.

Glasauer S, Schneider E, Grasso R, Ivanenko YP (2007) Spacetime relativity in self-motion reproduction. J Neurophysiol 97:451–461.

Glasser MF, Coalson TS, Robinson EC, Hacker CD, Harwell J, et al. (2016) A multimodal parcellation of human cerebral cortex. Nature 536:171–178.

Glimcher PW, Camerer C, Poldrack PA, Fehr E (2008) Neuroeconomics: Decision Making and the Brain. Cambridge, MA: Academic Press.

Glomm G, Ravikumar B (1992) Public versus private investment in human capital: endogenous growth and income inequality. J Polit Econ 100:818–834.

Gold JI, Shadlen MN (2000) Representation of a perceptual decision in developing oculomotor commands. Nature 404:390–394.

Gold JI, Shadlen MN (2007) The neural basis of decision making. Annu Rev Neurosci 30:535–574.

Goldman-Rakic PS, Funahashi S, Bruce CJ (1990) Neocortical memory circuits. Cold Spring Harb Symp Quant Biol 55:1025–1038.

Golub MD, Sadtler PT, Oby ER, Quick KM, Ryu SI, et al. (2018) Learning by neural reassociation. Nat Neurosci 21:607–616.

González-Forero M, Gardner A (2018) Inference of ecological and social drivers of human brain-size evolution Nature 557:554–557.

Goodale M, Milner A (1990) Separate visual pathways for perception and action. Trends Neurosci 15:20–25.

Goodale MA, Pelisson D, Prablanc C (1986) Large adjustments in visually guided reaching do not depend on vision of the hand or perception of target displacement. Nature 320:748–750.

Goodrich BG (2010) We do, therefore we think: time, motility, and consciousness. Rev Neurosci 21:331–361.

Google Inc. (2012). Google self-driving car project. http://googleblog.blogspot.com.

Gothard KM, Hoffman KL, Battaglia FP, McNaughton BL (2001) Dentate gyrus and CA1 ensemble activity during spatial reference frame shifts in the presence and absence of visual input. J Neurosci 21:7284–7292.

Gottlieb A (2009) A Nervous Splendor: The Wittgenstein Family Had a Genius for Misery. New York: The New Yorker. April issue.

Gould JL (1986) The locale map of honey bees: do insects have cognitive maps? Science 232:861–863.

Graf P, Schacter DL (1985) Implicit and explicit memory for new associations in normal and amnesic subjects. J Exp Psychol Learn Mem Cogn 11:501–518.

Granger CWJ (1969) Investigating causal relations by econometric models and cross-spectral methods. Econometria 37:424–438.

Grastyán E, Vereczkei L (1974) Effects of spatial separation of the conditioned signal from the reinforcement: a demonstration of the conditioned character of the orienting response or the orientational character of conditioning. Behav Biol 10:121–146.

Gray CM, König P, Engel A, Singer W (1989) Oscillatory responses in cat visual cortex exhibit inter-columnar synchronization which reflects global stimulus properties. Nature 338:334–337.

Graziano MSA (2013) Consciousness and the Social Brain. New York: Oxford University Press.

Graziano MSA, Hu TX, Gross CG (1997) Visuospatial properties of ventral premotor cortex. J Neurophysiol 77:2268–2292.

Graziano MSA, Yap GS, Gross CG (1994) Coding of visual space by premotor neurons. Science 266:1054–1057.

Greene B (2011) The Hidden Reality: Parallel Universes and the Deep Laws of the Cosmos. New York: Random House.

Greenfield P (1991) Language, tools and brain: the ontogeny of phylogeny of hierarchically organized sequential behavior. Behav Brain Sci 14:531–595.

Gregg J (2013) Are Dolphins Really Smart? The Mammal Behind the Myth. Oxford: Oxford University Press.

Gregory RL (1980) Perceptions as hypotheses. Philos Trans R Soc Lond B Biol Sci 290:181–197.

Grillner S (2006) Biological pattern generation: the cellular and computational logic of networks in motion. Neuron 52:751–766.

Groh A, Bokor H, Mease RA, Plattner VM, Hangya B, et al. (2014) Convergence of cortical and sensory driver inputs on single thalamocortical cells. Cereb Cortex 24:3167–3179.

Grosmark AD, Buzsáki G (2016) Diversity in neural firing dynamics supports both rigid and learned hippocampal sequences. Science 351:1440–1443.

Grosmark AD, Mizuseki K, Pastalkova E, Diba K, Buzsáki G (2012) REM sleep reorganizes hippocampal excitability. Neuron 75:1001–1007.

Gross CG, Graziano MSA (1995) Multiple representations of space in the brain. Neuroscientist 1:43–50.

Groth BHA (1914) The golden mean in the inheritance of size. Science 39:581–584.

Grush R (2004) The emulation theory of representation: motor control, imagery, and perception. Behav Brain Sci 27:377–442.

Gulyás AI, Miles R, Sík A, Tóth K, Tamamaki N, Freund TF (1993) Hippocampal pyramidal cells excite inhibitory neurons through a single release site. Nature 366:683–687.

Guo ZV, Inagaki HK, Daie K, Druckmann S, Gerfen CR, Svoboda K (2017) Maintenance of persistent activity in a frontal thalamocortical loop. Nature 545:181–186.

Gupta AS, van der Meer MA, Touretzky DS, Redish AD (2010) Hippocampal replay is not a simple function of experience. Neuron 65:695–705.

Guth A (1997) The Inflationary Universe. New York: Perseus Books Group.

Gutnisky DA, Dragoi V (2008) Adaptive coding of visual information in neural populations. Nature 452:220–224.

Hafting T, Fyhn M, Molden S, Moser MB, Moser EI (2005) Microstructure of a spatial map in the entorhinal cortex. Nature 436:801–806.

Hagoort P (2005) On Broca, brain, and binding: a new framework. Trends Cogn Sci 9:416–423.

Hahnloser RH, Kozhevnikov AA, Fee MS (2002) An ultra-sparse code underlies the generation of neural sequences in a songbird. Nature 419:65–70.

Haken H (1984) The Science of Structure: Synergetics. New York: Van Nostrand Reinhold.

Halpern BP (1983) Tasting and smelling as active, exploratory sensory processes. Am J Otolaryngol 4:246–249.

Hamad S (1990) The symbol grounding problem. Physica D 42:335–346.

Hämäläinen M, Hari R, Ilmoniemi RJ, Knuutila J, Lounasmaa OV (1993) Magnetoencephalography: theory, instrumentation, and applications to noninvasive studies of the working human brain. Rev Mod Phys 65:413–497.

Han Y, Kebschull JM, Campbell RAA, Cowan D, Imhof F, et al. (2018) The logic of single-cell projections from visual cortex. Nature 556:51–56.

Hannah R (2009a) Time in Antiquity. London: Routledge Press.

Hannah R (2009b) Timekeeping. In: The Oxford Handbook of Engineering and Technology of the Classical World (Oleson JP, ed), pp. 740–7158. Oxford: Oxford University Press.

Harari YV (2017) Homo Deus: A Brief History of Tomorrow. New York: Harper Publishing.

Hardcastle K, Maheswaranathan N, Ganguli S, Giocomo LM (2017) A multiplexed, heterogeneous, and adaptive code for navigation in medial entorhinal cortex. Neuron 94:375–387.

Hardy L (2007) Towards quantum gravity: a framework for probabilistic theories with non-fixed causal structure. J Phys A 40:3081–3099.

Harnad S (1990) The symbol grounding Problem. Physica D 42:335–346.

Harris JA, Mihalas S, Hirokawa, KE, Zeng H (2018) The organization of intracortical connections by layer and cell class in the mouse brain. bioRxiv. https://doi.org/10.1101/292961

Harris KD (2005) Neural signatures of cell assembly organization. Nat Rev Neurosci 6:399–407.

Harris KD, Csicsvari J, Hirase H, Dragoi G, Buzsáki G (2003) Organization of cell assemblies in the hippocampus. Nature 424:552–556.

Harris-Warrick R, Marder E, Selverston AI, Moulins M (1992) Dynamic Biological Networks: The Stomatogastric Nervous System. Cambridge, MA: MIT Press.

Harvey CD, Collman F, Dombeck DA, Tank DW (2012) Choice-specific sequences in parietal cortex during a virtual-navigation decision task. Nature 484:62–68.

Hassabis D, Kumaran D, Summerfield C, Botvinick M (2017) Neuroscience-inspired artificial intelligence. Neuron 95:245–258.

Hassabis D, Kumaran D, Vann SD, Maguire EA (2007) Patients with hippocampal amnesia cannot imagine new experiences. Proc Natl Acad Sci U S A 104:1726–1731.

Hassabis E, Maguire EA (2007) Deconstructing episodic memory with construction. Trends Cogn Sci 7:299–306.

Hasselmo ME (2012) How We Remember: Brain Mechanisms of Episodic Memory. Cambridge, MA: MIT Press.

Hasselmo ME, Stern CE (2015) Current questions on space and time encoding. Hippocampus 25:744–752.

Hatsopoulos NG. Suminski AJ (2011) Sensing with the motor cortex. Neuron 72:477–487.

Hayman CAG, Macdonald CA, Tulving E (1993) The role of repetition and associative interference in new semantic learning in amnesia. J Cogn Neurosci 5:375–389.

Hebb DO (1949) The Organization of Behavior: A Neuropsychological Theory. New York: Wiley.

Hechavarría JC (2013) Evolution of neuronal mechanisms for echolocation: specializations for target range computation in bats of the genus Pteronotus. J Acoustic Soc Amer 133:570.

Heidegger M (1977) The Question Concerning Technology. In: Martin Heidegger: Basic Writings (Krell DF, ed), pp. 287–317. New York: Harper & Row.

Heidegger M (1927/2002) Time and Being. Translated by Joan Stambaugh. Chicago, IL: University of Chicago Press.

Heidelberger M (2004) Life and Work. Nature from Within: Gustav Theodor Fechner and his Psychophysical Worldview. Pittsburgh, PA: University of Pittsburgh Press.

Heiligenberg W (1991) Neural Nets in Electric Fish. Cambridge, MA: MIT Press.

Heit G, Smith ME, Halgren E (1988) Neural encoding of individual words and faces by the human hippocampus and amygdala. Nature 333:773–775.

Helbing D (2013) Globally networked risks and how to respond. Nature 497:51–59.

Held R, Hein A (1983) Movement-produced stimulation in the development of visually guided behavior. J. Comp Physiol Psychol 56:872–876.

Helmholtz H (1866/1962) Treatise on Physiological Optics. New York: Dover Publications.

Hempel CG, Oppenheim P (1948). Studies in the logic of explanation. Philosoph Sci 15:135–175.

Hengen KB, Torrado Pacheco A, McGregor JN, Van Hooser SD, Turrigiano GG (2016) Neuronal firing rate homeostasis is inhibited by sleep and promoted by wake. Cell 165:180–191.

Henneberg M, Steyn M (1993) Trends in cranial capacity and cranial index in sub-Saharan Africa during the Holocene. Am J Human Biol 5:473–479.

Henson OW (1965) The activity and function of the middle-ear muscles in echo-locating bats. J Physiol 180:871–887.

Henze DA, Buzsáki G (2001) Action potential threshold of hippocampal pyramidal cells in vivo is increased by recent spiking activity. Neuroscience 105:121–130.

Henze DA, Wittner L, Buzsáki G (2002) Single granule cells reliably discharge targets in the hippocampal CA3 network in vivo. Nat Neurosci 5:790–795.

Herculano-Houzel S (2016) The Human Advantage: A New Understanding of How our Brains Became Remarkable. Boston, MA: MIT Press.

Herrnstein J, Charles M (1994) Bell Curve: Intelligence and Class Structure in American Life. New York: Simon and Schuster.

Hill AJ (1978) First occurrence of hippocampal spatial firing in a new environment. Exp Neurol 62:282–297.

Hinard V, Mikhail C, Pradervand S, Curie T, Houtkooper RH, et al. (2012) Key electrophysiological, molecular, and metabolic signatures of sleep and wakefulness revealed in primary cortical cultures. J Neurosci 32:12506–12517.

Hinman JR, Penley SC, Long LL, Escabi MA, Chrobak JJ (2011). Septotemporal variation in dynamics of theta: speed and habituation. J Neurophysiol 105:2675–2686.

Hinton G, Plaut D (1987) Using fast weights to deblur old memories. In: Proceedings of the Ninth Annual Conference of the Cognitive Science Society, pp. 177–186. New York: Erlbaum.

Hinton GE, Dayan P, Frey BJ, Neal R (1995) The wake-sleep algorithm for unsupervised Neural Networks. Science 268:1158–1161.

Hirabayashi T, Miyashita Y (2005) Dynamically modulated spike correlation in monkey inferior temporal cortex depending on the feature configuration within a whole object. J Neurosci 25:10299–10307.

Hirase H, Czurko A, Csicsvari J, Buzsáki G (1999) Firing rate and theta-phase coding by hippocampal pyramidal neurons during "space clamping." Eur J Neurosci 11:4373–4380.

Hirase H, Leinekugel X, Czurko A, Csicsvari J, Buzsáki G (2001) Firing rates of hippocampal neurons are preserved during subsequent sleep episodes and modified by novel awake experience. Proc Natl Acad Sci U S A 98:9386–9390.

Hochberg LR, Bacher D, Jarosiewicz B, Masse NY, Simeral JD, et al. (2012) Reach and grasp by people with tetraplegia using a neurally controlled robotic arm. Nature 485:372–375.

Hoerl C, McCormack T (eds) (2001) Time and Memory. Issues in Philosophy and Psychology. Oxford: Clarendon Press.

Hoffecker JF (2011) Landscape of the Mind: Human Evolution and the Archeology of Thought. New York: Columbia University Press.

Hoffman, DD (1998). Visual Intelligence: How We Create What We See. New York: W. W. Norton.

Hoffman DD, Singh M, Prakash C (2005) The interface theory of perception. Psychol Bull Rev 22:1480–1506.

Hofmann V, Sanguinetti-Scheck JI, Künzel S, Geurten B, Gómez-Sena L, Engelmann J (2013) Sensory flow shaped by active sensing: sensorimotor strategies in electric fish. J Exp Biol 216:2487–500.

Hollerman JR, Schultz W (1998) Dopamine neurons report an error in the temporal prediction of reward during learning. Nat Neurosci 1:304–309.

Holloway RL (1969) Culture: a human domain. Curr Anthropol 10:395–412.

Holmes G (1918) Disturbances of visual orientation. Br J Ophthalmol 2:449–468.

Honma M, Kuroda T, Futamura A, Shiromaru A, Kawamura M (2016) Dysfunctional counting of mental time in Parkinson's disease. Sci Rep 6:25421.

Hopfield JJ (1982) Neural networks and physical systems with emergent collective computational abilities. Proc Natl Acad Sci U S A 79:2554–2558.

Hopfinger JB, Buonocore MH, Mangun GR (2000) The neural mechanisms of top-down attentional control. Nat Neurosci 3:284–291.

Houde JF, Jordan MI (1998) Sensorimotor adaptation in speech production. Science 279:1213–1216.

Howard MF, Poeppel D (2010) Discrimination of speech stimuli based on neuronal response phase patterns depends on acoustics but not comprehension. J Neurophysiol 104:2500–2511.

Howard MW (2018) Memory as perception of the past: compressed time in mind and brain. Trends Cogn Sci 22:124–136.

Howard MW, Kahana MJ (2002) A distributed representation of temporal context. J Math Psychol 46:269–299.

Howard MW, MacDonald CJ, Tiganj Z, Shankar KH, Du Q, et al. (2014) A unified mathematical framework for coding time, space, and sequences in the hippocampal region. J Neurosci 34:4692–4707.

Hromádka T, Deweese MR, Zador AM (2008) Sparse representation of sounds in the unanesthetized auditory cortex. PLoS Biology 6:e16–137.

Hubel DH (1957) Tungsten microelectrode for recording from single units. Science 125:549–550.

Hubel DH, Wiesel TN (1962) Receptive fields, binocular interaction and functional architecture in the cat's visual cortex. J Physiol 160:106–154.

Hubel DH, Wiesel TN (1974) Uniformity of monkey striate cortex: a parallel relationship between field size, scatter, and magnification factor. J Compar Neurol 158:295–305.

Huber D, Gutnisky DA, Peron S, O'Connor DH, Wiegert JS, et al. (2012) Multiple dynamic representations in the motor cortex during sensorimotor learning. Nature 484:473–478.

Huber D, Petreanu L, Ghitani N, Ranade S, Hromádka T, et al. (2008) Sparse optical microstimulation in barrel cortex drives learned behaviour in freely moving mice. Nature 451:61–64.

Hublin JJ, Ben-Ncer A, Bailey SE, Freidline SE, Neubauer S, et al. (2017) New fossils from Jebel Irhoud, Morocco and the pan-African origin of Homo sapiens. Nature 546:289–292.

Hughes JR (1995) The phenomenon of travelling waves: a review. Clin Electroencephalogr 26:1–6.

Humphrey NK (1976) The social function of intellect. In: Growing Points in Ethology (PPG Bateson PPG, Hinde RA, eds), pp. 303–317. Cambridge: Cambridge University Press.

Hutcheon B, Yarom Y (2000) Resonance, oscillation and the intrinsic frequency preferences of neurons. Trends Neurosci 23:216–222.

Huxter J, Burgess N, O'Keefe J (2003) Independent rate and temporal coding in hippocampal pyramidal cells. Nature 425:828–832.

Hyde KL, Lerch J, Norton A, Forgeard M, Winner E, et al. (2009) Musical training shapes structural brain development. J Neurosci 29:3019–3025.

Iacoboni M, Dapretto M (2006) The mirror neuron system and the consequences of its dysfunction. Nat Rev Neurosci 7:942–951.

Iacoboni M, Woods RP, Brass M, Bekkering H, Mazziotta JC, Rizzolatti G (1999) Cortical mechanisms of human imitation. Science 286:2526–2528.

Ifrah G (1985) From One to Zero. New York: Viking.

Ikegaya Y, Aaron G, Cossart R, Aronov D, Lampel I, et al. (2004) Synfire chains and cortical songs: temporal modules of cortical activity. Science 304:559–564.

Ikegaya Y, Sasaki T, Ishikawa D, Honma N, Tao K, et al. (2013) Interpyramid spike transmission stabilizes the sparseness of recurrent network activity. Cereb Cortex 23:293–304.

Ingvar DH (1985) "Memory of the future": an essay on the temporal organization of conscious awareness. Hum Neurobiol 4:127-36.

Innocenti GM, Vercelli A, Caminiti R (2014) The diameter of cortical axons depends both on the area of origin and target. Cereb Cortex 24:2178–2188.

Insel T (2017) Join the disruptors of health science. Nature 551:23–26.

Isomura Y, Sirota A, Ozen S, Montgomery S, Mizuseki K, et al. (2006) Integration and segregation of activity in entorhinal-hippocampal subregions by neocortical slow oscillations. Neuron 52:871–882.

Ito HT, Zhang SJ, Witter MP, Moser EI, Moser MB (2015) A prefrontal-thalamo-hippocampal circuit for goal-directed spatial navigation. Nature 522:50–55.

Itskov V, Curto C, Pastalkova E, Buzsáki G (2011) Cell assembly sequences arising from spike threshold adaptation keep track of time in the hippocampus. J Neurosci 31:2828–2834.

Ivry RB, Spencer RM (2004) The neural representation of time. Curr Opin Neurobiol 14:225–232.

Jadhav SP, Kemere C, German PW, Frank LM (2012) Awake hippocampal sharp-wave ripples support spatial memory. Science 336:1454–1458.

Jafarpour A, Spiers H (2017) Familiarity expands space and contracts time. Hippocampus 27:12–16.

James W (1884) What is an emotion? Mind 9:188–205.

James W (1890) The Principles of Psychology, Volumes I and II. New York: Dover.

James W (1907) The Energies of Men. New York: Moffat, Yard and Company.

Janssen P, Shadlen MN (2005) A representation of the hazard rate of elapsed time in macaque area LIP. Nat Neurosci 8:234–241.

Järvilehto T (1999) The theory of the organism-environment system: III. Role of efferent influences on receptors in the formation of knowledge. Integr Physiol Behav Sci 34:90–100.

Jasper HH, Andrews HL (1938) Brain potentials and voluntary muscle activity in man. J Neurophysiol 1:87–100.

Jeannerod M (2001) Neural simulation of action: a unifying mechanism for motor cognition. Neuroimage 14:S103–S109.

Jeewajee A, Barry C, O'Keefe J, Burgess N (2008) Grid cells and theta as oscillatory interference: electrophysiological data from freely moving rats. Hippocampus 18:1175–1185.

Jensen O, Colgin LL (2007) Cross-frequency coupling between neuronal oscillations. Trends Cogn Sci 11:267–269.

Jensen O, Lisman JE (1996a) Hippocampal CA3 region predicts memory sequences: accounting for the phase precession of place cells. Learn Mem 3:279–287.

Jensen O, Lisman JE (1996b) Novel lists of 7±2 known items can be reliably stored in an oscillatory short-term memory network: interaction with long-term memory. Learn Mem 3:257–263.

Jensen O, Lisman JE (2000) Position reconstruction from an ensemble of hippocampal place cells: contribution of theta phase coding. J Neurophysiol 83:2602–2609.

Jensen O, Lisman JE (2005) Hippocampal sequence-encoding driven by a cortical multi-item working memory buffer. Trends Neurosci 28:67–72.

Jezek K, Henriksen EJ, Treves A, Moser EI (2011) Theta-paced flickering between place-cell maps in the hippocampus. Nature 478:246–249.

Ji D, Wilson MA (2007) Coordinated memory replay in the visual cortex and hippocampus during sleep. Nat Neurosci 10:100–107.

Jinno S, Klausberger T, Marton LF, Dalezios Y, Roberts JD, et al. (2007) Neuronal diversity in GABAergic long-range projections from the hippocampus. J Neurosci 27:8790–8804.

Jolliffe IT (1986) Principal component analysis. Springer Series in Statistics. New York: Springer.

John ER (1972) Switchboard versus statistical theories of learning and memory. Science 177:850–864.

John ER (1976) A model of consciousness. In: Consciousness and Self-Regulation (Schwartz GE, Shapiro DH, eds), pp. 6–50. New York: Plenum Press.

Johnson A, Redish AD (2007) Neural ensembles in CA3 transiently encode paths forward of the animal at a decision point. J Neurosci 27:12176–12189.

Johnson LA, Euston DR, Tatsuno M, McNaughton BL (2010) Stored trace reactivation in rat prefrontal cortex is correlated with down to-up state fluctuation density. J Neurosci 30:2650–2661.

Johnston D, Wu SM (1995) Foundations of Cellular Neurophysiology. Cambridge, MA: MIT Press.

Jones OP, Alfaro-Almagro F, Jbabdi S (2018) An empirical, 21st century evaluation of phrenology. Cortex 106:26–35.

Jones W, Klin A (2013) Attention to eyes is present but in decline in 2–6-month-old infants later diagnosed with autism. Nature 504:427–431.

Jordan KE, Maclean EL, Brannon EM (2008) Monkeys match and tally quantities across senses. Cognition 108:617–625.

Jørgensen CB (2003) Aspects of the history of the nerves: Bell's theory, the Bell-Magendie law and controversy, and two forgotten works by P. W. Lund and D. F. Eschricht. J Hist Neurosci 12:229–249.

Jortner RA, Farivar SS, Laurent G (2007) A simple connectivity scheme for sparse coding in an olfactory system. J Neurosci 27:1659–1669.

Josselyn SA, Köhler S, Frankland PW (2015) Finding the engram. Nat Rev Neurosci 16:521–534.

Jung C (1973) Synchronicity: An Acausal Connecting Principle. Princeton, NJ: Princeton University Press.

Kable JW, Glimcher PW (2009) The neurobiology of decision: consensus and controversy. Neuron 63:733–745.

Kaczmarek LK, Levitan IB (1986) Neuromodulation: The Biochemical Control of Neuronal Excitability. New York: Oxford University Press.

Kahneman D (2011) Thinking Fast and Slow. New York: Farrar, Straus and Giroux, Macmillan Publishers.

Kaila K (1994) Ionic basis of GABAA receptor channel function in the nervous system. Prog Neurobiol 42:489–537.

Kalueff AV, Stewart AM, Song C, Berridge KC, Graybiel AM, Fentress JC (2016) Neurobiology of rodent self-grooming and its value for translational neuroscience. Nat Rev Neurosci 17:45–59.

Kampis G (1991) Self-Modifying Systems in Biology and Cognitive Science. London: Pergamon Press.

Kandel ER, Schwartz JH, Jessell JM, Siegelbaum SA, Hudspeth HJ, Mack S. (2012) Principles of Neural Science (5th ed). New York: McGraw-Hill.

Kant I (1871) Critique of Pure Reason (Guyer P, Wood AW, translators). Cambridge Edition of the Works of Immanuel Kant.Cambridge: Cambridge University Press.

Karayannis T, Au E, Patel JC, Kruglikov I, Markx S, et al. (2014) Cntnap4/Caspr4 differentially contributes to GABAergic and dopaminergic synaptic transmission. Nature 511:236–240.

Karlsson MP, Frank LM (2009) Awake replay of remote experiences in the hippocampus. Nat Neurosci 12:913–918.

Karmos G, Martin J, Czopf J (1971) Jel-zaj viszony mértékének jelentősége agyi kiváltott potenciálsorozatok számítógépes értékelésénél. Mérés és Automatika, 19 (in Hungarian)
Katz B (1966) Nerve, Muscle and Synapse. New York: McGraw Hill.
Katz LC, Shatz CJ (1996) Synaptic activity and the construction of cortical circuits. Science 274:1133–1138.
Kawato M (1999) Internal models for motor control and trajectory planning. Curr Opin Neurobiol 9:718–727.
Kebschull JM, Garcia da Silva P, Reid AP, Peikon ID, Albeanu DF, Zador AM (2016) High-throughput mapping of single-neuron projections by sequencing of barcoded RNA. Neuron 91:975–987.
Kelemen E, Fenton AA (2010) Dynamic grouping of hippocampal neural activity during cognitive control of two spatial frames. PLoS Biol 8:e1000403.
Kelso JAS (1995) Dynamic Patterns: The Self-Organization of Brain and Behavior. Cambridge, MA: MIT Press.
Kenet T, Bibitchkov D, Tsodyks M, Grinvald A, Arieli A (2003) Spontaneously emerging cortical representations of visual attributes. Nature 425:954–956.
Kepecs A, Fishell G (2014) Interneuron cell types are fit to function. Nature 505:318–326.
Kéri S (2009) Genes for psychosis and creativity: a promoter polymorphism of the neuregulin 1 gene is related to creativity in people with high intellectual achievement. Psychol Sci 20:1070–1073.
Kerlin JR, Shahin AJ, Miller LM (2010) Attentional gain control of ongoing cortical speech representations in a "cocktail party." J Neurosci 30:620–628.
Keysers C, Wicker B, Gazzola V, Anton JL, Fogassi L, Gallese V (2004) A touching sight: SII/PV activation during the observation and experience of touch. Neuron 42:335–346.
Khazipov R, Sirota A, Leinekugel X, Holmes GL, Ben-Ari Y, Buzsáki G (2004) Early motor activity drives spindle bursts in the developing somatosensory cortex. Nature 432:758–761.
Khodagholy D, Gelinas JN, Buzsáki G (2017) Learning-enhanced coupling between ripple oscillations in association cortices and hippocampus. Science 358:369–372.
Kiebel SJ, Daunizeau J, Friston KJ (2008) A hierarchy of time-scales and the brain. PLoS Comput Biol 4:e1000209.
Kilner JM, Friston KJ, Frith CD (2007) Predictive coding: an account of the mirror neuron system. Cogn Processing 8:159–166.
Kim S, Sapiurka M, Clark RE, Squire LR (2013) Contrasting effects on path integration after hippocampal damage in humans and rats. Proc Natl Acad Sci U S A 110:4732–4737.
Kjelstrup KB, Solstad T, Brun VH, Hafting T, Leutgeb S, et al. (2008) Finite scale of spatial representation in the hippocampus. Science 321:140–143.
Klausberger T, Somogyi P (2008) Neuronal diversity and temporal dynamics: the unity of hippocampal circuit operations. Science 321:53–57.
Kleberg FI, Triesch J (2018) Neural oligarchy: how synaptic plasticity breeds neurons with extreme influence. BioRxiv http://dx.doi.org/10.1101/361394.

Knierim JJ, Neunuebel JP (2016) Tracking the flow of hippocampal computation: pattern separation, pattern completion, and attractor dynamics. Neurobiol Learn Mem 129:38–49.

Knierim JJ, Zhang K (2012) Attractor dynamics of spatially correlated neural activity in the limbic system. Annu Rev Neurosci 35:267–285.

Knill DC, Pouget A (2004) The Bayesian brain: the role of uncertainty in neural coding and computation. Trends Neurosci 27:712–719.

Knudsen, EI, Konishi M (1978) A neural map of auditory space in the owl. Science 200:795–797.

Ko H, Cossell L, Baragli C, Antolik J, Clopath C, et al. (2013) The emergence of functional microcircuits in visual cortex. Nature 496:96–100.

Koch AL (1966) The logarithm in biology I. Mechanisms generating the log-normal distribution exactly. J Theor Biol 12:276–290.

Koch C (2004) The Quest for Consciousness: A Neurobiological Approach. Englewood, CO: Roberts and Co.

Koch C, Rapp M, Segev I (1996) A brief history of time (constants). Cereb Cortex 6:93–101.

Koestler A (1973) The Roots of Coincidence. New York: Vintage.

Kolarik AJ, Scarfe AC, Moore BCJ, Pardhan S (2017) Blindness enhances auditory obstacle circumvention: assessing echolocation, sensory substitution, and visual-based. PLoS One 2017 Apr 13;12(4):e0175750.

Kolers PA, von Grünau M (1976) Shape and color in apparent motion. Vision Res 16:329–335.

Konorski J (1948) Conditioned reflexes and neuron organization. New York: Cambridge University Press.

Kopell N (2000) We got rhythm: dynamical systems of the nervous system. N Am Math Soc 47:6–16.

Kornhuber HH, Deecke L (1965) Hirnpotentialänderungen bei Willkurbewegungen und passiven Bewegungen des Menschen: Bereitschaftspotential und reafferente Potentiale Pflugers Archiv 284:1–17.

Kornmüller AE (1931) Eine experimentelle Anasthesie der ausseren Augenmuskeln am Menschen und ihre Auswirkungen. Journal fur Psychologie und Neurologie 41:354–366.

Koulakov AA, Hromádka T, Zador AM (2009) Correlated connectivity and the distribution of firing rates in the neocortex. J Neurosci 29:3685–3694.

Kozaczuk, W (1984) Enigma: How the German Machine Cipher Was Broken, and how it was read by the allies in World War Two. Kasparek C, ed. and translator (2nd ed.). Frederick, MD: University Publications of America (translation of the original Polish version in 1979, supplemented with appendices by Marian Rejewski).

König P, Luksch H (1998) Active sensing: closing multiple loops. Z Naturforsch C 53:542–549.

Kraus BJ, Brandon MP, Robinson RJ 2nd, Connerney MA, Hasselmo ME, Eichenbaum H (2015) During running in place, grid cells integrate elapsed time and distance run. Neuron 88:578–589.

Krakauer JW, Ghazanfar AA, Gomez-Marin A, MacIver MA, Poeppel D (2017) Neuroscience needs behavior: correcting a reductionist bias. Neuron 93:480–490.

Krauzlis RJ, Bollimunta A, Arcizet F, Wang L (2014). Attention as an effect not a cause. Trends Cogn Sci 18:457-464.

Kremer W (2012) Human echolocation: using tongue-clicks to navigate the world. BBC. Retrieved September 12, 2012.

Kremer Y, Léger JF, Goodman D, Brette R, Bourdieu L (2011) Late emergence of the vibrissa direction selectivity map in the rat barrel cortex. J Neurosci 31:10689-10700.

Kubie JL, Muller RU, Bostock E (1990) Spatial firing properties of hippocampal theta cells. J Neurosci 10:1110-1123.

Kubota K, Niki H (1971) Prefrontal cortical unit activity and delayed alternation performance in monkeys. J Neurophysiol 34:337-347.

Kudrimoti HS, Barnes CA, McNaughton BL (1999) Reactivation of hippocampal cell assemblies: effects of behavioral state, experience, and EEG dynamics. J Neurosci 19:4090-4101.

Kümmerle R, Ruhnke M, Steder B, Stachniss C, Burgard W (2014) Autonomous robot navigation in populated pedestrian zones. J Field Robotics 32:565-589.

Kupalov PS (1978) Mechanisms of the Establishments of Temporary Connections Under Normal and Pathologic Conditions (in Russian). Moscow: Meditsina.

Kushchayev SV, Moskalenko VF, Wiener PC, Tsymbaliuk VI, Cherkasov VG, et al. (2012) The discovery of the pyramidal neurons: Vladimir Betz and a new era of neuroscience. Brain 135:285-300.

Lakatos P, Karmos G, Mehta AD, Ulbert I, Schroeder CE (2008) Entrainment of neuronal oscillations as a mechanism of attentional selection. Science 320:110-113.

Lakatos P, Shah AS, Knuth KH, Ulbert I, Karmos G, Schroeder CE (2005) An oscillatory hierarchy controlling neuronal excitability and stimulus processing in the auditory cortex. J Neurophysiol 94:1904-1911.

Landau B, Spelke E, Gleitman H (1984) Spatial knowledge in a young blind child. Cognition 16:225-260.

Lange CG (1885/1912) The mechanisms of the emotions (B. Rand translation). In: The Classical Psychologists (Rand B, ed), pp. 672-684. Copenhagen (Original work published 1885, Om Sindsbevaegelser et Psyki-Fysiologist Studie).

Langguth B, Eichhammer P, Zowe M, Kleinjung T, Jacob P, et al. (2005) Altered motor cortex excitability in tinnitus patients: a hint at crossmodal plasticity. Neurosci Lett 380:326-329.

Langston RF, Ainge JA, Couey JJ, Canto CB, Bjerknes TL, et al. (2010) Development of the spatial representation system in the rat. Science 328:1576-1580.

Lansner A (2009) Associative memory models: from the cell-assembly theory to biophysically detailed cortex simulations. Trends Neurosci 32:178-186.

Lashley KS (1930) Basic neural mechanisms in behavior. Psychol Rev 30:237-2272 and 329-353.

Lashley KS (1951) The problem of serial order in behavior. In: Cerebral Mechanisms in Behavior: The Hixon Symposium (Jeffress LA, ed), pp. 112-136. New York: Wiley.

Lasztóczi B, Klausberger T (2016) Hippocampal place cells couple to three different gamma oscillations during place field traversal. Neuron 91:34-40.

Latchoumane CV, Ngo HV, Born J, Shin HS (2017) Thalamic spindles promote memory formation during sleep through triple phase-locking of cortical, thalamic, and hippocampal rhythms. Neuron 95:424-435.

Laurent G (1999) A systems perspective on early olfactory coding. Science 286:723–728.

Leaman O (1985) An Introduction to Medieval Islamic Philosophy. New York: Cambridge University Press.

Lebedev MA, O'Doherty JE, Nicolelis MA (2008) Decoding of temporal intervals from cortical ensemble activity. J Neurophysiol 99:166–186.

LeCun Y, Bengio Y, Hinton G (2015) Deep learning. Nature 521:436–444.

LeDoux J (2015) Anxious: Using the Brain to Understand and Treat Fear and Anxiety. New York: Penguin Random House.

LeDoux JE (2014) Coming to terms with fear. Proc Natl Acad Sci U S A 111:2871–2878.

LeDoux J, Daw ND (2018) Surviving threats: neural circuit and computational implications of a new taxonomy of defensive behaviour. Nat Rev Neurosci 19:269–282.

Lee AK, Wilson MA (2002) Memory of sequential experience in the hippocampus during slow wave sleep. Neuron 36:1183–1194.

Lee D (1950). Notes on the conception of the self among the Wintu Indians. J Abnorm Soc Psychol 45:538–543.

Lee D, Lin B-J, Lee AK (2012) Hippocampal place fields emerge upon single-cell manipulation of excitability during behavior. Science 337:849–853.

Lee DD, Seung SH (1999) Learning the parts of objects by non-negative matrix factorization. Nature 401:788–791.

Leinekugel X, Khazipov R, Cannon R, Hirase H, Ben-Ari Y, Buzsáki G (2002) Correlated bursts of activity in the neonatal hippocampus in vivo. Science 296:2049–2052.

Leon MI, Shadlen MN (2003) Representation of time by neurons in the posterior parietal cortex of the macaque. Neuron 38:317–327.

Leopold D, Murayama Y, Logothetis N (2003) Very slow activity fluctuations in monkey visual cortex: implications for functional brain imaging. Cereb Cortex 13:422–433.

Lerner Y, Honey CJ, Silbert LJ, Hasson U (2011) Topographic mapping of a hierarchy of temporal receptive windows using a narrated story. J Neurosci 31:2906–2915.

Lettvin JY, Maturana HR, McCulloch WS, Pitts WH (1959) What the frog's eye tells the frog's brain. Proc Inst Radio Engr 47:1940–1951.

Leutgeb S, Leutgeb JK, Treves A, Meyer R, Barnes CA, et al. (2005) Progressive transformation of hippocampal neuronal representations in "morphed" environments. Neuron 48:345–348.

Leutgeb S, Ragozzino KE, Mizumori SJ (2000) Convergence of head direction and place information in the CA1 region of hippocampus. Neuroscience 100:11–19.

Levenstein D, Watson BO, Rinzel J, Buzsáki G (2017) Sleep regulation of the distribution of cortical firing rates. Curr Opin Neurobiol 44:34–42.

Levinson SC (2003) Space in Language and Cognition: Explorations in Cognitive Diversity. New York: Cambridge University Press.

Lévi-Strauss C (1963) Structural Anthropology. New York: Basic Books.

Levy WB, Steward O (1983) Temporal contiguity requirements for long-term associative potentiation/depression in the hippocampus. Neurosci 8:791–797

Li X, Shu H, Liu Y, Li P (2006) Mental representation of verb meaning: behavioral and electrophysiological evidence. J Cogn Neurosci 18:1774–1787.

Li XG, Somogyi P, Ylinen A, Buzsáki G (1994) The hippocampal CA3 network: an in vivo intracellular labeling study. J Comp Neurol 339:181–208.

Libet B (1985) Unconscious cerebral initiative and the role of conscious will in voluntary action. Behav Brain Sci 8:529–566.
Libet B (2005) Mind Time: The Temporal Factor in Consciousness. Cambridge, MA: Harvard University Press.
Libet B, Wright EW, Feinstein B, Pearl DK (1979) Subjective referral of the timing for a conscious sensory experience: a functional role for the somatosensory specific projection system in man. Brain 102:193–224.
Liberman AM, Cooper FS, Shankweiler DP, Studdert-Kennedy M (1967) Perception of the speech code. Psychol Rev 74:431-461.
Limpert E, Stahel WA, Abbt W (2001) Log-normal distributions across the sciences: keys and clues. BioScience 51:341–352.
Lisman J, Redish AD (2009) Prediction, sequences and the hippocampus. Philos Trans R Soc Lond B Biol Sci 364:1193–1201.
Lisman JE (1997) Bursts as a unit of neural information: making unreliable synapses reliable. Trends Neurosci 20:38–43.
Lisman JE, Idiart MA (1995) Storage of 7 ± 2 short-term memories in oscillatory subcycles. Science 267:1512–1515.
Liu K, Sibille J, Dragoi (2018) Generative predictive codes by multiplexed hippocampal neuronal tuplets. Neuron XXX
Liu X, Ramirez S, Pang PT, Puryear CB, Govindarajan A, et al. (2012) Optogenetic stimulation of a hippocampal engram activates fear memory recall. Nature 484:381–385.
Livio M (2002) The Golden Ratio: The Story of Phi, the World's Most Astonishing Number. New York: Broadway Books.
Llinás R (1988) The intrinsic electrophysiological properties of mammalian neurons: insights into central nervous system function. Science 242:1654–1664.
Llinás R (2002) I of the Vortex: From Neurons to Self. Cambridge, MA: MIT Press.
Llinás R, Sugimori M (1980) Electrophysiological properties of in vitro Purkinje cell dendrites in mammalian cerebellar slices. J Physiol 305:197–213.
Locke J (1690) An essay concerning human understanding. London: T Basset.
Loewenstein Y, Kuras A, Rumpel S (2011) Multiplicative dynamics underlie the emergence of the log-normal distribution of spine sizes in the neocortex in vivo. J Neurosci 31:9481–9488.
Logothetis NK, Sheinberg DL (1996) Visual object recognition. Annu Rev Neurosci 19:577–621.
Long MA, Fee MS (2008) Using temperature to analyse temporal dynamics in the songbird motor pathway. Nature 456:189–194.
Losonczy A, Magee JC (2006) Integrative properties of radial oblique dendrites in hippocampal CA1 pyramidal neurons. Neuron 50:291–307.
Löwel S, Singer W (1992) Selection of intrinsic horizontal connections in the visual cortex by correlated neuronal activity. Science 255:209–212.
Lubenov EV, Siapas AG (2008) Decoupling through synchrony in neuronal circuits with propagation delays. Neuron 58:118–131.
Lubenov EV, Siapas AG (2009) Hippocampal theta oscillations are travelling waves. Nature 459:534–539.
Lucretius (2008) On the Nature of the Universe (translated by Melville, Robert). Oxford: Oxford University Press.

Luczak A, Barthó P, Harris KD (2009) Spontaneous events outline the realm of possible sensory responses in neocortical populations. Neuron 62:413–425.

Luczak A, Barthó P, Marguet SL, Buzsáki G, Harris KD (2007) Sequential structure of neocortical spontaneous activity in vivo. Proc Natl Acad Sci U S A 104:347–352.

Lundh L-G (1983) Mind and Meaning: Towards a Theory of the Human Considered As a System of Meaning Structures. Studia Psychologica-Upsaliensia 10: 1-208.

Luria AR (1966) Higher cortical functions in man. Tavistock, London.

Lutz J (2009) Music drives brain plasticity. F1000 Biol Rep 1:78.

Lyon A (2014) Why are normal distributions normal? British J Philos Sci 65:621–649.

MacDonald CJ, Lepage KQ, Eden UT, Eichenbaum H (2011) Hippocampal "time cells" bridge the gap in memory for discontiguous events. Neuron 71:737–749.

MacDonald S (1998) Aquinas's libertarian account of free will. Rev Int Philos 2:309–328.

Machens CK, Romo R, Brody CD (2005) Flexible control of mutual inhibition: a neural model of two-interval discrimination. Science 307:1121–1124.

MacKay DM (1956) The epistemological problem for automata. In: Automomata Studies (Shannon CE, McCarthy J, eds), pp. 235–251. Princeton, NJ. Princeton University Press.

Mackay DM (1963) Psychophysics of perceived intensity: a theoretical basis for Fechner's and Stevens' laws. Science 139:1213–1216.

MacKay DM (1967) Ways of looking at perception. In: Models for the Perception of Speech and Visual Form (Wathen-Dunn w, ed). Cambridge, MA: MIT Press.

Mackey M (1992) Time's Arrow: The Origins of Thermodynamic Behavior. Berlin: Springer-Verlag.

MacLean JN, Watson BO, Aaron GB, Yuste R (2005) Internal dynamics determine the cortical response to thalamic stimulation. Neuron 48:811–823.

MacLean PD (1970) The triune brain, emotion, and scientific bias. In: The Neurosciences (Schmitt FO, ed). New York: Rockefeller University Press.

MacLeod K, Bäcker A, Laurent G (1998) Who reads temporal information contained across synchronized and oscillatory spike trains? Nature 395:693–698.

Magee JC (2000) Dendritic integration of excitatory synaptic input. Nat Rev Neurosci 1:181-90.

Magee JC, Johnston D (1997) A synaptically controlled, associative signal for Hebbian plasticity in hippocampal neurons. Science 275:209–213.

Magendie F (1822) Expériences sur les fonctions des racines des nerfs rachidiens. Journal de physiologie expérimentale et de pathologie 276–279.

Maguire EA, Gadian DG, Johnsrude IS, Good CD, Ashburner J, et al. (2000) Navigation-related structural change in the hippocampi of taxi drivers. Proc Natl Acad Sci U S A 97:4398–4403.

Maguire EA, Nannery R, Spiers HJ (2006) Navigation around London by a taxi driver with bilateral hippocampal lesions. Brain 129:2894–907.

Maingret N, Girardeau G, Todorova R, Goutierre M, Zugaro M (2016) Hippocampo-cortical coupling mediates memory consolidation during sleep. Nat Neurosci 19:959–964.

Malafouris L (2009) "Neuroarchaeology": exploring the links between neural and cultural plasticity. Prog Brain Res 178:251–259.

Mannino M, Bressler SL (2015) Foundational perspectives on causality in large-scale brain networks. Phys Life Rev 15:107–23.

Manns JR, Hopkins RO, Reed JM, Kitchener EG, Squire LR (2003) Recognition memory and the human hippocampus. Neuron 37:171–180.

Mao T, Kusefoglu D, Hooks BM, Huber D, Petreanu L, Svoboda K (2011) Long-range neuronal circuits underlying the interaction between sensory and motor cortex. Neuron 72:111–123.

Maor E (1994) E: The Story of a Number. Princeton NJ: Princeton University Press.

Marcus G, Marblestone A, Dean T (2014) Neuroscience. The atoms of neural computation. Science 346:551–552.

Marder E, Goeritz ML, Otopalik AG (2015) Robust circuit rhythms in small circuits arise from variable circuit components and mechanisms. Curr Opin Neurobiol 31:156–163.

Marder E, Rehm KJ (2005) Development of central pattern generating circuits. Curr Opin Neurobiol 15:86–93.

Markov NT, Ercsey-Ravasz M, Van Essen DC, Knoblauch K, Toroczkai Z, Kennedy H (2013) Cortical high-density counterstream architectures. Science 342:1238406.

Markov NT, Misery P, Falchier A, Lamy C, Vezoli J, et al. (2011) Weight consistency specifies regularities of macaque cortical networks. Cereb Cortex 21:1254–1272.

Markram H, Lubke J, Frotscher M, Sakmann B (1997) Regulation of synaptic efficacy by coincidence of postsynaptic APs and EPSPs. Science 275:213–215.

Markram H, Rinaldi T, Markram K (2007) The intense world syndrome—an alternative hypothesis for autism. Front Neurosci 1:77–96.

Markram H, Tsodyks M (1996) Redistribution of synaptic efficacy between neocortical pyramidal neurons. Nature 382:807–810.

Marr D (1969) A theory of cerebellar cortex. J Physiol 202:437–470.

Marr D (1971) Simple memory: a theory for archicortex. Philos Trans R Soc Lond B Biol Sci 262:23–81.

Marr D (1982) Vision: A Computational Investigation into the Human Representation and Processing of Visual Information. New York: Freeman.

Martinez F (1971) Comparison of two types of tactile exploration in a task of a mirror-image recognition. Psychonom Sci 22:124–125.

Martinez-Conde S, Macknik SL, Hubel DH (2004) The role of fixational eye movements in visual perception. Nat Rev Neurosci 5:229–240.

Masquelier T, Guyonneau R, Thorpe SJ (2009) Competitive STDP-based spike pattern learning. Neural Comput 21:1259–1276.

Matell MS, Meck WH (2004) Cortico-striatalcircuits and interval timing: coincidence detection of oscillatory processes. Brain Res Cogn Brain Res 21:139–170.

Matsuzawa T (1985) Use of numbers by a chimpanzee. Nature 315:57–59.

Maturana HR, Varela FJ (1980) Autopoieis and Cognition: The Realization of the Living. D. Dordrecht, Netherlands: Reidel Publishing.

Mátyás F, Sreenivasan V, Marbach F, Wacongne C, Barsy B, et al. (2010) Motor control by sensory cortex. Science 330:1240–1243.

Mauk MD, Buonomano DV (2004) The neural basis of temporal processing. Annu Rev Neurosci 27:307–340.

Maurer AP, Burke SN, Lipa P, Skaggs WE, Barnes CA (2012) Greater running speeds result in altered hippocampal phase sequence dynamics. Hippocampus 22:737–747.

Maurer AP, Vanrhoads SR, Sutherland GR, Lipa P, McNaughton BL (2005) Self-motion and the origin of differential spatial scaling along the septo-temporal axis of the hippocampus. Hippocampus 15:841–852.

Mayer C, Bandler RC, Fishell G (2016) Lineage is a poor predictor of interneuron positioning within the forebrain. Neuron 92:45–51.

Mazor O, Laurent G (2005) Transient dynamics versus fixed points in odor representations by locust antennal lobe projection neurons. Neuron 48:661–673.

McAdams CJ, Maunsell JHR (1999) Effects of attention on orientation-tuning functions of single neurons in macaque cortical area V4. J Neurosci 19:431–441.

McBain CJ, Fisahn A (2001) Interneurons unbound. Nat Rev Neurosci 2:11–23.

McClelland JL, McNaughton BL, O'Reilly RC (1995) Why there are complementary learning systems in the hippocampus and neocortex: insights from the successes and failures of connectionist models of learning and memory. Psychol Rev 102:419–457.

McClelland JL, Rumelhart DE, the PDP Research Group (1986) Parallel Distributed Processing: Explorations in the Microstructure of Cognition. Volume 2: Psychological and Biological Models. Cambridge, MA: MIT Press.

McCloskey M, Cohen N (1989) Catastrophic interference in connectionist networks: The sequential learning problem. In: The Psychology of Learning and Motivation: Volume 24 (Bower GH, ed), pp. 109–164. Cambridge, MA: Academic Press.

McCormick DA, Thompson RF (1984) Cerebellum: essential involvement in the classically conditioned eyelid response. Science 223:296–299.

McElvain LE, Friedman B, Karten HJ, Svoboda K, Wang F, et al. (2017) Circuits in the rodent brainstem that control whisking in concert with other orofacial motor actions. Neuroscience 368:152–170.

McGregor RJ (1993) Composite cortical networks of multimodal oscillators. Biol Cybern 69:243–255.

McNaughton BL, Barnes CA, Gerrard JL, Gothard K, Jung MW, et al. (1996) Deciphering the hippocampal polyglot: the hippocampus as a path integration system. J Exp Biol 199:173–185.

McNaughton BL, Barnes CA, O'Keefe J (1983) The contributions of position, direction, and velocity to single unit activity in the hippocampus of freely-moving rats. Exp Brain Res 52:41–49.

McNaughton BL, Battaglia FP, Jensen O, Moser EI, Moser MB (2006) Path integration and the neural basis of the "cognitive map." Nat Rev Neurosci 7:663–678.

McNaughton BL, Morris RGM (1987) Hippocampal synaptic enhancement and information storage within a distributed memory system. Trends Neurosci 10:408–415.

Mehta MR (2015) From synaptic plasticity to spatial maps and sequence learning. Hippocampus 25:756–762.

Meister MLR, Buffalo EA (2016) Getting directions from the hippocampus: the neural connection between looking and memory. Neurobiol Learn Mem 134:135–144.

Mel BW (1999) Computational neuroscience. Think positive to find parts. Nature 401:759–760.

Merleau-Ponty M (1945/2005) Phenomenology of Perception (Smith C, translator). London: Routledge.

Mesgarani N, Chang EF (2012) Selective cortical representation of attended speaker in multi-talker speech perception. Nature 485:233–236.

Mesulam MM (1998) From sensation to cognition. Brain 121:1013–1052.

Micadei K, Peterson JPS, Souza AM, Sarthour RS, Oliveira IS, et al. (2017) Reversing the thermodynamic arrow of time using quantum correlations. arXiv:1711.03323.

Michelson AA, Morley EW (1887) On the relative motion of the earth and the luminiferous ether. Am J Sci 34:333–345.

Michon JA (1985) The complete time experiencer. In: Time, Mind and Behavior (Michon JA, Jackson JL, eds), pp. 21–52. Berlin: Springer.

Mickus T, Jung Hy, Spruston N (1999) Properties of slow, cumulative sodium channel inactivation in rat hippocampal CA1 pyramidal neurons. Biophys J 76:846–860.

Miesenbock G (2009) The optogenetic catechism. Science 326:395–399.

Miles R (1990) Synaptic excitation of inhibitory cells by single CA3 hippocampal pyramidal cells of the guinea-pig in vitro. J Physiol 428:61–77.

Milh M, Kaminska A, Huon C, Lapillonne A, Ben-Ari Y, Khazipov R (2007) Rapid cortical oscillations and early motor activity in premature human neonate. Cereb Cortex 17:1582–1594.

Miller G (1956). The magical number seven, plus or minus two: some limits on our capacity for processing information. Psychol Rev 63:81–97.

Miller R (1996) Neural assemblies and laminar interactions in the cerebral cortex. Biol Cybern 75:253–261.

Milner B, Corkin S, Teuber HL (1968) Further analysis of the hippocampal amnesic syndrome: 14-year follow-up study of H.M. Neuropsychologia 6:191–209.

Milner B, Squire LR, Kandel ER (1998) Cognitive neuroscience and the study of memory. Neuron 20:445–468.

Milner PM (1996). Neural representations: some old problems revisited. J Cogn Neurosci 8:69–77.

Minkowski H (1909) Raum und Zeit. Physikalische Zeitschrift 10:104–111. Reprinted and translated in Minkowski Spacetime: A Hundred Years Later (Vesselin P, ed), pp. xiv–xlii. Dordrecht: Springer 2010.

Mishkin M, Ungerleider L, Macko K (1983) Object vision and spatial vision: two cortical pathways. Trends Neurosci 6:414–417.

Mishkin M, Vargha-Khadem F, Gadian DG (1998) Amnesia and the organization of the hippocampal system. Hippocampus 8:212–216.

Mita A, Mushiake H, Shima K, Matsuzaka Y, Tanji J (2009) Interval time coding by neurons in the presupplementary and supplemental motor areas. Nat Neurosci 12:502–507.

Mitchell JF, Sundberg KA, Reynolds JH (2009) Spatial attention decorrelates intrinsic activity fluctuations in macaque area V4. Neuron 63:879–888.

Mitra A, Mitra SS, Tsien RW (2012) Heterogeneous reallocation of presynaptic efficacy in recurrent excitatory circuits adapting to inactivity. Nature Neurosci 15:250–257.

Mitra A, Snyder AZ, Hacker CD, Pahwa M, Tagliazucchi E, et al. (2016) Human cortical-hippocampal dialogue in wake and slow-wave sleep. Proc Natl Acad Sci U S A 113:E6868–E6876.

Mittelstaedt ML, Mittelstaedt H (1980) Homing by path integration in a mammal. Naturwissenschaften 67:566–567.

Mitzenmacher M (2003) A brief history of generative models for power law and lognormal distributions. Internet Math 1:226–251.

Miyashita Y (1993) Inferior temporal cortex: where visual perception meets memory. Ann Rev Neurosci 16:245–263.

Miyashita Y (2004) Cognitive memory: cellular and network machineries and their top-down control. Science 306:435–440.

Miyashita-Lin EM, Hevner R, Wassarman KM, Martinez S, Rubinstein JL (1999) Early neocortical regionalization in the absence of thalamic innervation. Science 285:906–909.

Mizuseki K, Buzsáki G (2013) Preconfigured, skewed distribution of firing rates in the hippocampus and entorhinal cortex. Cell Rep 4:1010–1021.

Mizuseki K, Diba K, Pastalkova E, Buzsáki G (2011) Hippocampal CA1 pyramidal cells form functionally distinct sublayers. Nat Neurosci 14:1174–1181.

Mizuseki K, Sirota A, Pastalkova E, Buzsáki G (2009) Theta oscillations provide temporal windows for local circuit computation in the entorhinal-hippocampal loop. Neuron 64:267–280.

Mnih V, Heess N, Graves A, Kavukcuoglu K (2014). Recurrent models of visual attention. arXiv:14066247.

Mölle M, Eschenko O, Gais S, Sara SJ, Born J (2009) The influence of learning on sleep slow oscillations and associated spindles and ripples in humans and rats. Eur J Neurosci 29:1071–1081.

Moore T, Armstrong KM, Fallah M (2003) Visuomotor origins of covert spatial attention. Neuron 40:671–683.

Moran J, Desimone R (1985) Selective attention gates visual processing in the extrastriate cortex. Science 229:782–784.

Morillon B, Hackett TA, Kajikawa Y, Schroeder CE (2015) Predictive motor control of sensory dynamics in auditory active sensing. Curr Opin Neurobiol 31:230–238.

Morris JS, Friston KJ, Büchel C, Frith CD, Young AW, et al. (1998) A neuromodulatory role for the human amygdala in processing emotional facial expressions. Brain 121, 47–57.

Morrone MC, Ross J, Burr D (2005) Saccadic eye movements cause compression of time as well as space. Nat Neurosci 8:950–954.

Moser EI, Kropff E, Moser MB (2008) Place cells, grid cells, and the brain's spatial representation system. Annu Rev Neurosci 31:69–89.

Moser EI, Moser MB, McNaughton BL (2017) Spatial representation in the hippocampal formation: a history. Nat Neurosci 20:1448–1464.

Moser EI, Roudi Y, Witter MP, Kentros C, Bonhoeffer T, Moser MB (2014) Grid cells and cortical representation. Nat Rev Neurosci 15:466–481.

Mosher CP, Zimmerman PE, Gothard KM (2014) Neurons in the monkey amygdala detect eye contact during naturalistic social interactions. Curr Biol 24:2459–2464.

Mountcastle VB (1957) Modality and topographic properties of single neurons of cat's somatic sensory cortex. J Neurophysiol 20:408–34.

Moyal JE (1949) Causality, determinism and probability. Philosophy 24:310–317.

Muessig L, Hauser J, Wills TJ, Cacucci F (2016) Place cell networks in pre-weanling rats show associative memory properties from the onset of exploratory behavior. Cereb Cortex 26:3627–3636.

Mukamel R, Ekstrom AD, Kaplan J, Iacoboni M, Fried I (2010) Single-neuron responses in humans during execution and observation of actions. Curr Biol 20:750–756.

Muller RA (2016) Now: The Physics of Time. New York: W. W. Norton and Company.

Muller RU, Kubie JL (1987) The effects of changes in the environment on the spatial firing of hippocampal complex-spike cells. J Neurosci 7:1951–1968.

Muller RU, Stead M, Pach J (1996) The hippocampus as a cognitive graph. J Gen Physiol 107:663–694.

Mumford L (1934) Technics and Civilization. New York: Harcourt, Brace & Company.

Murphy BK, Miller KD (2003) Multiplicative gain changes are induced by excitation or inhibition alone. J Neurosci 23:10040–10051.

Musall S, Kaufman MT, Gluf S, Churchland A (2018) Movement-related activity dominates cortex during sensory-guided decision making. BiorRxiV https://doi.org/10.1101/308288

Nadasdy Z, Hirase H, Czurko A, Csicsvari J, Buzsáki G (1999) Replay and time compression of recurring spike sequences in the hippocampus. J Neurosci 19:9497–9507.

Nadel L, Moscovitch M (1997) Memory consolidation, retrograde amnesia and the hippocampal complex. Curr Opin Neurobiol 7:217–227.

Navigli R, Lapata M (2010) An experimental study of graph connectivity for unsupervised word sense disambiguation. IEEE Trans Pattern Anal Mach Intell 32:678–692.

Nelson A, Schneider DM, Takatoh J, Sakurai K, Wang F, Mooney R (2013) A circuit for motor cortical modulation of auditory cortical activity. J Neurosci 33:14342–14353.

Newsome WT, Mikami A, Wurtz RH (1986) Motion selectivity in macaque visual cortex. III. Psychophysics and physiology of apparent motion. J Neurophysiol 55:1340–1351.

Newtson D, Engquist G, Bois, J (1977) The objective basis of behaviour units. J Personal Soc Psychol 35:847–862.

Nicolelis MA, Lebedev MA (2009) Principles of neural ensemble physiology underlying the operation of brain-machine interfaces. Nat Rev Neurosci 10:530–540.

Niell CM, Stryker MP (2010) Modulation of visual responses by behavioral state in mouse visual cortex. Neuron 65:472–479.

Nielsen JM (1958) Memory and Amnesia. Los Angeles, CA: San Lucas.

Niessing J, Friedrich RW (2010) Olfactory pattern classification by discrete neuronal network states. Nature 465:47–52.

Nigam S, Shimono M, Ito S, Yeh F-C, Timme NM, et al. (2016) Rich-club organization in effective connectivity among cortical neurons. J Neurosci 36:670–684.

Nobre AC, O'Reilly J (2004) Time is of the essence. Trend Cog Sci 8:387–389.

Noë A (2004) Action in Perception. Cambridge, MA: MIT Press.

Noë A (2009) Out of Our Heads: Why You Are Not Your Brain, and Other Lessons from the Biology of Consciousness. New York: Hill and Wang.

Norimoto H, Makino K, Gao M, Shikano Y, Okamoto K, et al. (2018) Hippocampal ripples down-regulate synapses. Science 358:1524–1527.

Normann RA, Perlman I (1979) The effects of background illumination on the photoresponses of red and green cones. J Physiol 286:491–507.

Nottebohm F, Stokes TM, Leonard CM (1976) Central control of song in the canary, Serinus canarius. J Comp Neurol 165:457–486.

O'Connor DH, Hires SA, Guo ZV, Li N, Yu J, et al. (2013) Neural coding during active somatosensation revealed using illusory touch. Nat Neurosci 16:958–965.

Oh SW, Harris JA, Ng L, Winslow B, Cain N, et al. (2014) A mesoscale connectome of the mouse brain. Nature 508:207–214.

Ohayon M, Zulley J, Guilleminault C, Smirne S (1999) Prevalence and pathologic associations of sleep paralysis in the general population. Neurology 52:1194–2000.

Ohki K Chung S, Ch'ng YH, Kara P, Reid RC (2005) Functional imaging with cellular resolution reveals precise micro-architecture in visual cortex. Natur 433:597–603.

O'Keefe J (1976) Place units in the hippocampus of the freely moving rat Exp. Neurol 51:78–109.

O'Keefe J (1991) An allocentric spatial model for the hippocampal cognitive map. Hippocampus 1:230–235.

O'Keefe J (1999) Do hippocampal pyramidal cells signal non-spatial as well as spatial information? Hippocampus 9:352–364.

O'Keefe J, Burgess N (1996) Geometric determinants of the place fields of hippocampal neurons. Nature 381:425–428.

O'Keefe J, Dostrovsky J (1971) The hippocampus as a spatial map. Preliminary evidence from unit activity in the freely-moving rat. Brain Res 34:171–175.

O'Keefe J, Nadel L (1978) The Hippocampus as a Cognitive Map. New York: Oxford University Press.

O'Keefe J, Recce ML (1993) Phase relationship between hippocampal place units and the EEG theta rhythm. Hippocampus 3:317–330.

Okun M, Steinmetz N, Cossell L, Iacaruso MF, Ko H, et al. (2015) Diverse coupling of neurons to populations in sensory cortex. Nature 521:511–515.

Ólafsdóttir HF, Barry C, Saleem AB, Hassabis D, Spiers HJ (2015) Hippocampal place cells construct reward related sequences through unexplored space. eLife 4:e06063.

Ólafsdóttir HF, Carpenter F, Barry C (2017) Task demands predict a dynamic switch in the content of awake hippocampal replay. Neuron 96:925–935.

Olsen SR, Bhandawat V, Wilson RI (2010) Divisive normalization in olfactory population codes. Neuron 66:287–299.

Olshausen BA, Anderson CH, Van Essen DC (1993). A neurobiological model of visual attention and invariant pattern recognition based on dynamic routing of information. J Neurosci 13:4700–4719.

Olton DS (1979) Mazes, maps, and memory. Am Psychol 34:583–596.

Omer DB, Maimon SR, Las L, Ulanovsky N (2018) Social place-cells in the bat hippocampus. Science 359:218–224.

Omura Y, Carvalho MM, Inokuchi K, Fukai T (2015) A lognormal recurrent network model for burst generation during hippocampal sharp waves. J Neurosc 35:14585–14601.

O'Neill J, Senior T, Csicsvari J (2006) Place-selective firing of CA1 pyramidal cells during sharp wave/ripple network patterns in exploratory behavior. Neuron 49:143–155.

O'Neill J, Senior TJ, Allen K, Huxter JR, Csicsvari J (2008) Reactivation of experience-dependent cell assembly patterns in the hippocampus. Nat Neurosci 11:209–215.

O'Neill J, Boccara CN, Stella F, Schoenenberger P, Csicsvari J (2017) Superficial layers of the medial entorhinal cortex replay independently of the hippocampus. Science 355:184–188.

O'Regan JK, Noë A (2001) A sensorimotor account of vision and visual consciousness. Beh Brain Sci 25:883–975.

Oscoz-Irurozqui M, Ortuño F (2016) Geniuses of medical science: friendly, open and responsible, not mad. Med Hypotheses 97:71–73.

Ossendrijver M (2016) Ancient Babylonian astronomers calculated Jupiter's position from the area under a time-velocity graph. Science 351:482–484.

Otero-Millan J, Troncoso XG, Macknik SL, Serrano-Pedraza I, Martinez-Conde S (2008) Saccades and microsaccades during visual fixation, exploration, and search: foundations for a common saccadic generator. J Vis 8:1–18.

Pagel M, Atkinson QD, Meade A (2007) Frequency of word-use predicts rates of lexical evolution throughout Indo-European history. Nature 449:717–720.

Paillard J (1991) Motor and representational framing of space. In: Brain and Space (Paillard J, ed), pp. 163–182. Oxford: Oxford University Press.

Palm G, Aertsen A (eds.) (1986) Brain Theory. Proceedings of the First Trieste Meeting on Brain Theory. Berlin: Springer Verlag.

Papale AE, Zielinski MC, Frank LM, Jadhav SP, Redish AD (2016) Interplay between hippocampal sharp-wave-ripple events and vicarious trial and error behaviors in decision making. Neuron 92:975–982.

Papez JW (1937) A proposed mechanism of emotion. Arch Neurol Psychiatry 38:725–744.

Parker ST, Gibson KR (1977) Object manipulation, tool use and sensimotor intelligence as feeding adaptations in cebus monkeys and great apes. J Hum Evol 6:623–641.

Parker Jones O, Alfaro-Almagro F, Jbabdi S (2018) An empirical, 21st century evaluation of phrenology. Cortex 106:26–35.

Parron C, Save E (2004) Evidence for entorhinal and parietal cortices involvement in path integration in the rat. Exp Brain Res 159:349–359.

Pasley BN, David SV, Mesgarani N, Flinker A, Shamma SA, Crone NE, Knight RT, Chang EF. (2012) Reconstructing speech from human auditory cortex. PLoS Biol 10:e1001251.

Pastalkova E, Itskov V, Amarasingham A, Buzsáki G (2008) Internally generated cell assembly sequences in the rat hippocampus. Science 321:1322–1327.

Patel J, Fujisawa S, Berényi A, Royer S, Buzsáki G (2012) Traveling theta waves along the entire septotemporal axis of the hippocampus. Neuron 75:410–417.

Patzke N, Spocter MA, Karlsson KÆ, Bertelsen MF, Haagensen M, et al. (2015) In contrast to many other mammals, cetaceans have relatively small hippocampi that appear to lack adult neurogenesis. Brain Struct Funct 220:361–383.

Paus T, Perry DW, Zatorre RJ, Worsley KJ, Evans AC (1996) Modulation of cerebral blood flow in the human auditory cortex during speech: role of motor-to-sensory discharges. Eur J Neurosci 8:2236–2246.

Payne K (2017) The broken ladder: how inequality affects the way we think, live and die. Viking.

Pearl J (1995) Causal diagrams for empirical research. Biometrika 82:669–709.

Pedroarena C, Llinás R (1997) Dendritic calcium conductances generate high-frequency oscillation in thalamocortical neurons. Proc Natl Acad Sci U S A 94:724–728.

Penrose R (2004) The Road to Reality: A Complete Guide to the Laws of the Universe. London: Jonathan Cape.

Penttonen M, Buzsáki G (2003) Natural logarithmic relationship between brain oscillators. Thalamus Related Systems 2:145–152.

Pepperberg I (1994) Numerical competence in an African gray parrot (Psittacus erithacus). J Comp Psychol 108:36–44.

Perbal S, Couillet J, Azouvi P, Pouthas V (2003) Relationships between time estimation, memory, attention, and processing speed in patients with severe traumatic brain injury. Neuropsychologia 41:1599–1610.

Perez-Orive J, Mazor O, Turner GC, Cassenaer S, Wilson RI, Laurent G (2002) Oscillations and sparsening of odor representations in the mushroom body. Science 297:359–365.

Perin R, Berger TK, Markram H (2011) A synaptic organizing principle for cortical neuronal groups. Proc Natl Acad Sci U S A 108:5419–5424.

Pesic P (2018) Polyphonic Minds: Music of the Hemispheres. Cambridge MIT Press.

Petreanu L, Gutnisky DA, Huber D, Xu Nl, O'Connor DH, et al. (2012) Activity in motor– sensory projections reveals distributed coding in somatosensation. Nature 489:299–302.

Petsche H, Stumpf C, Gogolák G (1962) The significance of the rabbit's septum as a relay station between midbrain and the hippocampus. I. The control of hippocampus arousal activity by the septum cells. Electroencephalogr Clin Neurophysiol 14:202–211.

Peyrache A, Battaglia FP, Destexhe A (2011) Inhibition recruitment in prefrontal cortex during sleep spindles and gating of hippocampal inputs. Proc Natl Acad Sci U S A 108:17207–17212.

Peyrache A, Lacroix MM, Petersen PC, Buzsáki G (2015) Internally organized mechanisms of the head direction sense. Nat Neurosci 18:569–575.

Peyrache A, Schieferstein N, Buzsáki G (2017) Transformation of the head-direction signal into a spatial code. Nat Commun 8:1752. doi:10.1038/s41467-017-01908-3.

Pezzulo G, Kemere C, van der Meer MAA (2017) Internally generated hippocampal sequences as a vantage point to probe future-oriented cognition. Ann N Y Acad Sci 1396:144–165.

Pfeiffer BE (2017) The content of hippocampal "replay." Hippocampus doi:10.1002/hipo.22824.

Pfeiffer BE, Foster DJ (2013) Hippocampal place-cell sequences depict future paths to remembered goals. Nature 497:74–79.

Pfungst O (2000) Clever Hans: The Horse of Mr. von Ostern. London: Thoemmes Press.

Piaget J (1946) Le D´eveloppement de la Notion de Temps chez l'Enfant. Paris: Presses Universitaires de France.

Piaget J (1957) The child and modern physics. Sci Am 196:46–51.

Pinker S (2003) The Blank Slate: The Modern Denial of Human Nature. New York: Viking.

Pizlo Z, Li Y, Sawada T, Steinman RM (2014) Making a Machine That Sees Like Us. New York: Oxford University Press.

Poincaré H (1905) La valeur de la science. Paris: Flammarion.

Poldrack RA (2010) Mapping mental function to brain structure: how can cognitive neuroimaging succeed? Perspect Psychol Sci 5:753–761.

Polyn SM, Natu VS, Cohen JD, Norman KA (2005) Category-specific cortical activity precedes retrieval during memory search. Science 310:1963–1966.

Popper K (1959) The Logic of Scientific Discovery. Abingdon-on-Thames, UK: Routledge.
Port RF, Van Gelder T (1995) Mind as Motion. Cambridge, MA: MIT Press.
Posamentier AS, Lehmann I (2007) The Fabulous Fibonacci Numbers. Amherst, NJ: Prometheus Books.
Pouget A, Sejnowski T (1994) A neural model of the cortical representation of egocentric distance. Cereb Cortex 4:314–329.
Pouget A, Sejnowski TJ (1997a) A new view of hemineglect based on the response properties of parietal neurones. Philos Trans R Soc Lond B Biol Sci 352:1449–1459.
Pouget A, Sejnowski TJ (1997b) Spatial tranformations in the parietal cortex using basis functions. J Cog Neurosci 9:222–237.
Poulet JF, Hedwig B (2006) The cellular basis of a corollary discharge. Science 311:518–522.
Power RA, Steinberg S, Bjornsdottir G, Rietveld CA, Abdellaoui A, et al. (2015) Polygenic risk scores for schizophrenia and bipolar disorder predict creativity. Nat Neurosci 18:953–955.
Prinz AA, Bucher D, Marder E (2004) Similar network activity from disparate circuit parameters. Nat Neurosci 7:1345–1352.
Prinz W, Beisert M, Herwig A (2013) Action Science: Foundations of an Emerging Discipline. Cambridge, MA: MIT Press.
Proffitt T, Luncz LV, Falótico T, Ottoni EB, de la Torre I, Haslam M (2016) Wild monkeys flake stone tools. Nature 539:85–88.
Pulvermüller F (2003) The Neuroscience of Language. Cambridge: Cambridge University Press.
Pulvermüller F (2010) Brain embodiment of syntax and grammar: discrete combinatorial mechanisms spelt out in neuronal circuits. Brain Lang 112:167–179.
Pulvermüller F (2013) Semantic embodiment, disembodiment or misembodiment? In search of meaning in modules and neuron circuits. Brain Lang 127:86–103.
Quian Quiroga R, Reddy L, Kreiman G, Koch C, Fried I (2005) Invariant visual representation by single neurons in the human brain. Nature 435:1102–1107.
Quilichini P, Sirota A, Buzsáki G (2010) Intrinsic circuit organization and theta-gamma oscillation dynamics in the entorhinal cortex of the rat. J Neurosci 30:11128–11142.
Quine WVO, Churchland PS, Føllesdal D (2013) Word and Object. Cambridge, MA: MIT Press
Quintana J, Fuster JM (1999) From perception to action: temporal integrative functions of prefrontal and parietal neurons. Cereb Cortex 9:213–221.
Rabinovich MI, Huerta R, Varona P, Afraimovich VS (2008) Transient cognitive dynamics, metastability, and decision making. PLoS Comput Biol 4:e1000072.
Radua J, Del Pozo NO, Gómez J, Guillen-Grima F, Ortuño F (2014) Meta-analysis of functional neuroimaging studies indicates that an increase of cognitive difficulty during executive tasks engages brain regions associated with time perception. Neuropsychologia 58:14–22.
Rall W (1964) Theoretical significance of dendritic trees for neuronal input-output relations. In: Neural Theory and Modeling (Reiss R, ed), pp. 73–97. Stanford, CA: Stanford University Press.

Ramachandran VS, Rogers-Ramachandran D, Cobb S (1995) Touching the phantom limb. Nature 377:489–490.

Ranck JB (1985) Head direction cells in the deep cell layer of dorsal presubiculum in freely moving rats. In: Electrical Activity of the Archicortex (Buzsáki G, Vanderwolf CH, eds), pp. 217–220. Budapest: Akadémiai Kiadó.

Rangel LM, Quinn LK, Chiba AA (2015) Space, time, and the hippocampus. In: The Neurobiological Basis of Memory (Jackson PA, Chiba AA, Berman RF, Ragozzino ME, eds), pp. 59–75. Berlin: Springer-Verlag.

Rao RP, Ballard DH (1999) Predictive coding in the visual cortex: a functional interpretation of some extra-classical receptive-field effects. Nat Neurosci 2:79–87.

Ratcliff R (1990) Connectionist models of recognition memory: constraints imposed by learning and forgetting functions. Psychol Rev 97:285–308.

Rauskolb FW, Berger K, Lipski C, Magnor M, Cornelsen K, et al. (2008) Caroline: an autonomously driving vehicle for urban environments. J Field Robotics 25:674–724.

Redican WK (1975) Facial expressions in nonhuman primates. In Primate Behavior (Rosenblum LA, ed), pp. 103–194. London: Academic Press.

Redish AD (2016) Vicarious trial and error. Nat Rev Neurosci 17:147–159.

Redish AD, Elga AN, Touretzky DS (1996) A coupled attractor model of the rodent head direction system. Netw Comput Neural Syst 7:671–685.

Redish AD, Rosenzweig ES, Bohanick JD, McNaughton BL, Barnes CA (2000) Dynamics of hippocampal ensemble activity realignment: time versus space. J Neurosci 20:9298–9309.

Redish AD, Touretzky DS (1997) Cognitive maps beyond the hippocampus. Hippocampus 7:15–35.

Reeves A (2017) The architecture of inequality. Nature 543:312–314.

Reid CR, Latty T, Dussutour A, Beekman M (2012) Slime mold uses an externalized spatial "memory" to navigate in complex environments. Proc Natl Acad Sci U S A 109:17490–17494.

Renart A, de la Rocha J, Bartho P, Hollender L, Parga N, et al. (2010) The asynchronous state in cortical circuits. Science 327:587–590.

Renfrew C, Frith C, Malafouris L (2009) The Sapient Mind: Archaeology Meets Neuroscience: Oxford: Oxford University Press.

Rescorla RA, Wagner AR (1972) A theory of Pavlovian conditioning: Variations in the effectiveness of reinforcement and non-reinforcement. In: Classical Conditioning II: Current Theory and Research (Black AH, Prokasy WF, eds), pp. 64–99. New York: Appleton-Century.

Reynolds JH, Heeger DJ (2009) The normalization model of attention. Neuron 61:168–185.

Rich PD, Liaw HP, Lee AK (2014) Place cells. Large environments reveal the statistical structure governing hippocampal representations. Science 345:814–817.

Rieke F, Bodnar DA, Bialek W (1995) Naturalistic stimuli increase the rate and efficiency of information transmission by primary auditory afferents. Proc Biol Sci 262:259–265.

Rieke F, Warland D, de Ruyter van Steveninck R, Bialek W (1997) Spikes: Exploring the Neural Code. Cambridge, MA: MIT Press.

Riggs LA, Ratliff F (1952) The effects of counteracting the normal movements of the eye. J Opt Soc Am 42:872–873.
Risold PY, Swanson LW (1996) Structural evidence for functional domains in the rat hippocampus. Science 272:1484–1486.
Rizzolatti G, Arbib MA (1998) Language within our grasp. Trends Cogn Sci 21:188–194.
Rizzolatti G, Craighero L (2004) The mirror-neuron system. Annu Rev Neurosci 27:169–192.
Rosenbaum P, Rubin DB (1983) The central role of the propensity score in observational studies for causal effects. Biometrika 70:41–55.
Rosenblum B, Kuttner F (2008) Quantum Enigma: Physics Encounters Consciousness. New York: Oxford University Press.
Ross J, Morrone MC, Goldberg ME, Burr DC (2001) Changes in visual perception at the time of saccades. Trends Neurosci 24:113–121.
Rothman JS, Cathala L, Steuber V, Silver RA (2009) Synaptic depression enables neuronal gain control. Nature 457:1015–1018.
Rothschild G, Nelken I, Mizrahi (2010) Functional organization and population dynamics in the mouse primary auditory cortex. Nat Neurosci 13:353–560.
Rothschild G, Eban E, Frank LM (2017) A cortical-hippocampal-cortical loop of information processing during memory consolidation. Nat Neurosci 20:251–259.
Roux L, Hu B, Eichler R, Stark E, Buzsáki G (2017) Sharp wave ripples during learning stabilize the hippocampal spatial map. Nat Neurosci 20:845–853.
Rovelli C (2016) Reality Is Not What It Seems: The Journey to Quantum Gravity. London: Allan Lane Publisher.
Roweis ST, Saul LK (2000) Nonlinear dimensionality reduction by locally linear embedding. Science 290:2323–2325.
Roxin A, Brunel N, Hansel D, Mongillo G, van Vreeswijk C (2011) On the distribution of firing rates in networks of cortical neurons. J Neurosci 31:16217–16226.
Royer S, Sirota A, Patel J, Buzsáki G (2010) Distinct representations and theta dynamics in dorsal and ventral hippocampus. J Neurosci 30:1777–1787.
Royer S, Zemelman BV, Losonczy A, Kim J, Chance F, et al. (2012) Control of timing, rate and bursts of hippocampal place cells by dendritic and somatic inhibition. Nat Neurosci 15:769–775.
Rudy B, Fishell G, Lee S, Hjerling-Leffler J (2011) Three groups of interneurons account for nearly 100% of neocortical GABAergic neurons. Dev Neurobiol 71:45–61.
Ruff CB, Trinkhous E, Holliday TW (1997) Body mass and encephalization in Pleistocene Homo. Nature 387:173–176.
Rumelhart DE, McClelland JL, the PDP Research Group (1986) Parallel Distributed Processing: Explorations in the Microstructure of Cognition. Volume 1: Foundations. Cambridge, MA: MIT Press.
Russell B (1992) On the notion of cause. In: The Collected Papers of Bertrand Russell v6: Logical and Philosophical Papers 1909–1913 (Slater J, ed), pp 193–210. London: Routledge Press.
Russell B, Slater JG, Frohmann B (1992) The Collected Papers of Bertrand Russell: Logical and Philosophical Papers, 1909–1913. New York: Routledge.
Rutishauser U, Tudusciuc O, Wang S, Mamelak AN, Ross IB, Adolphs R (2013) Single-neuron correlates of atypical face processing in autism. Neuron 80:887–899.

Sabbah S, Gemmer JA, Bhatia-Lin A, Manoff G, Castro G, et al. (2017) A retinal code for motion along the gravitational and bodyaxes. Nature 546:492–497.

Sajin SM, Connine CM (2014) Semantic richness: the role of semantic features in processing spoken words. J Mem Lang 70:13–35.

Sale K (1995) Rebels Against the Future: The Luddites and Their War on the Industrial Revolution: Lessons for the Computer Age. New York: Basic Books.

Salinas E, Abbott LF (1995) Transfer of coded information from sensory to motor networks. J Neurosci 15:6461–6474.

Salinas E, Sejnowski TJ (2001) Gain modulation in the central nervous system: where behavior, neurophysiology, and computation meet. Neuroscientist 7:430–440.

Salinas E, Thier P (2000) Gain modulation: a major computational principle of the central nervous system. Neuron 27:15–21.

Samsonovich A, McNaughton BL (1997) Path integration and cognitive mapping in a continuous attractor neural network model. J Neurosci 17:5900–5920.

Samsonovich AV, Ascoli GA (2005) A simple neural network model of the hippocampus suggesting its pathfinding role in episodic memory retrieval. Learn Mem 12:193–208.

Sanchez-Vives MV, McCormick DA (2000) Cellular and network mechanisms of rhythmic recurrent activity in neocortex. Nat Neurosci 3:1027–1034.

Sargolini F, Fyhn M, Hafting T, McNaughton BL, Witter MP, et al. (2006) Conjunctive representation of position, direction, and velocity in entorhinal cortex. Science 312:758–762.

Sarpeshkar R (2010) Ultra Low Power Bioelectronics: Fundamentals, Biomedical Applications, and Bio-inspired Systems. New York: Cambridge University Press.

Sawamura H, Shima K, Tanji J (2002) Numerical representation for action in the parietal cortex of the monkey. Nature 415:918–922.

Scarr S, McCartney K (1983) How people make their own environments: a theory of genotype → environment effects. Child Dev 54:424–435.

Schacter DL (2001) Forgotten Ideas, Neglected Pioneers: Richard Semon and the Story of Memory. Philadelphia: Psychology Press.

Schacter DL, Addis DR (2007) Constructive memory: the ghosts of past and future. Nature 445:27.

Schacter DL, Addis DR, Buckner RL (2007) Remembering the past to imagine the future: the prospective brain. Nat Rev Neurosci 8:657–661.

Schacter DL, Harbluk J, McLachlan D (1984) Retrieval without recollection: an experimental analysis of source amnesia. J Verb Learn Verb Behav 23:593–611.

Schaffer J (2016) The metaphysics of causation. In: The Stanford Encyclopedia of Philosophy (Zalta EN ed). Stanford, CA: Stanford University.

Scharnowski F, Rees G, Walsh V (2013) Time and the brain: neurorelativity: the chronoarchitecture of the brain from the neuronal rather than the observer's perspective. Trends Cogn Sci 17:51–52.

Scheidel W (2017) The Great Leveler: Violence and History of Inequality from the Stone Age to the Twenty-First Century. Princeton, NJ: Princeton University Press.

Schlegel AA, Rudelson JJ, Tse PU (2012) White matter structure changes as adults learn a second language. J Cogn Neurosci 24:1664–1670.

Schmajuk NA, Thieme AD (1992) Purposive behavior and cognitive mapping: a neural network model. Biol Cybern 67:165-174.
Schmitz TW, Duncan J (2018) Normalization and the cholinergic microcircuit: a unified basis for attention. Trend Cogn Sci 22:422-437.
Schneider DM, Nelson A, Mooney R (2014) A synaptic and circuit basis for corollary discharge in the auditory cortex. Nature 513:189-194.
Schneider F, Wildermuth D (2011) Results of the European land robot trial and their usability for benchmarking outdoor robot systems. Towards Autonomous Robotic Systems 408-409.
Scholz J, Klein MC, Behrens TE, Johansen-Berg H (2009) Training induces changes in white-matter architecture. Nat Neurosci 12:1370-1371.
Schomburg EW, Fernández-Ruiz A, Mizuseki K, Berényi A, Anastassiou CA, et al. (2014) Theta phase segregation of input-specific gamma patterns in entorhinal-hippocampal networks. Neuron 84:470-485.
Schroeder CE, Lakatos P (2009) Low-frequency neuronal oscillations as instruments of sensory selection. Trends Neurosci 32:9-18.
Schroeder CE, Lakatos P, Kajikawa Y, Partan S, Puce A (2008) Neuronal oscillations and visual amplification of speech. Trends Cogn Sci 12:106-113.
Schultz W (1998) Predictive reward signal of dopamine neurons. J Neurophysiol 80:1-27.
Schultz W (2015) Neuronal reward and decision signals: from theories to data. Physiol Rev 95:853-951.
Scoville WB, Milner B (1957) Loss of recent memory after bilateral hippocampal lesions. J Neurol Neurosurg Psychiat 20:11-21.
Seetharaman G, Lakhotia A, Blasch E (2006) Unmanned vehicles come of age: the DARPA grand challenge. Computer 39:26-29.
Sejnowski TJ (2018) The Deep Learning Revolution. Cambridge, MA: MIT Press.
Seligman MEP (1971) Phobias and preparedness. Behavior Ther 2:307-320.
Seligman MEP (1975) Helplessness: On Depression, Development, and Death. San Francisco, CA: W. H. Freeman.
Senzai Y, Buzsáki G (2017) Physiological properties and behavioral correlates of hippocampal granule cells and mossy cells. Neuron 93:691-704.
Seth AK (2005) Causal connectivity analysis of evolved neuronal networks during behavior. Netw Comput Neural Syst 16:35-55.
Shadlen MN, Kiani R (2013) Decision making as a window on cognition. Neuron 80:791-806.
Shafer G (1996) The Art of Causal Conjecture. Cambridge, MA: MIT Press.
Shankar KH, Howard MW (2012) A scale-invariant representation of time. Neural Computation 24:134-193.
Shannon CE (1948) A mathematical theory of communication. Bell System Technical Journal 623-656.
Shannon CE (1956) The bandwagon. IRE Trans InformTheory 2:3.
Shannon RV, Zeng FG, Kamath V, Wygonski J, Ekelid M (1995) Speech recognition with primarily temporal cues. Science 270:303-304.
Sharma J, Angelucci A, Sur M (2000) Induction of visual orientation modules in auditory cortex. Nature 404:841-847.

Shaw GL, Silverman DJ, Pearson JC (1985) Model of cortical organization embodying a basis for the theory of information processing and memory recall. Proc Natl Acad Sci.82:2364–2368.

Sheer DE, Grandstaff NW, Benignus VA (1966) Behavior and 40-c-sec. electrical activity in the brain. Psychol Rep 19:1333–1334.

Shenoy KV, Sahani M, Churchland MM (2013) Cortical control of arm movements: a dynamical systems perspective. Annu Rev Neurosci 36:337–359.

Shepard RN, Kilpatric DW, Cunningham JP (1975) The internal representation of numbers. Cogn Psychol 7:82–138.

Sherrington CS (1942) Man on His Nature. Cambridge: Cambridge University Press.

Shilnikov AL, Maurer AP (2016) The art of grid fields: geometry of neuronal time. Front Neural Circuits 10:12.

Shimamura AP, Squire LR (1987). A neuropsychological study of fact memory and source amnesia. J Exp Psychol Learn Mem Cogn 13:464–473.

Shipley TF, Zacks JM (eds) (2008) Understanding Events: From Perception to Action. Oxford: Oxford University Press, 2008.

Shmueli G (2010) To explain or to predict? Statist Sci 25:289–310.

Shumway-Cook A, Woollacott MH (1995) Motor Control: Theory and Practical Applications. Philadelphia, PA: Lippincott Williams & Wilkins.

Siapas AG, Wilson MA (1998) Coordinated interactions between hippocampal ripples and cortical spindles during slow-wave sleep. Neuron 21:1123–1128.

Siegler RS, Opfer JE (2003) The development of numerical estimation: evidence for multiple representations of numerical quantity. Psychol Sci 14:237–243.

Silva D, Feng T, Foster DJ (2015) Trajectory events across hippocampal place cells require previous experience. Nat Neurosci 18:1772–1779.

Silver D, Huang A, Maddison CJ, Guez A, Sifre L, et al. (2016) Mastering the game of Go with deep neural networks and tree search. Nature 529:484–489.

Silver RA (2010) Neuronal arithmetic. Nat Rev Neurosci 11:474–489.

Singer AC, Carr MF, Karlsson MP, Frank LM (2013) Hippocampal SWR activity predicts correct decisions during the initial learning of an alternation task. Neuron 77:1163–1173.

Singer W (1999) Neuronal synchrony: a versatile code for the definition of relations? Neuron 24:49–65.

Singh N, Theunissen F (2003) Modulation spectra of natural sounds and ethological theories of auditory processing. J Acoust Soc Am 114:3394–3411.

Sinha C, Da Silva Sinha V, Zinken J, Sampaio W (2011) When time is not space: the social and linguistic construction of time intervals and temporal event relations in an Amazonian culture. Lang Cogn 3:137–169.

Sinnot EW (1937) The relation of gene to character in quantitative inheritance. Proc Natl Acad Sci USA 23:224–227.

Sirota A, Csicsvari J, Buhl D, Buzsáki G (2003) Communication between neocortex and hippocampus during sleep in rodents. Proc Natl Acad Sci U S A 100:2065–2069.

Sirota A, Montgomery S, Fujisawa S, Isomura Y, Zugaro M, Buzsáki G (2008) Entrainment of neocortical neurons and gamma oscillations by the hippocampal theta rhythm. Neuron 60:683–697.

Skaggs WE, McNaughton BL (1996) Replay of neuronal firing sequences in rat hippocampus during sleep following spatial experience. Science 271:1870–1873.

Skaggs WE, McNaughton BL, Wilson MA, Barnes CA (1996) Theta phase precession in hippocampal neuronal populations and the compression of temporal sequences. Hippocampus 6:149–172.

Skarda CA, Freeman WJ (1987) How brains make chaos in order to make sense of the world. Behav Brain Sci 10:161–173.

Skeide MA, Kumar U, Mishra RK, Tripathi VN, Guleria A, et al. (2017) Learning to read alters cortico-subcortical cross-talk in the visual system of illiterates. Sci Adv 3:e1602612.

Skinner BF (1938) The Behavior of Organisms: An Experimental Analysis. Boston, MA: D. Appleton & Company.

Slotine JJE, Li W (1991) Applied Nonlinear Control. Engelwood, NJ: Prentice-Hall.

Smear M, Resulaj A, Zhang J, Bozza T, Rinberg D (2013) Multiple perceptible signals from a single olfactory glomerulus. Nat Neurosci 16:1687–1691.

Smolin L (2013) Time Reborn: From the Crisis of Physics to the Future of the Universe. Boston, MA: Houghton Mifflin Harcourt.

Soares S, Atallah BV, Paton JJ (2016) Midbrain dopamine neurons control judgment of time. Science 354:1273–1277.

Sobel D (1995) Longitude: The True Story of a Lone Genius Who Solved the Greatest Scientific Problem of His Time. New York: Walker Publishing Company, Inc.

Sokolov EN (1960) Neuronal models and the orienting reflex. In: The Central Nervous System and Behavior (Brazier MAB, ed), pp. 187–276. New York: Josiah Macy, Jr. Foundation.

Sokolov EN (1963) Perception and the Conditioned Reflex. New York: Pergamon Press.

Soltesz I (2005) Diversity in the Neuronal Machine: Order and Variability in Interneuronal Microcircuits. New York: Oxford University Press.

Soltesz I, Deschênes M (1993) Low- and high-frequency membrane potential oscillations during theta activity in CA1 and CA3 pyramidal neurons of the rat hippocampus under ketamine-xylazine anesthesia. J Neurophysiol 70:97–116.

Sommer MA, Wurtz RH (2006) Influence of the thalamus on spatial visual processing in frontal cortex. Nature 444:374–377.

Song S, Sjöström PJ, Reigl M, Nelson S, Chklovskii DB (2005) Highly nonrandom features of synaptic connectivity in local cortical circuits. PLoS Biology 3:e68.

Sperry RW (1950) Neural basis of the spontaneous optokinetic response produced by visual inversion. J Compar Physiol Psychol 43:482–489.

Spirtes P, Glymour C, Scheines R (2000) Causation, Prediction, and Search (2nd ed). Cambridge, MA: MIT Press.

Sporns O (2010) Networks of the Brain. Cambridge, MA: MIT Press.

Spruijt BM, van Hooff JA, Gispen WH (1992) Ethology and neurobiology of grooming behavior. Physiol Rev 72:825–852.

Squire LR (1992a) Declarative and nondeclarative memory: multiple brain systems supporting learning and memory. J Cogn Neurosci 4:232–243.

Squire LR (1992b) Memory and the hippocampus: a synthesis from findings with rats, monkeys, and humans. Psychol Rev 99:195–231.

Squire LR, Alvarez P (1995) Retrograde amnesia and memory consolidation: a neurobiological perspective. Curr Opin Neurobiol 5:169–177.

Squire LR, Slater PC, Chace PM (1975) Retrograde amnesia: temporal gradient in very long term memory following electroconvulsive therapy. Science 187:77–79.

Squire LR, Stark CE, Clark RE (2004) The medial temporal lobe. Annu Rev Neurosci 27:279–306.

Srinivasamurthy A, Subramanian S, Tronel G, Chordia P (2012) A beat tracking approach to complete description of rhythm in Indian classical music. Proc 2nd Comp Music Workshop, 72–78.

Staddon JE (2005) Interval timing: memory, not a clock. Trends Cogn Sci 9:312–314.

Staddon JER (1978) Theory of behavioral power functions. Psychol Rev 85:305–320.

Staddon JER, Simmelhag VL (1971) The "superstition" experiment: a reexamination of its implications for the principles of adaptive behavior. Psychol Rev 78:3–43.

Stark E, Roux L, Eichler R, Buzsáki G (2015) Local generation of multineuronal spike sequences in the hippocampal CA1 region. Proc Natl Acad Sci U S A 112:10521–10526.

Steinhardt PJ, Turok N (2002) A cyclic model of the universe. Science 296:1436–1439

Steriade M, Contreras D, Curró Dossi R, Nuñez A (1993a) The slow (< 1 Hz) oscillation in reticular thalamic and thalamocortical neurons: scenario of sleep rhythm generation in interacting thalamic and neocortical networks. J Neurosci 13:3284–3299.

Steriade M, Nunez A, Amzica F (1993b) A novel slow (~1 Hz) oscillation of neocortical neurons in vivo: depolarizing and hyperpolarizing components. J Neurosci 13:3252–3265.

Steriade M, Nuñez A, Amzica F (1993c) Intracellular analysis of relations between the slow (< 1 Hz) neocortical oscillation and other sleep rhythms of the electroencephalogram. J Neurosci 13:3266–3283.

Stevens SS (1961) To honor Fechner and repeal his law: a power function, not a log function, describes the operating characteristic of a sensory system. Science 133:80–86.

Stone JV (2013) Bayes' Rule: A Tutorial Introduction to Bayesian Analysis. London: Sebtel Press.

Stringer C, Pachitariu M, Steinmetz N, Reddy CB, Carandini M, Harris KD (2018) Spontaneous behaviors drive multidimensional, brain-wide population activity. BiorRxiV. https://doi.org/10.1101/306019.

Sturm I, Blankertz B, Potes C, Schalk G, Curio G (2014) ECoG high gamma activity reveals distinct cortical representations of lyrics passages, harmonic and timbre-related changes in a rock song. Front Human Neurosci 8:798.

Suddendorf T, Corballis MC (2007) The evolution of foresight: what is mental time travel, and is it unique to humans? Behav Brain Sci 30:299–313.

Suga N, Schlegel P (1972) Neural attenuation of responses to emitted sounds in echolocating bats. Science 177:82–84.

Sullivan D, Mizuseki K, Sorgi A, Buzsáki G (2014) Comparison of sleep spindles and theta oscillations in the hippocampus. J Neurosci 34:662–674.

Sussillo D, Abbott LF (2009) Generating coherent patterns of activity from chaotic neural networks. Neuron 63:544–557.

Sutton RS, Barto AG (1998) Reinforcement Learning: An Introduction (Adaptive Computation and Machine Learning). Cambridge, MA: MIT Press.

Swadlow HA (2000) Information flow along neocortical axons. In: Time and the Brain (Miller R, ed). Reading, UK: Harwood Academic Publishers.

Szent-Györgyi A (1951) Nature of the contraction of muscle. Nature 167:380–381.

Takahashi N, Sasaki T, Matsumoto W, Matsuki N, Ikegaya Y (2010) Circuit topology for synchronizing neurons in spontaneously active networks. Proc Natl Acad Sci U S A 107:10244–10249.

Takehara-Nishiuchi K, McNaughton BL (2008) Spontaneous changes of neocortical code for associative memory during consolidation. Science 322:960–963.

Talmy L (1985) Lexicalization patterns: semantic structure in lexical forms. In: Language Typology and Semantic Description (Shopen T, ed), pp. 57–149. Cambridge: Cambridge University Press.

Taube JS (2007) The head direction signal: origins and sensory-motor integration. Annu Rev Neurosci 30:181–207.

Tegmark M (2014) Our Mathematical Universe: My Quest for the Ultimate Nature of Reality. New York: Vintage Books, Random House.

Tenenbaum JB, de Silva V, Langford JC (2000) A global geometric framework for non-linear dimensionality reduction. Science 290:2319–2323.

Tennesen M (2009) More Animals Seem to Have Some Ability to Count. Sci Am. September. http://www.scientificamerican.com/article/how-animals-have-the-ability-to-count/.

Terada S, Sakurai Y, Nakahara H, Fujisawa S (2017) Temporal and rate coding for discrete event sequences in the hippocampus. Neuron 94:1248–1262.

Teyler TJ, DiScenna P (1986) The hippocampal memory indexing theory. Behav Neurosci 100:147–154.

Thaler L, Arnott SR, Goodale MA (2011) Neural correlates of natural human echolocation in early and late blind echolocation experts. PLoS One 6:e20162.

Thelen E (1989) Self-organization in developmental processes: Can systems approaches work? In: Systems and Development: The Minnesota Symposia on Child Psychology, Vol. 22 (Gunnar M, Thelen E, eds), pp. 17–171. Hillsdale, NJ: Erlbaum.

Thomas L (1972) Antaeus in Manhattan. N Engl J Med 286:1046–1047.

Thompson RF (2005) In search of memory traces. Annu Rev Psychol 56:1–23.

Thomson AM, Deuchars J (1994) Temporal and spatial properties of local circuits in neocortex. Trends Neurosci 17:119–126.

Thomson AM, West DC (2003) Presynaptic frequency filtering in the gamma frequency band; dual intracellular recordings in slices of adult rat and cat neocortex. Cereb Cortex 13:136–143.

Thorpe S, Delorme A, Van Rullen R (2001) Spike-based strategies for rapid processing. Neural Netw 14:715–725.

Tiganj Z, Cromer JA, Roy GE, Miller EK, Howard MW (2017) Compressed timeline of recent experience in monkey lPFC. bioRxiv 126219.

Tingley D, Buzsáki G (2018) Transformation of a spatial map across the hippocampal-lateral septal circuit. Neuron 98:1229–1242.

Tinbergen N (1951) The Study of Instinct. New York: Oxford University Press.

Tolman EC (1948) Cognitive maps in rats and men. Psychol Rev 55:189–208.

Tomasino B, Fink GR, Sparing R, Dafotakis M, Weiss PH (2008) Action verbs and the primary motor cortex: a comparative TMS study of silent reading, frequency judgments, and motor imagery. Neuropsychologia 46:1915–1926.
Tonegawa S, Liu X, Ramirez S, Redondo R (2015) Memory engram cells have come of age. Neuron 87:918–931.
Tononi G (2012) Phi: A Voyage from the Brain to the Soul. New York: Pantheon Books.
Tononi G, Cirelli C (2003) Sleep and synaptic homeostasis: a hypothesis. Brain Res Bull 62:143–150.
Tononi G, Cirelli C (2014) Sleep and the price of plasticity: from synaptic and cellular homeostasis to memory consolidation and integration. Neuron 81:12–34.
Tononi G, Sporns O, Edelman GM (1994) A measure for brain complexity: relating functional segregation and integration in the nervous system. Proc Natl Acad Sci U S A 91:5033–5037.
Toth N (1985) The Oldowan reassessed: a close look at early stone artifacts. J Archeol Sci 12:101–120.
Toulmin S, Goodfield J (1982) The Discovery of Time. Chicago: The University of Chicago Press.
Treue S (2003) Visual attention: the where, what, how and why of saliency. Curr Opin Neurobiol 13:428–432.
Treue S, Martínez-Trujillo JC (1999) Feature-based attention influences motion processing gain in macaque visual cortex. Nature 399:575–579.
Treves A, Rolls ET (1994) Computational analysis of the role of the hippocampus in memory. Hippocampus 4:374–391.
Trivers RL (2011) The Folly of Fools. New York: Basic Books.
Trope Y, Liberman N (2010) Construal-level theory of psychological distance. Psychol Rev 117:440–463.
Troxler IPV (1804) Über das Verschwinden gegebener Gegenstände innerhalb unseres Gesichtskreises [On the disappearance of given objects from our visual field]. Ophthalmologische Bibliothek (in German) 2:1–53.
Truccolo W, Hochberg LR, Donoghue JP (2010) Collective dynamics in human and monkey sensorimotor cortex: predicting single neuron spikes. Nat Neurosci 13:105–111.
Tse PU, Intriligator J, Rivest J, Cavanagh P (2004) Attention and the subjective expansion of time. Percept Psychophys 66:1171–1189.
Tsodyks M, Kenet T, Grinvald A, Arieli A (1999) Linking spontaneous activity of single cortical neurons and the underlying functional architecture. Science 286:1943–1946.
Tsodyks MV, Skaggs WE, Sejnowski TJ, McNaughton BL (1996) Population dynamics and theta rhythm phase precession of hippocampal place cell firing: a spiking neuron model. Hippocampus 6:271–280.
Tulving E (1972) Episodic and semantic memory. In: Organization of Memory (Tulving E, Donaldson W, eds), pp. 381–402. New York: Academic Press.
Tulving E (1983) Elements of Episodic Memory. Oxford: Clarendon Press.
Tulving E (2002) Episodic memory: from mind to brain. Ann Rev Psych 53:1–15.
Tulving E, Schacter DL (1990) Priming and human memory systems. Science 247:301–306.
Tversky A (1977) Features of similarity. Psychological Review 84:327–352.

Tversky A, Kahneman D (1974) Judgment under uncertainty: heuristics and biases. Science 185:1124–1131.

Tversky A, Kahneman D (1981) The framing of decisions and the psychology of choice. Science 211:453–4458.

Ullman S (1980) Against direct perception. Behav Brain Sci 3:373–416.

Umeno MM, Goldberg ME (1997) Spatial processing in the monkey frontal eye field. I. Predictive visual responses. J Neurophysiol 78:1373–1383.

Urmson C, Anhalt J, Bae H, Bagnell A, Baker CR, et al. (2008) Autonomous driving in urban environments: boss and the urban challenge. J Field Robotics 25:425–466.

Valeeva G, Janackova S. Nasretdinov A, Richkova V, Makarov R, et al. (1998) Coordinated activity in the developing entorhinal-hippocampal network. Proc Natl Acad Sci (USA) in press

van den Bos E, Jeannerod M (2002) Sense of body and sense of action both contribute to self-recognition. Cognition 85:177–187.

van den Heuvel MP, Sporns O (2011) Rich-club organization of the human connectome. J Neurosci 31:15775–15786.

Vandecasteele M, Varga V, Berenyi A, Papp E, Bartho P, et al. (2014) Optogenetic activation of septal cholinergic neurons suppresses sharp wave ripples and enhances theta oscillations in the hippocampus. Proc Natl Acad Sci U S A 111:13535–13540.

Vanderwolf CH (2007) The Evolving Brain: The Mind and the Neural Control of Behavior. Berlin: Springer.

van de Ven GM, Trouche S, McNamara CG, Allen K, Dupret D (2016) Hippocampal offline reactivation consolidates recently formed cell assembly patterns during sharp wave-ripples. Neuron 92:968–974.

VanRullen R, Guyonneau R, Thorpe SJ (2005) Spike times make sense. Trends Neurosci 28:1–4.

Varela F, Lachaux JP, Rodriguez E, Martinerie J (2001) The brainweb: phase synchronization and large-scale integration. Nat Rev Neurosci 2:229–239.

Varela FJ, Thompson E, Rosch E (1991) The Embodied Mind: Cognitive Science and Human Experience. Cambridge, MA: MIT Press.

Vargha-Khadem F, Gadian DG, Watkins KE, Connelly A, Van Paesschen W, Mishkin M (1997) Differential effects of early hippocampal pathology on episodic and semantic memory. Science 277:376–380.

Verhage M, Maia AS, Plomp JJ, Brussaard AB, Heeroma JH, et al. (2000) Synaptic assembly of the brain in the absence of neurotransmitter secretion. Science 287:864–869.

Vida I, Bartos M, Jonas P (2006) Shunting inhibition improves robustness of gamma oscillations in hippocampal interneuron networks by homogenizing firing rates. Neuron 49:107–117.

Villette V, Malvache A, Tressard T, Dupuy N, Cossart R (2015) Internally recurring hippocampal sequences as a population template of spatiotemporal information. Neuron 88:357–366.

von der Malsburg C (1994). The correlation theory of brain function. In: Models of Neural Networks II: Temporal Aspects of Coding and Information (Domany E, van Hemmen JL, Schulten K, eds). New York: Springer.

von Economo C, Koskinas GN (1929) The Cytoarchitectonics of the Human Cerebral Cortex. London: Oxford University Press.

von Holst E, Mittelstaedt H (1950) Das reafferezprincip Wechselwirkungen zwischen Zentralnerven-system und Peripherie. Naturwissenschaften 37:467–476, 1950. [The reafference principle. In: *The Behavioral Physiology of Animals and Man. The Collected Papers of Erich von Holst*, translated by Martin R. Coral Gables, FL: Univ. of Miami Press, 1973, p. 139–173, 176–209].

Von Neumann J (1956) Probabilistic logics and synthesis of reliable organisms from unreliable components. In: Automata Studies. Annals of Mathematical Studies, No. 34 (Shannon CE, McCarthy J, eds), pp. 43–98. Princeton, NJ: Princeton University Press.

Von Neumann J (1958) The Computer and the Brain. New Haven, CT: Yale University Press.

von Uexküll J (1934/2011) A Foray into the Worlds of Animals and Humans, with a Theory of Meaning. Minneapolis: University of Minnesota Press.

Vyazovskiy VV, Harris KD (2013) Sleep and the single neuron: the role of global slow oscillations in individual cell rest. Nat Rev Neurosci 14:443–451.

Wachowiak M (2011) All in a sniff: olfaction as a model for active sensing. Neuron 71:962–973.

Wallenstein GV, Hasselmo ME (1997) GABAergic modulation of hippocampal population activity: sequence learning, place field development, and the phase precession effect. J Neurophysiol 78:393–408.

Walsh V (2003) A theory of magnitude: common cortical metrics of time, space and quantity. Trends Cogn Sci 7:483–488.

Walter WG (1950) An imitation of life. Sci Am 182:42—45.

Walter WG, Cooper R, Aldridge VJ, McCallum WC, Winter AL (1964) Contingent negative variation: an electric sign of sensorimotor association and expectancy in the human brain. Nature 203:380–384.

Wang HP, Spencer D, Fellous JM, Sejnowski TJ (2010) Synchrony of thalamocortical inputs maximizes cortical reliability. Science 328:106–109.

Wang Q, Sporns O, Burkhalter A (2012) Network analysis of corticocortical connections reveals ventral and dorsal processing streams in mouse visual cortex. J Neurosci 32:4386–4399.

Wang RF, Spelke ES (2000) Updating egocentric representations in human navigation. Cognition 77:215–250.

Wang SS, Shultz JR, Burish MJ, Harrison KH, Hof PR, et al. (2008) Functional trade-offs in white matter axonal scaling. J Neurosci 28:4047–4056.

Wang X, Merzenich MM, Beitel R, Schreiner CE (1995) Representation of a species-specific vocalization in the primary auditory cortex of the common marmoset: temporal and spectral characteristics. J Neurophysiol 74:2685–2706.

Wang XJ (2010) Neurophysiological and computational principles of cortical rhythms in cognition. Physiol Rev 90:1195–1268.

Wang Y, Romani S, Lustig B, Leonardo A, Pastalkova E (2015) Theta sequences are essential for internally generated hippocampal firing fields. Nat Neurosci 18:282–288.

Ward HC, Kotthaus S, Grimmond CS, Bjorkegren A, Wilkinson M, et al. (2015) Effects of urban density on carbon dioxide exchanges: observations of dense urban, suburban and woodland areas of southern England. Environ Pollut 198:186–200.

Ward LM (2002) Dynamical Cognitive Science. Cambridge, MA: MIT Press.
Watson BO, Levenstein D, Greene JP, Gelinas JN, Buzsáki G (2016) Network homeostasis and state dynamics of neocortical sleep. Neuron 90:839–852.
Watson JB (1930) Behaviorism. Chicago, IL: University of Chicago Press.
Watson RA, Szathmáry E (2016) How can evolution learn? Trends Ecol Evol 31:147–157.
Watts DJ, Strogatz SH (1998) Collective dynamics of "small-world" networks. Nature 393:440–442.
Wearden JH (2015) Passage of time judgements. Conscious Cogn 38:165–171.
Wearden JH, Lejeune H (2008) Scalar properties in human timing: conformity and violations. Quart J Exp Psychol 61:569–587.
Weatherall JO (2016) Void: The Strange Physics of Nothing (Foundational Questions in Science). New Haven, CT: Yale University Press.
Wehner R, Menzel R (1990) Do insects have cognitive maps? Ann Rev Neurosci 13:403–414.
Weng J (2004) Developmental robotics: theory and experiments. Int J Humanoid Robot 1:199–236.
Wennekers T, Sommer F, Aertsen A (2003) Neuronal assemblies. Theory Biosci 122:1–104.
Werner, G. (1988) Five decades on the path to naturalizing epistemology. In: Sensory Processing in the Mammalian Brain (J. S. Lund, ed), pp. 345–359. New York: Oxford University Press.
Wertheimer M (1912) Experimentelle Studien über das Sehen von Bewegung. Zeitschrift für Psychologie 61:161–265.
Whishaw IQ, Brooks BL (1999) Calibrating space: exploration is important for allothetic and idiothetic navigation. Hippocampus 9:659–667.
Whishaw IQ, Hines DJ, Wallace DG (2001) Dead reckoning (path integration) requires the hippocampal formation: evidence from spontaneous exploration and spatial learning tasks in in light (allothetic) and dark (idiothetic) tests. Behav Brain Res 127:49–69.
Whittington MA, Traub RD, Kopell N, Ermentrout B, Buhl EH (2000) Inhibition-based rhythms: experimental and mathematical observations on network dynamics. Int J Psychophysiol 38:315–336.
Wickelgren WA (1999) Webs, cell assemblies, and chunking in neural nets: introduction. Can J Exp Psychol 53:118–131.
Wicker B, Keysers C, Plailly J, Royet JP, Gallese V, Rizzolatti G (2003) Both of us disgusted in my insula: the common neural basis of seeing and feeling disgust. Neuron 40:655–664.
Wiener N (1956) The theory of prediction. In: Modern Mathematics for Engineers, Series 1 (Beckenback EF, ed). New York: Dover Publications.
Wikenheiser AM, Redish AD (2013) The balance of forward and backward hippocampal sequences shifts across behavioral states. Hippocampus 23:22–29.
Wills TJ, Lever C, Cacucci F, Burgess N, O'Keefe J (2005) Attractor dynamics in the hippocampal representation of the local environment. Science 308:873–876.
Wills TJ, Muessig L, Cacucci F (2014) The development of spatial behaviour and the hippocampal neural representation of space. Philos Trans R Soc Lond B Biol Sci 369:20130409.
Willshaw DJ, Buneman OP, Longuet-Higgins HC (1969) Non-holographic associative memory. Nature 222:960–962.

Wilson MA, McNaughton BL (1993) Dynamics of the hippocampal ensemble code for space. Science 261:1055–1058.

Wilson MA, McNaughton BL (1994) Reactivation of hippocampal ensemble memories during sleep. Science 265:676–679.

Wilson NR, Runyan CA, Wang FL, Sur M (2012) Division and subtraction by distinct cortical inhibitory networks in vivo. Nature 488:343–348.

Winfree AT (1980) The Geometry of Biological Time. (Biomathematics, Vol. 8.). Berlin-Heidelberg-New York: Springer-Verlag.

Winter SS, Clark BJ and Taube JS (2015) Disruption of the head direction cell network impairs the parahippocampal grid cell signal. Science 347:870–874.

Wittgenstein L (1973) Philosophical Investigations. (3rd edition). London: Pearson.

Wolff SBE, Ölveczky BP (2018) The promise and perils of causal circuit manipulations. Curr Opin Neurobiol 49:84–94.

Wolpert DM, Ghahramani Z, Jordan MI (1995) An internal model for sensorimotor integration. Science 269:1880–1882.

Wood E, Dudchenko PA, Robitsek RJ, Eichenbaum H (2000) Hippocampal neurons encode information about different types of memory episodes occurring in the same location. Neuron 27:623–633.

Woolf J (2010) The Mystery of Lewis Carroll. New York: St. Martin's Press.

Wootton D (2015) The Invention of Science. A New History of Scientific Revolution. New York: HarperCollins Publishers.

Wositsky J, Harney BY (1999) Born Under the Paperbark Tree. Marlston, SA: JB Books.

Wu X, Foster DJ (2014) Hippocampal replay captures the unique topological structure of a novel environment. J Neurosci 34:6459–6469.

Yarbus AL (1967) Eye Movements and Vision (Haigh B, translator). Original Russian edition published in Moscow, 1965. New York: Plenum Press.

Yarrow K, Haggard P, Heal R, Brown P, Rothwell JC (2001) Illusory perceptions of space and time preserve cross-saccadic perceptual continuity. Nature 414:302–305.

Yassin L, Benedetti BL, Jouhanneau JS, Wen JA, Poulet JFA, Barth AL. (2010) An embedded subnetwork of highly active neurons in the neocortex. Neuron 68:1043–1050.

Yasumatsu N, Matsuzaki M, Miyazaki T, Noguchi J, Kasai H (2008) Principles of long-term dynamics of dendritic spines. J Neurosci 28:13592–13608.

Yoshimura Y, Dantzker JL, Callaway EM (2005) Excitatory cortical neurons form fine-scale functional networks. Nature 433:868–873.

Yuste R, MacLean JN, Smith J, Lansner A (2005) The cortex as a central pattern generator. Nat Rev Neurosci 6:477–483.

Zatorre RJ, Chen JL, Penhune VB (2007) When the brain plays music: auditory-motor interactions in music perception and production. Nat Rev Neurosci 8:547–558.

Zeh HD (2002) The Physical Basis of the Direction of Time (4th ed.). Berlin: Springer.

Zhang Q, Li Y, Tsien RW (2009) The dynamic control of kiss-and-run and vesicular reuse probed with single nanoparticles. Science 323:1448–1453.

Zion Golumbic EM, Ding N, Bickel S, Lakatos P, Schevon CA, et al. (2013) Mechanisms underlying selective neuronal tracking of attended speech at a "cocktail party." Neuron 77:980–991.

Zipser D, Andersen RA (1988) A back-propagation programmed network that simulates response properties of a subset of posterior parietal neurons. Nature 331:679–684.

Ziv Y, Burns LD, Cocker ED, Hamel EO, Ghosh KK, et al. (2013) Long-term dynamics of CA1 hippocampal place codes. Nat Neurosci 16:264–266.

Zohary E, Shadlen MN, Newsome WT (1994) Correlated neuronal discharge rate and its implications for psychophysical performance. Nature 370:140–143.

Zorzi M, Priftis K, Umiltà C (2002) Brain damage: neglect disrupts the mental number line. Nature 417:138–139.

Zucker RS, Regehr WG (2002) Short-term synaptic plasticity. Annu Rev Physiol 64:355–405.

Zugaro MB, Monconduit L, Buzsáki G (2004) Spike phase precession persists after transient intrahippocampal perturbation. Nature Neurosci 8:67–71.

INDEX

Note: Page references followed by an "*f*" indicate figure.

Abramovic, Marina, 53
abstraction, 280n.5
   reference frame, 291*f*, 292
   translation invariance, 292–93
accidentalism, 38–40n.12
actin, 154–55
action
   body maps influenced by, 79
   in body sensation, 76
   motor output and, 54–55
   perception separated from, 54
   primacy of, 60n.19, 114–15
   retinal stabilization and, 57
   speed of, 58
   thought as, 55
   Troxler effect and, 57, 57n.13
action-induced corollary signaling, 75
action paths, inactivation of, 67–68
action-perception loop, 21, 59. *See also*
     corollary discharge mechanism
   body motion and, 61
   in evolution, 60
   externalization of thought as, 228
   habituation, 62n.23
   in motor regions, 61
   movement control mechanisms, 60
   neuronal networks in, 62–63
   orienting reflex, 62n.23
   patellar reflex, 41
   sender-receiver partnership in, 193
   in sensory cortex, 61, 61n.22
   technology and, 221–22
action plans, 7–8n.11
action science, 21n.45
active sensing, 69, 70*f*
   affordance theory and, 69n.33
   among blind, 71, 71n.36
   body sensation, 75
   as closed loop system, 69n.33
   definition of, 69
   echolocation, 71, 71n.36
   electrolocation in, 70–71
   epithelium stimulation, 72–73n.38
   forms of, 69–70
   hearing, 74
   jamming avoidance, 70–71n.34
   mechanisms of, 70
   observation and, 69
   olfaction, 72
   vision, 73. *See also* saccades
adaptive filtering, 65–66
Adey, W. Ross, 142n.4
affordance theory, 69n.33
Ahissar, Ehud, 69n.33
AI. *See* artificial intelligence
allocentric map-based navigation model, 175n.14
allocentric map representation, 111–12, 117
American Standard Code for Information Exchange (ASCII), 26–27n.53
amnesia. *See* retrograde amnesia

amnesic syndrome, 123n.40
amygdala, 48–49
  behavioral function of, 134
  eye movements and, 132n.61
  facial cues and, 133–34
  neuronal recordings of, 133n.66
Anderson, Richard, 289
animal vocalizations, speech and, 159n.29
anticipation, 35–36
antonym pairs, 87n.11
anxiety, 60–61n.20
Aquinas, Thomas. *See* Thomas Aquinas
arcopallium, birdsongs and, 168–69
Aristotle, 21n.45, 32, 106n.9
  on causality, 40–41
  on cause-and-effect, 40–41
  on dual role of mind, 3, 3–4n.4
  on explanandum, 3
  on explanans, 3
  on logic, 47n.29, 91–92n.18
  tabula rasa, 10–11, 23–25, 23n.48
artificial intelligence (AI), 19n.40
  externalization of thought, 235, 238–39
  information coding in, 28
  inside-out framework and, 360
ASCII. *See* American Standard Code for Information Exchange
Ashby, W. Ross, 47n.28
association cortex, 101n.38
associations
  action plans and, 7–8n.11
  causation and, 7–8
  contiguity in time and place and, 7–8
  contraprepared, 23–24
  correlations and, 49–50
  in empiricism model of brain function, 8
  memories and, 7–8n.11
  principles of, 7–8
  resemblance and, 7–8
association theories, 7–8n.11
associative strength, 9n.19
asymmetric transformation, 255n.44
asymmetry, in episodic memory, 176–77
attention
  autonomic correlates of, 294
  BOLD signals, 294–95n.31
  focal, 294–95n.31
  global, 294–95n.31
  pupil diameter, 294n.29, 294–95
  vasoactive intestine peptide and, 299n.40
  velocity to, 298
attentional mechanisms, 151
auditory feedback, 74–75, 137–38
auditory neurons, 20n.42, 65–66
autism
  eye movements and, 132
  grooming syntax and, 171–72n.11
  mirror neurons system and, 131n.58
  perceptual models of, 133n.65
autonomous robots, 119–20
Autrey, Wesley, 338

Bach-y-Rita, Paul, 76
Bacon, Francis, 23n.48, 35–36n.6
Barabási, László, 83
Barnes, Carol, 114, 114n.23, 115
Bayes, Thomas, 341–42
Bayesian brain model, 341
behavior, foundations of, 7
behavioral psychology, 23–24n.51
behaviorism, 10–11
Being and Time (Heidegger), 245–46n.15
Berger, Hans, 147–48n.15, 147–48n.115
Bergson, Henri, 250, 358
Bernstein, Nicolai, 85n.5, 253n.38
Betz, Vladimir, 54n.5
Betz cells, 54n.5, 223
Big Bang theory, 248
biophysical adaptation mechanisms, 188–89
bio-sonar. *See* echolocation
birdsongs
  arcopallium, 168–69
  neural syntax of, 168
  neuronal trajectories and, 197
  self-generated syllables in, 169–70
  semantic composition in, lack of, 169–70n.7
  syrinx, 168–69
  VTA and, 170n.9
blank slate. *See* tabula rasa
Blank Slate (Pinker), 23n.48
blind persons
  active sensing among, 71, 71n.36
  body maps among, 79–80n.50

INDEX

blindsight, 79–80n.50
Bliss, Tim, 201n.7
blocking, 9n.19
blood oxygen level-dependent (BOLD) signals, 294–95n.31
body-centered space, 253
body-derived signals, 114n.23
body maps
  among blind persons, 79–80n.50
  in conjoined twins, 80–81n.53, 80
  illusions and, 80
  somatosensory cortex model and, 79–80n.49
  ventral premotor cortex and, 79–80
body motion, 61
body sensation
  action in, 76
  as active sensing, 75
  in calibrated brain, 76–77
  sensory-to-motor communication in, 75–76, 75–76n.44
  in somatosensory cortex, 75–76
body teaching brain, 77
  delta-brush oscillations in, 78–79n.48
  muscle jerks and, 78–79
  proprioceptive information in, 78–79
  spindles in, 78–79
  spontaneous retinal waves, 78–79n.47
  stretch sensors in, 77–78
  tracé alternans in, 78–79n.48
BOLD signals. *See* blood oxygen level-dependent signals
Boolean logic, 41n.18
Borges, Jose Luis, 199
Boyden, Ed, 48–49n.30
Braille, Louis, 68–69n.31
brain. *See also* body teaching brain; inside-out framework; outside-in framework; *specific topics*
  Bayesian model, 341–42
  Broca's area, 225–26n.20
  calibrated, 76–77
  cell assembly in, 95
  cytoarchitectural organization of, 4–5
  genetic guidance for, 20, 20n.43
  good-enough, 337–38, 350

  information transfer in, 194*f*
  knowledge base for, 65–66
  lateral geniculate body, 16, 67–68n.30
  medial prefrontal cortex, 222–23n.10, 223
  motor arc, 1–2
  motor effectors, 1–2
  MRI imaging of, 4–5
  network dynamics, 37–38n.8
  perception, location of, 1–2
  plasticity in, 26
  preconfigured, 25–26, 25n.52, 32
  primary motor cortex, 1–2
  protomaps, 20n.43
  pyramidal cells, 1–2
  regional division of, 3–5
  as self-organized system, xiii
  space in, 251, 265
  superior colliculus, 67–68n.30
  time in, 251, 257, 265
  triune brain, 21n.45
brain activity, 16
  action-inducing neurons in, 19
  applied stimuli, 17–18
  closed-loop systems, 19
  grounding and, 17–18, 17–18n.38
  higher order sensory areas, 20
  internal models for, 138–39n.76
  motor inputs, 20
  of musicians, 33n.2
  neuronal code, 16
  neuronal responses, 17–18
  neuronal spike patterns, 16
  neurons, 17–20
  patterns in, 21n.47
  phylogenetic experience, 16n.35
  primary visual cortex, 16–17
  representations in, 19n.41
  self-organized, 138–39n.76
  sensory inputs, 20
  in sensory region, 17–18
  single protein deletions, 20n.43
  supplementary motor areas, 20
  synaptic activity in, absence of, 20n.43
  thalamic lateral geniculate body, 16
  visual systems, 16–17, 19n.40
brain diseases, x–xi

brain dynamics, preformed, 340, 344, 346f, 349–50. *See also* tabula rasa
connectionism and, 340, 340n.9
neural words and sentences in, construction of, 343–44
restored associations in, 343n.15
brain function. *See also* empiricism model; inside-out framework; outside-in framework
externalization of thought of, 239–40
learning and, 10
map construction as externalization of, 109
Marr model of, 10
for memory, 128n.52
principles of, xi
Brain Initiative, x–xi
brain-machine interface, 349–50n.21
brain networks
interactions in, 47
self-organized, 47, 47n.29
brain organization, 103f, 105f
cognition and, 104–5, 104–5n.7, 106
cognitive map theory, 101–2n.3
corollary discharge mechanisms and, 104
disengagement from environment and, 102
input-dependence in, 106
memory and, 103–4
in multiple loop patterns, 103–4
ontogenetic accumulation of experience and, 104
phylogenetic accumulation of experience and, 104
prediction and, 103–4
through self-organized activity, 104
sensory inputs, 104–5
small nervous systems and, 102
stimulus response and, 102n.4
brain patterns, 21n.47
brain research, 4f, 4–5
brain rhythms, 152–53n.21, 154f
actin and, 154–55
coincidence detectors, 161n.31
electric fields generated by, 163n.35

ephaptic effects, 163n.35
inhibition by, 150–51
myosin and, 154–55
preservation of, 156–57
speech rhythms as, 157
temporal organization of, 153–54
timing information and, 161n.31
Braitenberg, Valentino, 95–96n.27
Brecht, Michael, 79–80n.49
Brenner, Sydney, 30–31n.65
Broca's area, 225–26n.20
Brodmann, Korbinian, 4–5
Brunelleschi, Filippo, 63
Buddhism
causation in, 38–40
decision-making in, 8n.16, 8
externalization of thought in, 222n.9
freedom in, 8n.16
Kundalini yoga and, 3–4n.6
psychological centers of the body in, 3–4n.6
Troxler effect and, 57n.13
Bullock, Theodore, xi–xii
Bürgi, Jost, 302–3n.6

calibrated brain, 76–77
Cannon, Walter, 134–35n.68
Carroll, Lewis, 199
catastrophic interference, 189n.38
causality, 40–41, 248–49
analysis of, 44n.23, 44–45
deterministic, 40, 42–43
in entorhinal-hippocampal circuits, 44n.23
as inference tool, 43n.22
in relationships, 45–46
standard modeling, 42–43n.20
testing for, 43n.22
causation, 37
accidentalism and, 38–40n.12
association and, 7–8
in Buddhism, 38–40
classical, 40n.17
conflation of statistical data and, 35n.5
correlations compared to, 48
cultural differences in, 38–40
definition of, 38–40

Descartes on, 38–40
determinism and, 40, 42–43
explanadum and, 37–38
explanans and, 37–38
explanation and, 38–40n.14
Hume on, 40
immediate causes, 40
Jung on, 38–40n.13
laws of, 38, 38n.11
logical inference in, 41–42
through manipulation of physical objects, 37–38
in Muslim philosophy, 38–40, 41–42n.19
noncausal agents, 37–38n.10
origination and, 38–40n.12
in outside-in framework, 37–38n.8
permissiveness and, 44–45n.24
philosophers on, 41–42
predisposing causes, 40
principles of, 38–40n.12
probabilistic, 45, 46n.27
problems with, 43
reciprocal, 47n.29
scientific thinking and, 46
self-causation, 47
subjectivism and, 41–42
substrate-dependent, 46
sustaining causes, 40
cause-and-effect, 40
Aristotle on, 40–41
through imaging, 43
knowledge and, 7–8
logic of, 42–43n.21
in neuroscience, 41
cell assembly, 84–85, 93f
antonym pairs and, 87n.11
in brain vocabulary, 95
conceptual development of, 85–86, 85–86n.7, 87n.10
conditional signals in, 86–87n.9
definition of, 95–96n.27, 97
embedded neurons in, 93–94n.22
engrams, 86–87, 86–87n.8
gamma wave frame of, 95
in Gestalt psychology, 85–86, 85–86n.6
Hebb on, 85–88, 85n.5, 85–86n.7

hippocampal neurons and, 92–93, 97n.28
in hippocampus, 184n.28
Hopfield attractor network, 87n.10
identification of, 91–92
inhibitory neurons, 94–95n.23
internally-generated sequences, 184
interneurons and, 97n.29
IPSPs in, 95n.24
LFP in, 86–87n.9
as meaningful events, 91
membrane time constant in, 93–94n.22
population vector and, relationship with, 90
in prefrontal cortex, 184–ixn.28
pyramidal cells and, 97n.29, 98–99
reader mechanisms in, 91–92
as representational concept, 91
self-organized, 172, 173f
sharp wave ripples and, 216–17
for short-term memory, 85–86
during sleep, 90–91
spike communication in, 98–99n.32
spike timing-dependent plasticity and, 86–87n.9–10
subjective idealism and, 91–92n.18
synapse strength in, 98–99n.31
synchrony of, 97–98n.30
temporal relationship of, 173–74
trajectories of, 166
unconditional signals in, 86–87n.9
upstream neuron discharge, 93–94
utility of, 91
velocity and, 298n.38
Central Limit Theorem, 307n.20
cerebral nerve fibers, 54n.5
chakras, 3–4n.6
Champollion, Jean-François, 29–30
Childe, Gordon, 233n.35
China Brain Project, x–xi
Chomsky, Noam, 141, 350n.22
Christianity, tabula rasa as influence on, 23
chunking, 142–43n.5, 354–55
Churchill, Winston, ix
cipher, 29–30n.60

Clarke, Arthur C., 239n.43
classical causation, 40n.17
classical conditioning, 9n.19
clocks, 230. *See also* time
   circadian, 272n.81
   internal, 259–60n.51
   neuronal, 258
closed-loop systems, for brain activity, 19
coactivations, 208–9
code-breaking, 28–29n.59, 29
   through ciphers, 29–30n.60
   with Enigma machines, 29n.60, 30
   for genetic code, 30–31
   for languages, 29–30
   of neuronal codes, 31
   Rosetta Stone, 29–30
   Turing and, 30, 30n.62
coding. *See* information coding; neuronal code
cognition
   brain organization and, 104–5, 104–5n.7, 106
   embodied, 9n.17, 104–5n.7
   foundations of, 7
   internally organized, 101–2
   internal neuronal sequences and, 176
   neuronal networks and, 185–86, 188–89n.37
   in outside-in framework, 13–14
cognitive effort, 267–68n.72
cognitive map theory, 21n.45, 123, 124n.45
   brain organization and, 101–2n.3
   hippocampal, 117, 117n.29
   memory and, 127
   mental navigation and, 123
cognitive psychology, 23–24n.51
   philosophical roots of, 5
cognitive representation of space, 110n.16
cognitive revolution, 227n.23, 227
cognitive science
   framework for, 6–7
   James on, 5–6
   philosophical roots of, 5
   tabula rasa in, 23
cognitivism, 9
coincidence. *See* meaningful coincidences

coincidence detectors, 161n.31
Columbus, Christopher, 115n.26
comparators, 63–64, 63–64n.27
computational modeling
   for head direction sense, 107n.11
   of mind, 3–4n.4
   of neuronal trajectories, 190–91n.41, 191n.44
computer science
   decision-making and, 8n.15
   neuroscience and, as metaphor for, 12–13
conditioning
   associative strength and, 9n.19
   blocking and, 9n.19
   classical, 9n.19
   conditioned suppression and, 9n.19
   extinction and, 9n.19
   higher order, 9n.19
   latent learning and, 9n.19
   masking and, 9n.19
   Pavlovian, 9, 9n.18, 9n.19
   stimulus generalization and discrimination, 9n.19
The Conduct of Life (Emerson), 53n.3
Confucianism, 55n.8
Confucius, 101, 101n.2, 337
connectionism, 23, 23n.50, 340, 340n.9
consciousness
   embodied cognition and, 9n.17
   in empiricism model, of brain function, 8
   externalization of thought and, 228n.26
   Gestalt psychology and, 9n.17
   Global Workspace version, 16n.36
   goal-directed behavior and, 9n.17
   Multiple Drafts model, 16n.36
   perception-action cycle and, 9n.17
   umwelt and, 9n.17
conservation of timing, across species, 155
   local interneuron networks, 155–56n.24
   long-range interneurons, 155–56n.24
   synaptic path length and, 155
constructive memory, 275n.88
contextual mechanisms, 151
contiguity in time and place, 7–8

contingency, 35–36
continuous vision, 68
　intermittent sampling and, 68–69
　saccades, 68
contraprepared associations, 23–24
co-occurrences, 34
corollary discharge mechanism, 62, 63, 65f, 70f, See also eye movements
　action path inactivation and, 67–68
　auditory neurons, 65–66
　brain organization and, 104
　Brunelleschi and, 63
　comparators, 63–64, 63–64n.27
　emotions and, 135
　loop organization and, 63–64n.27
　mirror neurons system and, 130, 131
　motor control theory and, 63–64n.26
　optokinetic reflex and, 54
　parietal cortex neurons, 67–68n.30
　predictive coding and, 63–64n.26
　from saccades, 73, 73–74n.41
　self-motion and, 115
　sensory physiology and, 63–64n.26
　time divisions in, 66–67
　in vertebrates, 67
　visual information in, 67–68, 67–68n.30
corporeal substance. See res extensa
correlations, 34
　associations and, 49–50
　in brain functions, 48–49
　causation compared to, 48
　challenges to, 46, 48
　as inference tool, 50
　perturbations, 48–49, 48–49n.30, 50
cortical interneurons, 150n.17
corticospinal tract, 54n.5
credit assignment problem, 11–12n.24
　with grounding, 17–18n.38
Crick, Francis, 30–31n.65
cross-frequency phase coupling, inhibition and, 151

DARPA. See Defense Advanced Research Projects Agency
dead reckoning, 115, 221–22n.8
De Anima (Aristotle), 106n.9

deciphering. See code-breaking
decision-making
　in Buddhism, 8n.16, 8
　as computer science metaphor, 8n.15
　in empiricism model, of brain function, 8
declarative memory, 123
deduction, 35
deep learning, 28, 28n.56
deep Q-networks, 215n.29
Defense Advanced Research Projects Agency (DARPA), 119n.33
degeneracy, 302–3n.7, 304–5n.11
Dehaene, Stanislas, 351n.26
Deisseroth, Karl, 48–49n.30
Delage, Yves, 85n.5
delta-brush oscillations, 78–79n.48
Dennett, Daniel, 55n.8
deSacy, Silvestre, 29–30n.61
Descartes, René, 21n.45, 35–36n.6
　on causation, 38–40
　dualist view of mind, 54–55n.8
　representations for, 12–13n.25
deterministic causality, 40, 42–43
Diagnostic and Statistical Manual of Mental Disorders, 358–59
directed notion of time, 247n.20
discrimination. See stimulus generalization and discrimination
distance calibration, in physical navigation, 110
distance-duration issue, 268–69
distance representation, 297
distance-time compression, 175, 187f
　allocentric map-based navigation model, 175n.14
　oscillation frequency of neurons in, 175–76n.15
　phase precession in, 175n.14
　sharp wave ripples and, 206–7n.14
　theta cycle in, 175
Donoghue, John, 88–90n.14
dopamines, 12n.24
　signaling, 262–63, 262–63n.60, 263–64n.61
　time and, 262–63, 262–63n.60, 263–64n.61

Dragoi, George, 327–28n.56, 344–45n.16
Dyson, Freeman, 219

echolocation, 71, 71n.36
  saccades and, 73–74n.41
ECoG. *See* electrocardiogram
Economo, Constantin von, 223n.11
Eddington, Arthur, 247
Edelman, Gerald, 87n.11
EEG. *See* electroencephalograms
egocentric episodic memory, 125–26
egocentric path integration, 117
Eichenbaum, Howard, 261
Einstein, Albert, 249, 262–63
electrocardiogram (ECoG), 144–45
  speech experiments, 158–59, 161–62
electroencephalograms (EEG), 43, 144–45, 144n.10, 158–59, 161–62
electrolocation, 70–71
embodied cognition, 9n.17, 104–5n.7
Emerson, Ralph Waldo, 53n.3, 53
emotions
  corollary discharge systems and, 135
  facial cues and, 133–35, 133n.64
  James-Lange theory of emotion, 134–35, 134–35n.68
empiricism model, for brain function
  association in, 8
  brain signals, 7–9
  consciousness in, 8
  decision-making in, 8
  executive function in, 8n.13
  foundations of, 7–8
  free will in, 8
  Pavlovian conditioning in, 9, 9n.18, 9n.19
  perceptual processing in, 8
  sensory inputs in, 8–9
  stimulus associations in, 8–9
  tabula rasa as influence on, 23
  top-down functions in, 8n.13
encryption, 28–29
engrams, 86–87, 86–87n.8
Enigma machines, 29n.60, 30
entorhinal cortex, 110, 111–12n.18
  declarative memory in, 127

entorhinal-hippocampal circuits, 44n.23
entropy, 46n.27, 247, 247n.20
ephaptic effects, 163n.35
episodic memory, 125*f*
  asymmetry and, 176–77
  boosting effect of, 126n.47
  cognitive space in, 124
  egocentric, 125–26
  evolutionary roots of, 221–22n.8
  in hippocampal system, 124n.45
  internal neuronal sequences and, 176–77
  planning in, 124
  recall in, 124
  semantic memory compared to, 123n.43, 126n.47
  temporal contiguity and, 176–77
epithelium stimulation, 72–73n.38
EPSPs. *See* excitatory postsynaptic potentials
error trials, 180–81n.20
Euler, Leonhard, 145n.14
evoked neuronal responses, 13–14
evolutionary science
  action-perception loop and, 60
  biological coding in, 28–29
  Lamarck and, 3–4
  perception-action loop in, 60
evolvability, 302–3n.7
excitatory gain
  long-range interneurons, 201–3n.10
  during sharp wave ripples, 201–3n.8, 201, 202*f*
  spike transfers in, 203–4n.11
excitatory postsynaptic potentials (EPSPs), 95n.24
executive function, 8n.13
executive mechanisms, 151
experience, 22
  internalization of, 105–6n.8
  as matching process, 344
  ontogenetic accumulation of, 104
  phylogenetic accumulation of, 104
explanandum (things to-be-explained), 3
  causation and, 37–38
explanans (things that explain), 3
  causation and, 37–38

explicit knowledge, 127n.48
externalization of thought
    as action-perception loop, 228
    AI, 235, 238–39
    of brain function, 239–40
    in Buddhism, 222n.9
    clocks as, 230
    consciousness and, 228n.26
    by homo sapiens, 224–26
    imagination and, 222–23
    information revolution as, 234
    internalization of thought as complementary processes for, 228
    neurotechnology as revolution, 236
    numbers, 228
    primary motor cortex and, 222–23
    thoughts as actions, 222
extinction, 9n.19
extrapolation, 35–36
    in head direction sense, 109
eye movements, 131
    amygdala hyperactivity and, 132n.61
    autism and, 132
    patterns of, alteration of, 132
    saccades and, 73–74n.41, 267
eye-tracking techniques, 57

facial cues, 133
    amygdala and, 133–34
    emotions and, 133–35, 133n.64
    imaging studies of, 133n.64
    as internal experience, 133n.64
    by primates, 134n.67
familiar-novel continuum, 324
Fechner, Gustav Theodor, 306
Fee, Michael, 168–69
feedback
    auditory, 74–75, 137–38
    loops, 47
    sensory, 58–59n.18
feed-forward inhibitory mechanisms, 75
Feynman, Richard, 46n.27, 245n.13, 249
filler terms, x–xin.4
fixed action patterns, 171, 171n.10
fMRI. *See* functional magnetic resonance imaging

focal attention, 294–95n.31
forecasting, 35–36
forward signaling, 67–68
foveal fields, 253
freedom, in Buddhism, 8n.16
free energy, 27n.54
Freeman, Walter, 147–48n.115
free will, 8
Fried, Itzhak, 182
Friston, Karl, 274n.87
frontal structures, 54
functional magnetic resonance imaging (fMRI), 6, 7n.9
    cause-and-effect through, 43
    for memory, 127–28
    of neuronal trajectories, 183
Fuster, Joaquín, 9n.17

GABA receptors. *See* gamma aminobutyric acid receptors
Gagarin, Yuri, 142n.4
gain, 281f
    attention-induced, 293–94n.28, 293
    control and, 292–93n.26
    coordinate system transformation, 287, 288f
    corollary discharge mechanisms and, 292n.25
    electrical engineering and, 280–81n.6
    entorhinal cortex and, 292–93n.27
    GABA receptors and, 283
    inhibition and, 283
    input magnitude normalization, 285, 286f
    intensity-invariant sensory neurons, 297
    internalization of, 298
    mechanisms of, 280
    movement coordination and, 290
    neural implementation of, 281
    neuromodulation and, 284–85n.15, 284, 292–93n.26
    norepinephrine and, 284–85n.16, 284
    normalization and, 281–82n.7, 282–83n, 282–83n.17, 285, 286–87n.17, 286–87n.18, 300
    under Ohm's law, 282–83n.8

gain (*cont.*)
　principles of, 280
　receptive fields and, 287–88n.20, 287, 293–94
　retinal view and, 290
　short-term plasticity, 283
　shunting and, 282, 283n.11
　STP and, 283–84n.13
　by velocity, 295–96n.32, 295, 296f, 297n.35
gain fields, 287
Galilei, Galileo, 91–92n.18, 337
Gall, Franz Joseph, 3–4, 3–4n.6
gamma aminobutyric acid (GABA) receptors, 148–50
　gain and, 283
　grooming syntax and, 171–72n.11
gamma waves, 95
　for neural syntax, in language, 147–48n.115
Gamow, Georgiy, 30–31n.65
Gastaut, Hans, 147–48n.115
Gauss, Karl Friedrich, 301
Gehry, Frank, 165–66
Geller, Uri, 301n.1
gene expression, 6
genetic code, code-breaking of, 30–31
geometry, 231
Georgopolous, Apostolos, 88–90
Gergonne, Joseph-Diez, ix
Gestalt psychology
　cell assembly in, 85–86, 85–86n.6
　cognitivism and, 9
　consciousness and, 9n.17
　outside-in framework and, 9
gestural communication hypothesis, 136
Gibson, James, 69n.33
global attention, 294–95n.31
global positioning system (GPS), 109n.15, 119–20, 243, 243n.7
Global Workspace model, of consciousness, 16n.36
goal-directed behavior, consciousness and, 9n.17
Gödel, Kurt, 251n.34
Goldwater, Barry, 142n.4

good-enough brain, 337–38, 350
GPS. *See* global positioning system
grammar. *See also* Universal Grammar
　neural syntax and, 142–43n.7
grandmother cells, 255n.44
Granger, Clive, 43
Grastyán, Endre, 101–iiin.3, 106–7n.10
grid cells, 111–12, 111–12n.18
grid maps
　allocentric representation in, 111–12
　in entorhinal cortex, 110, 111–12n.18
　grid cells, 111–12, 111–12n.18
　spatial offset for, 111–12
grooming syntax, 171
　autism and, 171–72n.11
　behavioral sequences in, 171n.10
　fixed action patterns in, 171, 171n.10
　neuronal mechanisms of, 171–72n.11
　neuronal trajectories and, 197
　obsessive-compulsive disorder and, 171–72n.11
　physiological correlates of, 171–72
　for sexual behavior, 171n.10
grounding, 17–18, 17–18n.38
guessing, 35
Gutenberg, Johannes, 227–28

habituation, 62n.23
Haken, Hermann, 47n.29
hallucinations, 60–61
　mirror neurons system and, 131n.58
haptic flow, 256n.47
Harnad, Steve, 17–18n.38
Harris, Brian, 165
Harris, Ken, 344–45n.16
Harrison, John, 231–32n.32
Häusser, Michael, 183–84n.27
head direction neurons, 110
head direction sense, internalization of, 106, 107f, 138–39, 353–54
　cells in, 106–8
　computational models for, 107n.11
　extrapolations in, 109
　interpolation in, 109
　neuronal code of, 108
　during non-REM sleep, 108, 108n.13

during REM, 108, 108n.13
  as self-organized mechanism, 108n.14
hearing
  action-induced corollary signaling
    and, 75
  as active sensing, 74
  auditory feedback and, 74–75
  feed-forward inhibitory mechanisms
    in, 75
Hebb, Donald O., 23n.50, 85–88,
  85n.5
  on inhibitory neurons, 94–95n.23
  plasticity rule for, 86–87
  spike timing-dependent plasticity,
    86–87n.9
Hegel, Georg W. F., 334n.66
Heidegger, Martin, 219,
  245–46n.15, 250–51
Heiligenberg, Walter, 70–71n.34
hemi-neglect patients, 254
heteroclinic attractor, 186, 186n.33
higher order conditioning, 9n.19
hippocampus, 48–49
  cell assembly in, 184n.28
  cognitive maps, 117, 117n.29
  computations by, 274n.86
  damage to, 119–20n.35
  episodic memory in, 124n.45
  internal neuronal sequences in,
    177–78, 180
  learned experience and, 207–ivn.15
  mental navigation and, 120–23, 120n.36,
    120–22n.38
  place cells in, 112–13, 112–13n.19
  place fields in, 112–13
  sender-receiver partnership
    and, 193–94
  as sequence generator, 272, 273f
  sharp wave ripples and, 212n.24,
    214n.27, 216
  space and, 256
Hofstadter, Douglas, 279
Holey, Robert W., 30–31n.65
Holst, Erich von, 63–64, 63–64n.26
homeostasis, 302–3n.7
homo Erectus, 224–25
homo habilis, 224–25

homo sapiens
  agricultural development by, 226
  Broca's area in, 225–26n.20
  cognitive revolution for, 227n.23, 227
  early archaeological remains, 223–24n.13
  externalization of thought by, 224–26
  mathematics for, 227–28n.22
  superiority of, 223
  toolmaking by, 223–24n.14, 224–25n.16,
    224–25n.18, 225–26
  urban revolution for, 226
Hopfield, John, 142n.4
Hopfield attractor network, 87n.10
Hubel, David, 14–15, 72–73n.38
Hughes, Howard, 142n.4
Human Brain Project, x–xi
Hume, David, 7–8, 32, 333n.65
  on logic, 47n.29
  on representation, 42
  subjectivism of, 41–42
Hunyadi, János, 231–32n.31
Hussein (King), 142n.4
hypothesized consolidation model, 208

ideal forms, 342
illusions, 56–57n.10
  body maps and, 80
imagination, externalization of thought
    and, 222–23
The Imitation Game, 30n.62
immediate causes, 40
implicit memory, 216n.31
indigenous peoples, time and space for,
  243–45, 244–45n.11
inductive reasoning, knowledge and, 7–8
information coding, 26
  in AI programs, 28
  ASCII rules for, 26–27n.53
  in biology, 28–29
  code-breaking, 28–29n.59, 29
  in deep learning processes, 28, 28n.56
  definition of, 28–29
  encryption in, 28–29
  predictive, 63–64n.26
  representations in, 26–27n.53, 28n.55
  Shannonian theory for, 27n.54
  types of, 26–27n.53

information processing, definition of, 10
information revolution
  externalization of thought and, 234
  Internet and, 236
  social media, 236–37
  technology and, 234
information theory, 26–27
  coding in, 26–27
  definition of, 28
  free energy and, 27n.54
  prediction error and, 27n.54
information transfer, in brain, 194f
inhibition
  in brain rhythms, 150–51
  cortical interneurons, 150n.17
  cross-frequency phase coupling, 151
  GABA and, 148–50
  gain and, 283
  long-range interneurons, 148–51, 152–53
  oscillations created by, 150–51
  principal cells and, 148–50
  punctuation by, 148
inhibitory corollary discharge, 65–66
inhibitory neurons, 94–95n.23
input magnitude normalization, 285, 286f
input-output coordination, 152
inputs
  motor, 20
  in outside-in framework, 11–12
  reinforcement of, 11–12
  sensory, 104–5
  through sensory channels, 11–12
insects, neuronal trajectories in, 168–69n.5, 190–91n.40
inside-out framework, of brain systems, xiii, 6–7, 20–25
  action-perception loop in, 21
  AI and, 360
  complementary strategies in, 21
  correspondences in, 20–21
  experience and, 22
  learning through action, 22n.47
  neural observer strategy, 22
  outside-in framework compared to, 21n.45
  preconfigured brain in, 25–26, 25n.52

preexisting constraints in, 23
reader-centric view, 22
representations in, 20–21
signals in, 21
instrumental space, 253
interface model, of perception, 56n.9
interictal spikes, 211n.20
internal clocks, 259–60n.51
internalization of thought
  externalization of thought as complementary process of, 228
  for time, 232f, 232
internal neuronal sequences, 179f
  cognition and, 176
  episodic memory and, 176–77
  error trials and, 180–81n.20
  in hippocampus, 177–78, 180
  prediction in, 180n.18
  theta oscillation cycles and, 180n.19, 180–82, 180–viin.21, 181–82n.24
internal space, 253
Internet, 236
interneurons, 97n.29
  cortical, 150n.17
  gain and, 283–84n.13
  local networks, 155–56n.24
  long-range, 148–51, 152–53, 155–56n.24
interpolation, 35–36
  in head direction sense, 109
I of the Vortex (Llinás), 58–59n.18
IPSPs. See inhibitory postsynaptic potentials
is-ought problem, 333n.65

James, William, 1, 5–6, 133n.64, 241–42, 276
  on emotions, 134–35
  empiricism as influence on, 7–9
James-Lange theory of emotion, 134–35, 134–35n.68
jamming avoidance, 70–71n.34
Jasper, Herbert, 75–76n.44
John, E. Roy, 16n.36
Jung, Carl, 38–40n.13
just noticeable difference, 305–6

Kaczynski, Theodore, 220n.4
Kahneman, Daniel, 338–39, 339n.8
Kant, Immanuel, 41–42, 242–44, 242–43n.4
Kassai, Lajos, 279–80
Katz, Bernard, 98–99n.31
Kepler, Johannes, 302–3n.6
Khorana, Har Gobind, 30–31n.65
Kish, Daniel, 71n.36
knowledge
    cause-and-effect relationships and, 7–8
    explicit, 127n.48
    inductive reasoning and, 7–8
    loss of, 237
    perception and, associations as part of, 7–8
    semantic, 123n.42, 125–26, 126n.47
    sensory inputs as source of, 8–9
    stimulus associations as source of, 8–9
Knox, Dillwyn, 29n.60
Koestler, Arthur, 38–40n.14
Konishi, Mark, 161n.31
Korsakoff's syndrome, 127n.49
Kundalini yoga, 3–4n.6
Kupalov, Piotr Stefanovitch, 101–iiin.3

Lamarck, Jean-Baptiste, 3–4
Lampedusa, Giuseppe di, 301
Lange, Carl, 134–35, 134–35n.68
language. *See also* neural syntax; space; speech; time
    auditory feedback and, 137–38
    constraints of, 350
    expansion to spoken language, 138
    gestural communication hypothesis, 136
    as internalized action, 135
    mirror neuron system, 135, 138
    motor theory of speech perception, 136n.69
    neural mechanisms for, 137
    through phonetic gestures, 136n.69
    pointing and gazing as, 137n.73
    semantic grounding, 136
    space in, 242
    syntactical rules of, 158–59n.28
    time in, 242
Laplace, Pierre-Simon, 341n.12

Lashley, Karl, 86–87n.8, 171n.10, 270n.80
latent learning, 9n.19
lateral geniculate body, 67–68n.30
    thalamic, 16
lateral intraparietal (LIP) cortex, 254–55n.42
Laurent, Gilles, 168–69n.5
learned experience
    behavioral event-based template-matching methods, 208–9n.17
    coactivations, 208–9
    hippocampal networks and, 207–ivn.15
    hypothesized consolidation model, 208
    during non-REM sleep, 207
    sharp wave ripples and, 210
    two-stage model of, 209
learning, 302–3n.7
    through action, 22n.47
    brain function and, 10
    deep, 28, 28n.56
    latent, 9n.19
    neuronal trajectories during, 182
    post-learning sleep, 211–12n.23
    second languages, 33n.2
    species-specific, 23–24
Le Corbusier, 357
Ledoux, Joe, 23–24n.51, 135
Levithan, David, 279
LFP. *See* local field potential
LIP cortex. *See* lateral intraparietal cortex
Liszt, Franz, 58–59
Llinás, Rodolfo, 58–59n.18–19, 283–84n.12
local field potential (LFP), 86–87n.9, 144–45, 144n.10
local interneurons, 155–56n.24
Locke, John, 8, 23n.48
    representations for, 12–13n.25
logic
    Aristotelian, 47n.29, 91–92n.18
    Boolean, 41n.18
    of cause-and-effect, 42–43n.21
    Humean, 47n.29
    mathematics and, 38
    scientific thinking and, 38
logos, 302–3n.6
logos, multiple meanings of, 302–3n.6

log rule, of perception, 305
log scales, 352
long-range interneurons, 148–51, 152–53, 155–56n.24
    excitatory gain and, 201–3n.10
long-term synaptic plasticity, 173–74n.13
loop quantum gravity, 250
Lorentz, Henrik, 241
loss of knowledge, 237
Luddites, 219–20

Mach, Ernst, 241
Machina speculatrix, 19n.41, 119
MacLean, Paul, 21n.45
magnetic resonance imaging (MRI)
    brain imaging with, 4–5
    fMRI, 6, 7n.9
magnetoencephalographic (MEG) data, 43, 144–45
    in speech experiments, 158–59, 161–62
magnocellular pathways, 67–68n.30
map-based navigation hypothesis, 180n.19
map construction. *See also* grid maps
    allocentric map-based navigation model, 175n.14
    as externalization of brain function, 109
    GPS and, 109n.15
    physical navigation through, 109
    SLAM, 114n.32, 117–19
Marr, David, 10, 10n.21
Marxism, tabula rasa as influence on, 23
masking, 9n.19
mathematics
    geometry, 231
    for homo sapiens, 227–28n.22
    as language-dependent, 351n.26
    logic and, 38
    logos in, 302–3n.6
    proportion in, 302–3n.6
    ratio in, 302–3n.6
McNaughton, Bruce, 114, 114n.23, 208–9
meaning
    through exploration, 353
    representation as distinct from, 12–13n.26
meaningful coincidences, 38–40n.13

medial prefrontal cortex, 222–23n.10, 223
MEG data. *See* magnetoencephalographic data
Memento, 127n.49
memory. *See also* episodic memory; semantic memory
    in association theories, 7–8n.11
    brain organization and, 103–4
    chunking and, 142–43n.5
    cognitive map and, 127
    constructive, 275n.88
    declarative, 123
    fMRI imaging of, 127–28
    for future, 275n.88, 275
    implicit, 216n.31
    internal neuronal sequences and, 180n.18
    Korsakoff's syndrome and, 127n.49
    migration of, 127
    navigation, 260
    neuronal trajectories during, 182
    segmentation and, 142–43n.5
    semantic proximity and, 127
    short-term, 85–86, 306
    spatial, 128n.51, 211–12n.23
    synaptic plasticity and, 201n.6
    two-trace model, 127n.49, 127–28
memory fields, 255
Menendez de la Prida, Liset, 222n.9
Meno (Plato), 357
mental navigation, 120
    association cortex, 101n.38
    cognitive maps, 123
    declarative memories, 123
    dorsomeidal entorhinal cortex, 101n.38
    hippocampal system and, 120–23, 120n.36, 120–22n.38
    nonsensory representations, 120–22
    semantic knowledge and, 123n.42
    temporal components of, 123
Menuhin, Yehudi, 141
Mertonian mean speed theorem, 246–47n.17
microsaccades, 73–74
Mill, John Stuart, 23n.48

mind
  computational model, 3–4n.4
  dual role of, 3, 3–4n.4
mind time, 55n.7, 215n.28
mind-world connection, James on, 5–6
Minkowski, Hermann, 246–47, 248–49n.24
mirror neurons system, 128
  in action conditions, 130, 131
  autism and, 131n.58
  corollary discharge and, 130, 131
  delusion of agent influence and, 131n.58
  extracellular recordings of, 130
  hallucinations and, 131n.58
  language as internalized action and, 135, 138
  motor systems and, 129–xvn.54
  in premotor cortex, 129
  in theory of mind, 129n.53
mirror symmetry, 255n.44
Mittelstaedt, Horst, 63–64, 63–64n.26
mobile robots, 119
morse code, 142, 142n.4
Moser, Edvard, 110n.16
Moser, May-Britt, 110n.16
motion, time and space and, 271
motor arc, 1–2
motor cells, 54n.5
motor control theory, 63–64n.26
motor cortex. *See* primary motor cortex
motor-driven path integration, 117–19
motor effectors, 1–2
motor inputs, 20
motor outputs, 54–55
motor theory of speech perception, 136n.69
Mountcastle, Vernon, 88–90n.14
movement control mechanisms, 60
MRI. *See* magnetic resonance imaging
Multiple Drafts model, of consciousness, 16n.36
multiple time scale representations, 107*f*, 172
  long-term synaptic plasticity, 173–74n.13
  theta cycles and, 172–73n.12, 173–75
multivoxel pattern analysis, 183
Murakami, Haruki, 101

muscle receptors, 59–60
musicians, brain activity of, 33n.2
Muslim philosophy, causation in, 38–40, 41–42n.19
myosin
  brain rhythms and, 154–55
  perception and, 58–59n.17

Napier, Scot John, 302–3n.6
Napier's constant, 145n.14
natural numbers, 352–53
navigation memory systems, 260
neocortical events, 211
Neumann, John von, 12–13
neural information, richness of, 191, 192*f*
neural syntax
  of birdsongs, 168
  brain hierarchies for, 143–44n.9, 143
  as coded information, 142–43
  cross-frequency coupling in, 148, 148n.16
  cross-frequency phase modulation, 147–48
  definition of, 142–43n.7
  ECoG and, 144–45
  EEG and, 144–45, 144n.10
  framework for, 143
  gamma waves, 147–48n.115
  grammar and, 142–43n.7
  LFP and, 144–45, 144n.10
  MEG and, 144–45
  oscillatory ripple waves, 148
  punctuation by inhibition, 148
  rhythms in, 145n.12, 145
neural words and sentences, construction of, 343–44
neuroarcheology, 224–25n.18
neuromodulation, gain and, 284–85n.15, 284, 292–93n.26
neuronal activity, 48–49n.30
neuronal clocks, 258
neuronal code, 16
  code-breaking of, 31
neuronal connections, 6
neuronal letter, 24

neuronal networks, 62–63, 187f
  biophysical adaptation mechanisms and, 188–89
  catastrophic interference and, 189n.38
  cognition and, 185–86, 188–89n.37
  as heteroclinic attractor, 186, 186n.33
  multiple trajectories in, 187–88
  pyramidal neurons in, 188
  segmentation for, 195
  self-organized sequences in, 185
  sender-receiver partnership with, 193–94
  synaptic plasticity of, 188
  synfire chains in, 185–86n.32
  as two-dimensional, 186
neuronal sequences, 7–8n.11
neuronal spiking, 58n.15
neuronal trajectories
  birdsongs and, 197
  computational modeling of, 190–91n.41, 191n.44
  in fish, 189–90n.39
  grooming and, 197
  in insects, 190–91n.40
  during learning, 182
  during memory recall, 182
  multivoxel pattern analysis, 183
  NMDA activation and, 190–91n.41
  place cell assembly and, 191
  in primates, 182–83n.25
  reading, 189
  sharp wave ripples and, 213n.26
neurons. *See also* population vector; pyramidal neurons
  action-inducing, 19
  assembly of. *See* cell assembly
  biophysical properties, 1–2
  brain activity and, 17–20
  comparators, 63–64, 63–64n.27
  in distance-time compression, 175–76n.15
  firing patterns for, 6
  head direction, 110
  hippocampal, 92–93, 97n.28
  inferior temporal, 292–93n.27
  inhibitory, 94–95n.23
  intensity-invariant sensory, 297

interneurons, 97n.29
nonegalitarian votes of, 325
plastic, 329
single neuron recording, 14f, 14–15, 83–84, 85–86n.6, 157n.25
specialist, 84–85n.4
spiking activity, 16, 84–85, 96
upstream discharge of, 93–94
von Economo, 54n.5
neuroscience. *See also* causation; cognitive science; logic
  cause-and-effect in, 41
  contemporary framework for, 3, 6–7
  cytoarchitectural organization in, 4–5
  fundamental goals of, xi
  Gall and, 3–4, 3–4n.6
  James on, 5–6
  objectivity of, 3
  observation in, 16–17
  philosophical roots of, 3, 5
  psychology and, 3
  representations in, 12–13, 12–13n.25
  subjectivity of explanations in, xii
  veridicality of, 3
neuroscientist, as term, 1–2n.3
neurotechnology, as revolution, 236
Nicolelis, Miguel, 88–90n.14
Nirenberg, Marshall, 30–31n.65
N-methyl-D-aspartate (NMDA) receptors, 190–91n.41
Nolan, Christopher, 127n.49
noncausal agents, 37–38n.10
nonegalitarian votes of neurons, 325, 327f
non-REM sleep, 108, 108n.13, 200–1
  learned experience during, 207
nonsensory representations, 120–22
Noor (Queen), 142n.4
norepinephrine, 284–85n.16, 284
normal distributions, 307
numbers. *See also* mathematics
  externalization of thought and, 228
  natural, 352–53

objective reality, 125–26n.46
observations, 34

INDEX

active sensing and, 69
conflation of statistical data, 35n.5
deduction from, 35
extrapolation from, 35–36
induction, 35
interpolation in, 35–36
mirror neurons system and, 130
in neuroscience, 16–17
passive, 56–57
postdiction from, 35, 35n.5
predictions from, 35–36
obsessive-compulsive disorder, 171–72n.11
offline brain states
non-REM sleep, 108, 108n.13, 200–1
sharp wave ripples in, 200–1n.4–5, 201
Ohm's law, 282–83n.8
O'Keefe, John, 101–2n.3, 106–7n.10, 110n.16, 112–13, 256
cognitive map theory and, 21n.45, 123, 124n.45
olfaction
as active sensing, 72
implant odorant, 72–73n.38
oliviocerebellar system, 58–59n.18
optic flow, 66–67, 256n.47
optogenetics, 48–49n.30
optokinetic reflex, 54
The Organization of Behavior (Hebb), 85
organized spike sequences, 204, 205f
orienting reflex, 62n.23
orthogonalization, 112–13n.20, 325–26n.52
oscillations, 150–51
Broca's area and, 160n.30
in distance-time compression, frequency of neurons in, 175–76n.15
fMRI imaging studies of, 160n.30
grouping by, 160
parsing by, 160
speech and, 159
syntactic segmentation by, 160
temporal dynamics of speech, 157n.25
theta cycle, in internal neuronal sequences, 180n.19, 180–82, 180–viin.21, 181–82n.24
velocity-controlled, 268–69
voltage-controlled, 268–69n.74

oscillatory ripple waves, 148
outside-in framework, of brain systems, 6–7. *See also* empiricism model
behaviorism in, 10–11
causation in, 37–38n.8
cognition in, 13–14
discovery in, 13–14
evoked neuronal responses in, 13–14
Gestalt psychology and, 9
inside-out framework compared to, 21n.45
Marr model of, 10
memory research and, 340
perception mechanisms in, 13–14
reinforcement in, 10
selective associations as part of, 7n.9
signaling in, 10
stimulus in, 10
tabula rasa, brain as, 10–11

Paillard, Jacques, 253n.38
palm space, 253
panic attack, 60–61n.20
Pareto, Vilfredo, 331n.60
parietal cortex neurons, 67–68n.30
internal sequences for, 185n.30
parvocellular pathways, 67–68n.30
passive observers, 56–57
Pastalkova, Eva, 177–78n.17
patellar reflex, 41
path integration, 115–16
egocentric, 117
motor-driven, 117–19
in physical navigation, 115, 115n.26, 116, 116n.28, 117
Pauli, Wolfgang, 38–40n.13
Pavlov, Ivan Petrovich, 9, 9n.18, 9n.19, 62n.23, 102n.4
Pavlovian conditioning, 9, 9n.18, 9n.19
Pei, I. M., 165
Penfield, Wilder G., 75–76n.44
Penrose, Roger, 230n.28
perception. *See also* action-perception loop; active sensing; corollary discharge mechanism
development of, 62n.23

perception (*cont.*)
  eye-tracking techniques, 57
  interface model of, 56n.9
  just noticeable difference, 305–6
  knowledge and, associations as part of, 7–8
  log rule of, 305
  motor outputs, 55
  myosin and, 58–59n.17
  in outside-in framework, 13–14
  phi phenomenon, 56–57n.10, 56–57n.11, 56
  sensation as distinct from, 54–55
  sensory feedback, 58–59n.18
  sensory inputs, 55
  separation from action, 54
  speed of, 58
  Troxler effect, 57, 57n.13
  veridicality and, 58
  visual artists' exploitation of, 58n.16
  under Weber-Fechner law, 305–6
perception-action loop, 9n.17, 55, 57, 59. *See also* action-perception loop
  in evolution, 60
perceptual processing, 8
perceptual world. *See* umwelt
peripheral fields, 253
permissiveness, 44–45n.24
perpetual speech area, 136
Persian philosophy, tabula rasa as influence on, 23
perturbations, 48–49, 48–49n.30, 50
Phi (Tononi), 56–57n.10
Philo of Alexandria, 276
philosophy
  causation and, 41–42
  cognitive psychology and, 5
  cognitive science and, 5
  neuroscience and, 3, 5
  Persian, 23
  sensors and, 3
phi phenomenon
  causation and, 56–57n.10, 56–57n.11, 56
  neuronal spiking, 58n.15
  visual areas in, 58n.15
phobias, 23–24n.51

phonetical syntax, of speech, 158
phonetic gestures, 136n.69
phrenology, 3–4
phylogenetic experience, 16n.35
physicalism, 252
physical navigation, 118f. *See also* robots
  allocentric map representation and, 117
  body-derived signals in, 114n.23
  through cognitive representation of space, 110n.16
  dead reckoning in, 115
  distance calibration in, 110
  egocentric path integration, 117
  through experience, 109
  head direction neurons and, 110
  hippocampal cognitive map and, 117, 117n.29
  with hippocampal damage, 119–20n.35
  through map construction, 109
  path integration in, 115, 115n.26, 116, 116n.28, 117
  primacy of action in, 114–15
  self-localization in, 114n.23
  self-motion and, 114
physics, 245, 248, 276n.89, 277–78
Pinker, Steven, 23n.48
place cells, 110, 112
  hippocampal, 112–13, 112–13n.19
  inputs in, 113
  neuronal trajectories and, 191
  orthogonalization in, 112–13n.20
  pyramidal neurons in, 113
  sharp wave ripples and, 206–7n.13
  spatial layout of, 113
place fields, 112
  hippocampal maps in, 112–13
  modifications of, 114–15n.24
  orthogonalization in, 112–13n.20
planning, in episodic memory, 124
plasticity
  in brain, 26
  gain and, 283
  Hebbian rule on, 86–87
  long-term synaptic, 173–74n.13
  in neuronal networks, 188

offline brain states and, 201n.6
spike timing-dependent, 86–87n.9–10
synaptic, 188, 201n.6
plastic neurons, 329
Plato, 357
Poe, Edgar Allan, 141, 241
Poincaré, Henri, 21n.45, 307n.20
population vector, 88, 89f
  brain-machine interface experiments, 88–90, 88–90n.14
  cell assembly and, relationship with, 90
  cortical columns and, 88–90n.14
  definition of, 90
  hypothesis for, 88–90, 88–90n.13
  multiple neuron activity networks, 87–88n.12
postdiction, 35, 35n.5
post-learning sleep, 211–12n.23
precipitating causes, 40
preconfigured brain, 25–26, 25n.52
  tabula rasa and, 32
prediction. *See also* extrapolation
  brain organization and, 103–4
  error, 27n.54
  in internal neuronal sequences, 180n.18
  from observations, 35–36
predictive coding, 63–64n.26
predisposing causes, 40
preformed brain dynamics. *See* brain dynamics
prefrontal cortex, 48–49
  cell assembly in, 184–ixn.28
  sharp wave ripples and, 211n.20
premotor cortex, 129
preplay sequences, 327–28n.55
primacy of action, 60n.19, 114–15
primary motor cortex, 1–2
  Betz cells, 54n.5
  externalization of thought and, 222–23
  mirror neurons system and, 129–xvn.54
  supplementary motor areas, 20
primary visual cortex
  brain activity and, 16–17
  neurons in, 14f, 16–17
primates
  facial cues by, 134n.67

neuronal trajectories in, 182–83n.25
priming, 216n.31
principal cells, 148–50
principle of causality, 248–49
The Principles of Psychology (James), 5–6, 241–42
printing revolution, 227–28
probability
  causation and, 45, 46n.27
  in metaphysical disputes, 46
proprioceptive information, 78–79
prosodic features, of speech, 157–58, 157–58n.27
protomaps, 20n.43, 354–55
proximal causes, 40
pupil diameter, attention and, 294n.29, 294–95
Purkinje cells, 98–99n.31
pyramidal cells, 1–2, 97n.29, 98–99
  gain and, 283–84n.13
pyramidal neurons, 113
  internal neuronal sequences and, 178, 181n.23
  local activation of, 348–49n.19
  in neuronal networks, 188
  sharp wave ripples and, 201–3, 203–4n.11

quantities, sense of, 351

ramping timekeepers, 258–59
Ranck, James, 106–7, 106–7n.10
rapid eye movement (REM), 60–61
  head direction sense during REM sleep, 108, 108n.13
reader mechanisms
  in cell assembly, 91–92
  information synthesis in, 195–96
readiness potential, 264–65n.66
reality. *See* objective reality
recall, in episodic memory, 124
receptive fields, 73–74, 287–88n.20, 287, 293–94
reciprocal causation, 47n.29
recurrent loops, 63–64n.27, 64–65
redundancy, 302–3n.7

reflexes
   conditioned, 9
   unconditioned, 9
reinforcement
   credit assignment problem, 11–12n.24
   definition of, 11–12
   dopamines, 12n.24
   in outside-in framework, 10
reinforcement learning, 12n.24
Rejewski, Marian, 30n.64
relationships, 351
relativity theory, 248–49n.24
REM. *See* rapid eye movement
representationism, 340n.10
representations
   allocentric, in grid maps, 111–12
   in brain activity, 19n.41
   in digital technology, 13n.27
   Hume on, 42
   in information coding, 26–27n.53, 28n.55
   in inside-out framework, 20–21
   meaning as distinct from, 12–13n.26
   in neuroscience, 12–13, 12–13n.25
   nonsensory, 120–22
   as Platonic concept, 12–13n.25
   in sensory inputs, 17
   space as, 251
   time as, 251
res cogitans (thinking soul), 21n.45
res extensa (corporeal substance), 21n.45
resilience, 302–3n.7
restored associations, 343n.15
retinal stabilization, 57
retrograde amnesia, 127n.49
rigid-to-plastic continuum, 326
Rinberg, Dmitri, 72–73n.38
Rizzolatti, Giacomo, 129
robots, physical navigation by, 117, 121*f*
   autonomous, 119–20
   DARPA, 119n.33
   GPS and, 119–20
   Machina speculatrix, 119
   mobile, 119
   through motor-driven path integration, 117–19

   through sensor-dependent landmark detection, 117–19
   SLAM and, 114n.32, 117–19
   spring map model for, 117–19n.31
robustness, 302–3n.7
rodents, grooming syntax in, 171
Roosevelt, Franklin, 220
The Roots of Coincidence (Koestler), 38–40n.14
Rosetta Stone, 29–30
rotational symmetry, 255n.44
Russell, Bertrand, 33, 46

saccades, 68
   corollary discharge from, 73, 73–74n.41
   echolocation and, 73–74n.41
   eye movements and, 73–74n.41, 267
   microsaccades, 73–74
   neuronal response, 73n.40
   receptive fields, 73–74
saddle point attractor-repellor state vector, 270n.79
Schatz, Carla, 86–87n.9
Schwartz, Andrew, 88–90n.14
science. *See also* neuroscience
   professional writers of, xii–
   purpose and function of, xi
   simplicity of language in, xii–xiii
scientific thinking, 15–16n.30
   causation and, 46
   logic and, 38
segmentation, 142–43n.5
   brain solutions for, 196
   for neuronal networks, 195
   theta waves and, 195–96n.48
Sejnowski, Terry, 142n.4
selective associations, in outside-in framework, 7n.9
self-causation, 47
   in complex systems, 47
   feedback loops and, 47
   self-organization and, 47n.28
self-localization, 114n.23
self-motion
   corollary discharge from, 115
   path integration and, 115n.26

INDEX

physical navigation and, 114
Seligman, Martin, 23–24n.51
semantic features, of speech, 161–62n.32
semantic grounding, 136
semantic knowledge, 123n.42, 125–26, 126n.47
semantic memory
  episodic memory compared to, 123n.43, 126n.47
  evolutionary roots of, 221–22n.8
  objective reality and, 125–26n.46
semantic proximity, 127
Semon, Richard, 86–87n.8
sender-receiver partnership, in brain, 192
  action-perception cycle and, 193
  hippocampus and, 193–94
  in neuronal networks, 193–94
sensation, perception as distinct from, 54–55
sensors
  higher order areas, 20
sensory analyzers, 1–2
sensory channels, 11–12
sensory cortex, 61, 61n.22
sensory feedback, 58–59n.18
sensory inputs, 104–5
  brain activity and, 20
  in empiricism model, of brain function, 8–9
  as knowledge source, 8–9
  representations in, 17
sensory stimuli
  in peripheral sensors, 1–2
  transduction of, as electricity, 1–2
sensory-to-motor communication, 75–76, 75–76n.44
serial syntax, 185n.31
sexual behavior, grooming syntax for, 171n.10
Shannon, Claude, 27n.54, 28
sharp wave ripples
  animal awareness of, 215
  cell assembly sequences, 216–17
  construction functions of, 212
  deep Q-networks, 215n.29
  distance-time compression and, 206–7n.14
  entorhinal cortex, 215n.30
  excitatory gain during, 201–3n.8, 201, 202f
  hippocampal outputs, 212n.24, 214n.27, 216
  interictal spikes and, 211n.20
  learned experience and, 210
  mind time, 215n.28
  with neocortical events, 211
  neuronal sequences, 204–6
  neuronal trajectories and, 213n.26
  in offline brain states, 200–1n.4–5, 201
  organized spike sequences of, 204, 205f
  place cells and, 206–7n.13
  post-learning sleep and, 211–12n.23
  prefrontal cortex and, 211n.20
  priming and, 216n.31
  pyramidal neurons, 201–3, 203–4n.11
  replay sequences and, 213
  as self-organized events, 207
  semantic information with, 216
  spatial memory and, 211–12n.23
  theta sequences and, 212–13n.25
  two-stage model for, 212
Sheen, Fulton J., 279
Sheer, Daniel, 147–48n.115
short-term memory
  cell assembly in, 85–86
  Weber-Fechner law and, 306
short-term potentiation (STP), 283–84n.13
shunting, 282, 283n.11
sigmoid curve, 34–35n.3
signals, signaling and
  action-induced corollary, 75
  body-derived, 114n.23
  conditional, in cell assembly, 86–87n.9
  dopamines, 262–63, 262–63n.60, 263–64n.61
  in empiricism model, of brain function, 7–9
simultaneous localization and mapping (SLAM), 114n.32, 117–19
single-neuron recordings, 14–15
single neurons, 14–15, 85–86n.6, 157n.25

single protein deletions, 20n.43
Sirota, Anton, 192–93n.45
skewed distributions, 307
Skinner, B. F., 10–11
SLAM. *See* simultaneous localization and mapping
sleep. *See also* rapid eye movement
  cell assembly during, 90–91
  non-REM, 108, 108n.13
  post-learning, 211–12n.23
  REM, 60–61, 108, 108n.13
sleep paralysis, 60–61, 60–61n.20
small nervous systems, 102
small-world networks, 155–56n.24
Smetana, Bedrich, 137–38n.75
Smolen, Lee, 249n.28
social media, 236–37
social sciences, tabula rasa in, 23
Society for Neuroscience, 165–66, 183–84n.27
Socrates, 357
Sokolov, Evgeny, 62n.23
somatosensory cortex, 75–76
  body maps and, 79–80n.49
  model of, 79–80n.49
Sommer, Fritz, 142n.4
space
  Big Bang theory and, 248
  body-centered, 253
  in brain, 251, 265
  as construction, 251
  distance-duration issue, 268–69
  distortion of, 267–68n.72
  external, 253
  grounding problem of, 251
  for hemi-neglect patients, 254
  hippocampal system, 256
  for indigenous peoples, 243–45, 244–45n.11
  inflation of, 248
  inside-out approach to, 276
  instrumental, 253
  internal, 253
  Kant on, 242–43n.4
  in language, 242
  loop quantum gravity, 250
  Mertonian mean speed theorem for, 246–47n.17
  Minkowski on, 248–49n.24
  motion and, 271
  navigation in, 256
  palm, 253
  physicalism and, 252
  in physics, 245, 248, 277–78
  principle of causality, 248–49
  relativity theory for, 248–49n.24
  as representation, 251
  spatial map theory and, 256
  velocity-controlled oscillators, 268–69
  warping of, 267
SPACE coding, 26–27n.53
spacetime, 248
Spanish Civil War, 29n.60
spatial map theory, 256
spatial memory, 128n.51, 211–12n.23
specialist neurons, 84–85n.4
species-specific learning, 23–24
speech
  acoustic features of, 161–62n.32
  animal vocalizations and, 159n.29
  brain rhythms and, 157
  ECoG experiments, 158–59, 161–62
  EEG experiments, 158–59, 161–62
  MEG experiments, 158–59, 161–62
  oscillations and, 159
  phonetical syntax of, 158
  prosodic features of, 157–58, 157–58n.27
  semantic features of, 161–62n.32
  temporal dynamics of, 157n.25
  variance of syllables in, 170n.8
Sperry, Roger, 63–64
spike timing-dependent plasticity, 86–87n.9–10, 324–25n.49
spike transfers, in excitatory gain, 203–4n.11
spiking patterns, for neurons, 16, 84–85, 96
spindle cells, 54n.5, 223, 223n.11
splitter cells, 172–73n.12
spoken language, 138

INDEX

spontaneous alteration, 23–24
spontaneous retinal waves, 78–79n.47
spring map model, for robots, 117–19n.31
stimulus
    brain organization and, 102n.4
    in information coding, 27
    in outside-in framework, 10
stimulus associations, as knowledge source, 8–9
stimulus generalization and discrimination, 9n.19
stimulus-response strategy, 20–21
Stojanovic, Dejan, ix
STP. *See* short-term potentiation
stretch sensors, 77–78
Strogatz, Steve, 155–56n.24
subjective feedback, 56–57n.10
subjective idealism, cell assembly and, 91–92n.18
subjective time compression, 264
subjectivism, 41–42
substrate-dependent causation, 46
sulcus of Rolando, 54n.5
superior colliculus, 67–68n.30
supervisory functions, 8n.13
supplementary motor areas, 20
Sur, Mrganka, 20n.42
sustaining causes, 40
symbol grounding, 17–18n.38
symbols, 12–13, *See also* representations
    semantic meaning of, 17–18n.38
synapses
    in brain activity, 20n.43
    in cell assembly, strength of, 98–99n.31
    long-term plasticity of, 173–74n.13
synaptic path length, 155
synchronicity, 38–40n.13
synergetics, 47n.29
synfire chains, 185–86n.32
syntax. *See* birdsongs; grooming syntax; neural syntax
syrinx, birdsongs and, 168–69
systems properties
    component diversity of, 304

distribution of creativity, 303–4n.9
diversity of, 303

tabula rasa (blank slate), 10–11, 23–25, 23n.48, 340n.10
    conceptual criticism of, 23–24
    preconfigured brain and, 32
tactile writing system, 68–69n.31
technology
    action-perception loop and, 221–22
    complaints about, 221n.7
    digital, 13n.27
    driving force of, 236–37n.42
    historical development of, 221
    hostility towards, 220n.4
    implications of, 220–21n.6
    information revolution and, 234
    Luddites and, 219–20
    social media, 236–37
Tegmark, Max, 230n.28
template-matching methods, 208–9n.17
temporal contiguity, episodic memory and, 176–77
territory, distributions of, 329, 330*f*
thalamic lateral geniculate body, 16
theory of mind, 129n.53
theory of relativity, 249n.28, 250
theta cycle, 34–35n.3
    in distance-time compression, 175
    internal neuronal sequences, 180n.19, 180–82, 180–viin.21, 181–82n.24
    multiple time scale representations and, 172–73n.12, 173–75
    segmentation and, 195–96, 195–96n.48
    sharp wave ripples and, 212–13n.25
things that explain. *See* explanans
thing to-be-explained. *See* explanandum
Thinking, Fast and Slow (Kahneman), 338–39
thinking soul. *See* res cogitans
Thomas, Lewis, 199, 200
Thomas Aquinas, 8
thought. *See also* externalization of thought
    as action, 55, 222
threat behavior, 23–24n.51

time
  accumulation mechanisms, 259–60n.53
  as arrow, 247
  behavioral mediation of, 260
  in brain, 251, 257, 265
  computation of, 257n.49
  as construction, 251
  contingent negative variation, 259–60n.53
  as coordination of velocities, 233–34, 233–34n.38
  directed notion of, 247n.20
  distortion of, 267–68n.72
  dopamine signaling, 262–63, 262–63n.60, 263–64n.61
  elapsed, estimation of, 257
  elapsed duration of, 260–61
  future research on, 275
  grounding problem of, 251
  for Heidegger, 245–46n.15
  for indigenous peoples, 243–45, 244–45n.11
  inflation of, 248
  inside-out approach to, 276
  internalization of thought for, 232f, 232
  Kant on, 242–43n.4
  in language, 242
  measurement of, 245–47, 245n.13
  Minkowski on, 248–49n.24
  motion and, 271
  physicalism and, 252
  in physics, 245, 248, 276n.89, 277–78
  processing pathways for, 245–46n.14
  readiness potential, 264–65n.66
  as redundant, 250
  relativity theory for, 248–49n.24
  as representation, 251
  subjective time compression, 264
  theory of relativity, 249n.28, 250
  uncertainty principle for, 246n.16
  warping of, 262, 267
  in world, 234
time cells, 260
timekeeping, 258f, 258, 261
time-symmetry axiom, 43
tissue culture circuit, 18–19

TMS studies. *See* transcranial magnetic stimulation studies
Tonegawa, Susumu, 86–87n.8, 183–84
Tononi, Giulio, 56–57n.10
toolmaking, by homo sapiens, 223–24n.14, 224–25n.16, 224–25n.18, 225–26
  externalization of thought through, 229
top-down control mechanisms, 151
top-down functions, 8n.13
tracé alternans, 78–79n.48
trajectories. *See* neuronal trajectories
transcranial magnetic stimulation (TMS) studies, 137
translational symmetry, 255n.44
triune brain, 21n.45
Trivers, Robert, 53, 53n.1
Troxler, Ignatz, 57n.13
Troxler effect, 57, 57n.13
truth, objectiveness of, 341–42n.14
Tulving, Endel, 124, 187–88n.35
Turing, Alan, 30, 30n.62
Tversky, Amos, 338–39
two-trace model, for memory, 127n.49, 127–28

Uexküll, Jacob von, 9n.17
umwelt (perceptual world), consciousness and, 9n.17
uncertainty principle, 246n.16
unconditional signals, in cell assembly, 86–87n.9
unconditioned reflexes, 9
Universal Grammar, 350n.22
upstream neuron discharge, 93–94
urban revolution, for homo sapiens, 226

Vanderwolf, Cornelius, 85n.5
Vanini, Giulio Cesare, 233n.33
Vasarely, Victor, 58n.16
vasoactive intestinal peptide, 299n.40
velocity
  to attention, 298
  cell assembly sequence compression and, 298n.38
  distance representation and, 297
  gain by, 295–96n.32, 295, 296f, 297n.35

velocity-controlled oscillators, 268–69
Venter, Craig, 165
ventral premotor cortex, 79–80
ventral tegmental area (VTA), 170n.9
veridicality, 53n.1, 58
vertebrates, corollary discharge mechanism in, 67
Vinci, Leonardo da, 33
vision. *See also* saccades
    as active sensing, 73
    continuous, 68–69
    spontaneous retinal waves in, 78–79n.47
visual cortex. *See* primary visual cortex
visual systems
    brain activity and, 16–17, 19n.40
visuo-cephalo motor space, 253
visuo-locomotor space, 253
visuo-ocular motor space, 253
voltage-controlled oscillations, 268–69n.74
Voltaire, 337
von Economo neurons, 54n.5, 223, 223n.11
VTA. *See* ventral tegmental area

Walter, William Grey, 19n.41, 119, 264n.64
warping
    of space, 267
    of time, 262, 267
Watson, James, 30–31n.65
Watson, John B., 10–11
Weber, Ernst Heinrich, 305–6
Weber-Fechner law, 305–6, 355–56
Weber fraction, 305–6
Weber law, 305–6
Wertheimer, Max, 56–57n.10
white matter, 33n.2
Wiesel, Thorsten, 14–15
Wilkins, Maurice, 30–31n.65
Wilson, Matt, 208–9
Wittgenstein, Ludwig, 1
word naming, 160n.30
World War II, code-breaking during, 29n.60, 30

Yarbus, Alfred, 57
Young, Thomas, 29–30